AF589461

Key Notes on
PLANT BIOTECHNOLOGY

The Author

Venkata R. Prakash Reddy received his M.Sc. (Ag.) in Genetics and Plant Breeding from University of Agricultural Sciences, Dharwad, Karnataka where he received University Gold Medal. He is a recipient of Inspire Fellowship of Department of Science and Technology for Ph.D. programme and doing Ph.D. in Acharya NG Ranga Agricultural University. Currently, working as Assistant Professor at Acharya NG Ranga Agricultural University, Agricultural College, Mahanandi. He has also published several research, review papers in various national and international journals of repute. He has already authored two books titled Heterosis and Combining Ability Studies in Sunflower and Key Notes on Genetics and Plant Breeding. His research interests include Plant Breeding and Molecular Biology.

Key Notes on
PLANT BIOTECHNOLOGY

Venkata R. Prakash Reddy
Assistant Professor,
Department of Genetics and Plant Breeding,
Acharya N G Ranga Agricultural University,
Agricultural College – Mahanandi

2017
Daya Publishing House®
A Division of
Astral International Pvt. Ltd.
New Delhi – 110 002

Cataloging in Publication Data--DK
Courtesy: D.K. Agencies (P) Ltd. <docinfo@dkagencies.com>

Prakash Reddy, Venkata R., author.
Key notes on plant biotechnology / Venkata R. Prakash Reddy.
pages cm
Includes bibliographical references.

ISBN 978-93-86071-32-3 (International Edition)

1. Plant biotechnology. I. Title. II. Title: Plant biotechnology.

SB106.B56R43 2017 DDC 630 23

Published by : **Daya Publishing House®**
A Division of
Astral International Pvt. Ltd.
– ISO 9001:2008 Certified Company –
4760-61/23, Ansari Road, Darya Ganj
New Delhi-110 002
Ph. 011-43549197, 23278134
E-mail: info@astralint.com
Website: www.astralint.com

Preface

This book has been written primarily for undergraduate and post graduate students of Agriculture, Biotechnology, Genetics and Plant Breeding. Biotechnology is the use of living systems and organisms to develop or make products, or "any technological application that uses biological systems, living organisms or derivatives thereof, to make or modify products or processes for specific use". The wide concept of this technology encompasses a wide range of procedures for modifying living organisms according to human purposes, going back to domestication of animals, cultivation of plants, and improvements to these through breeding programmes that employ artificial selection and hybridization. Modern usage also includes genetic engineering as well as cell and tissue culture technologies. Students of these disciplines have to answer both descriptive and objective questions in various competitive examinations. This book will help the students in the development of this faculty through the presentation of typical questions and answers and help them gain an understanding of basic biotechnology principles. I am conscious that all attempts leave scope for further improvements and shall be most grateful to the readers to give me the benefit of their advice and suggestions so that these can be incorporated in a later edition.

Venkata R. Prakash Reddy

Contents

Abbreviations

1. **ACC**: Aminocyclopropane-2-carboxylate
2. **AChE: A**cetylcholinesterase
3. **AES:** American Electrophoresis Society
4. **AFLP:** Amplified Fragment Length Polymorphism
5. **AIDS**: Acquired immune deficiency syndrome
6. **AMOVA:** Analysis of Molecular Variance
7. **ARS**: Autonomous replication sequence
8. **ASB**: American Society for Biotechnology
9. **AS-PCR**: Allele specific polymerase chain reaction
10. **ATP**: Adenosine triphosphate
11. **BAC**: Bacterial artificial chromosome
12. **BADH**: Betaine aldehyde dehydrogenase
13. **BAP**: Bacterial alkaline phosphatase
14. **BARC:** Baba Atomic Research Center
15. **BILs:** Backcross Inbred Lines
16. **BLAST** - Basic Local Alignment Search Tool
17. **Bp:** Base pair(s)
18. **Bt**: *Bacillus thuringiensis*
19. **CAE**: Capillary array electrophoresis
20. **cAMP**: Cyclic adenosine-3',5'-monophosphate

21. **CCMB:** Center for Cellular and Molecular Biology
22. **CDFD:** Center for DNA Finger Printing and Diagnostics
23. **cDNA:** Complementary DNA
24. **CE:** Capillary electrophoresis
25. **CHA:** Chemical Hybridizing Agent
26. **cM:** Centimorgan
27. **CMS:** Cytoplasmic Male Sterility
28. **CPMB:** Center for Plant Molecular Biology
29. **dATP:** Deoxyadenosine triphosphate
30. **dCTP:** Deoxycytidine triphosphate
31. **ddNTP**: Dideoxynucleoside triphosphate
32. **DGGE:** Denaturing gradient gel electrophoresis
33. **dGTP:** Deoxyguanine triphosphate
34. **DH:** Doubled haploid
35. **DMSO** - Dimethyl sulfoxide
36. **DNA**: Deoxyribonucleic acid
37. **Dnase**: Deoxyribonuclease
38. **dNTP**: Deoxynucleoside triphosphate (dNTP)
39. **dUTP**: Deoxuridine triphosphate
40. **EDTA:** Ethylene Diamine Tetra Acetic acid
41. **ELISA:** Enzyme-Linked Immunosorbent Assay
42. **EMS:** Ethyl Methane Sulfonate
43. **EST:** Expressed Sequence Tag
44. **FAD**: Flavin-adenine nucleotide
45. **FISH**: Fluorescence in situ hybridization
46. **FMN:** Flavin mononucleotide
47. **FNP:** Functional Nucleotide Polymorphism
48. **GDB/OMIM**: Genome Database/Online Mendelian Inheritance in Man
49. **GDB**: Genome Database
50. **gDNA**: guide DNA
51. **GEAC:** Genetic Engineering Approval Committee
52. **GISH** - Genomic *in situ* hybridization
53. **GMO**: Genetically modified organism
54. **HAC**: Human artificial chromosome
55. **HAEC**: Human artificial episomal chromosome
56. **h-DNA:** Heteroduplex DNA

57. **HEPA** - High Efficiency Particulate Air
58. **HEPA:** Higrowth hormone Efficiency Particulate Air
59. **HFCS:** Higrowth hormone Fructose Corn Syrup
60. **HGP**: Human Genome Project
61. **HPLC:** High-Performance Liquid Chromatography
62. **IEF** - Isoelectric Focusing
63. **IR**: inverted repeat (as in inverted repeat sequence)
64. **IRAP:** Inter-Retrotransposon Amplified Polymorphism
65. **ISSR:** Inter-Simple Sequence Repeat
66. **Kb:** Kilobase
67. **Kb:** Kilobases
68. **LCR**: Ligase chain reaction
69. **LD50:** Lethal dose/fifty LD50
70. **MALDI** - matrix Assisted Laser Desorption/Ionization
71. **MAS**: Marker Assisted Selection
72. **Mb:** Mega basepairs
73. **miRNA**: microRNA
74. **mRNA:** messenger RNA
75. **mt DNA:** Mitochondrial DNA
76. **NADP+**: Nicotinamide adenine dinucleotide phosphate (oxidized)
77. **NADPH**: Nicotinamide adenine dinucleotide phosphate (reduced)
78. **NCBI** - National Centre of Biotechnology Information
79. **NILs:** Near-isogenic lines
80. **PAC**: P1-derived artificial chromosome
81. **PAGE:** Polyacrylamide gel electrophoresis
82. **PCR:** Polymerase Chain Reaction
83. **PEG:** Polyethylene Glycol
84. **PFGE:** Pulsed-field gel electrophoresis
85. **QDFM:** Quantitative DNA Fiber Mapping
86. **QTL:** Quantitative Trait Loci
87. **RAPD:** Random Amplified Polymorphic DNA
88. **rDNA: R**ecombinant DNA
89. **REMAP:** Retrotransposon-Microsatellite Amplified Polymorphism
90. **RFLP:** Restriction Fragment Length Polymorphism
91. **RH:** Radiation hybrid
92. **RILs:** Recombinant Inbred Lines

93. **RLGS**: Restriction Landmark Genomic Scanning
94. **RNA**: Ribonucleic acid
95. **RNAi**: RNA interference
96. **RNase:** Ribonuclease
97. **rRNA**: Ribosomal RNA
98. **RT-PCR** - Reverse transcriptase Polymerase Chain Reaction
99. **RT-PCR:** Real-time Polymerase-Chain Reaction
100. **SAMPL:** Selective Amplification of Microsatellite Polymorphic Loci
101. **SBH**: Sequencing by hybridization
102. **SCAR:** Sequence-Characterized Amplified Region
103. **SDS:** Sodium Dodecyl Sulfate
104. **SDS-PAGE:** Sodium dodecyl sulfate Polyacrylamide Gel Electrophoresis
105. **SNP:** Single Nucleotide Polymorphism
106. **snRNA**: Small nuclear RNA
107. **snRNP:** Small nuclear ribonucleo protein particle
108. **SSLP:** Simple Sequence Length Polymorphism
109. **SSR**: Simple sequence repeat
110. **STMS:** Sequence Tagged Microsatellite Site
111. **STRP**: Short tandem repeat polymorphism
112. **STS:** Sequence tagged site
113. **Taq:** Thermus aquaticus
114. **t-DNA:** Transfer DNA
115. **Ti plasmid:** Tumer inducing
116. **Tm:** Melting temperature of DNA
117. **TRIM:** Terminal-repeat Retrotransposon in Miniature
118. **tRNA**: Transfer RNA
119. **UNEP:** United Nations Environment Programme
120. **VIGS:** Virus-Induced Gene Silencing
121. **VNTR:** Variable-number-of-tandem-repeats
122. **XIC:** X-inactivation center
123. **XIST:** X-inactivation specific transcript
124. **YAC:** Yeast Artificial Chromosome

1

Cell Structure and Function

Define cell? And write the features of cell theory?

- ☆ Cell is the basic unit of structure of all living matter.
- ☆ It was first discovered by Robert Hook in 1665 in cork tissue.
- ☆ Loewy and Sickevitz in 1963 defined a cell as "a unit of biological activity de-limited by a semi-permeable membrane and capable of self-reproduction in a medium free of other living system".
- ☆ The cell has also been defined as "a unit of life that is the smallest unit which can carry on the activities indispensable to life to grow, to synthesize new living material and to produce new cells".

Features of cell theory

- ☆ The cell theory or cell doctrine was formulated by independently by M.J. Schleiden and Theodor Schwann in 1838-39.
- ☆ It states that the animals and plants differ from each other superficially. But they have same pattern of organization and construction.
- ☆ It further states that the bodies of both animals and plants are composed of cells and that each cell can not only act independently but also function as an integral part of complete organism.
- ☆ Thus the cell is considered as morphological and physiological unit of living organisms or in words of Schwann and Schleiden as "functional and biological unit".

Write the difference between prokaryotic cell and eukaryotic cell?

Sl.No.	*Feature*	*Prokaryotic Cell*	*Eukaryotic Cell*
1.	Meaning	Prokaryotes are primitive organisms (Pro – primitive and karyon - nucleus)	Eukaryotes are higher organisms (Eu – good or true and karyon - nucleus)
2.	Origin	Fossil prokaryote as old as 3.3 billion years are known	Oldest eukaryotic fossils are less than 1 billion years old
3.	Number of cells	Generally uni cellular	Generally multi cellular
4.	Diameter of cell	Average diameter of prokaryotic cell ranges from 5–10 μm	Average diameter of eukaryotic cell ranges from 10–100 μm
4.	Envelope system	Posses only one envelope system.	Posses two envelop system.
5.	Cell wall	Cell wall is composed of peptidoglycan, a complex of oligosaccharides and proteins. The oligosaccharides component consists of linear chains of alternating N-acetyl glucosamine (GlcNAC) and N-acetyl muramic acid (NAM) through β(1-4) glycosidic linkage	Cell wall is made up of cellulose, hemicellulose and pectinsIn plants, cell wall consists of three layers *viz.*, primary cell wall, secondary cell wall and middle lamella
	Plasmodesmata	Absent	Present in plant cells
6.	Cell organelles	Don't posses well defined cytoplasmic organelles.	Posses well defined cytoplasmic organelles like endosplasmic reticulum, golgi bodies, chloroplast, mitochondria.
	Chloroplast	Pigments are distributed throughout the cytoplasm.	Pigments are present in plastids
	Mitochondria	Absent: Blue green algae have smaller photosynthetic structures, but they do not have grana	All green [plants have typical chloroplasts which have typical grana
		Absent, oxidative phosphorylation is associated with plasmalemma	Cells contain mitochondria, which are the sites of oxidative phosphorylation
7.	Nucleus condition	Lack nucleus and chromosomes. Chromosomes are attached to plasmalemma Nucleolus and nuclear membrane are absent. Spindle fibres are absent Histone proteins are absent	Well developed nucleus and chromosomes. Chromosome ends are attached to the nuclear membrane Nucleolus and nuclear membrane are present. Spindle fibres are present Histone proteins are present

Contd...

Contd...

Sl.No.	*Feature*	*Prokaryotic Cell*	*Eukaryotic Cell*
8.	DNA	DNA is circular and lies free in the cytoplasm The amount of DNA per cell is a small fraction of that found in eukaryotic cells	DNA is linear and lies within the nucleus The amount of DNA per cell is manifold greater than that in prokaryotic cell
9.	Cell division	Cell division is by amitosis (binary fission) Spindle fibres absent; Chromosome movement is produced by an enlargement of the plasmalemma to which they are attached.	Cell division is by mitosis and meiosis Spindle apparatus is organized; Spindle fibres are associated with the movement of chromosomes/chromatids to the opposite poles of a cell.
10.	Ribosomes	Posses ribosomes of 70 S type; They dissociate into 50S and 30S subunits; All ribosomes are free in the cytoplasm	Posses ribosomes of 80 S type; They dissociate into 60S and 40S subunits; Most of ribosomes are attached to endoplasmic reticulum, but some are free in the cytoplasm
11.	Protein synthesis	Transcription and translation takes place simultaneously	Transcription takes place in nucleus and translation takes place in cytoplasm
12.	Respiration by	Mesosomes support respiration	Mitochondria support respiration
13.	Examples	*e.g.* Bacteria, blue green algae, *E. coli,* PPLOs (Pleuropheumonia like organisms)	*e.g.* Plant and animal cells

- Cells reproduce, assimilate, respire, respond to changes in the environment and absorb water and other materials from internal or external courses.
- R. Virchow formulated cell lineage theory *i.e.* cell is hereditary unit of an organism.

Explain in brief about the structure and functions of cell wall in plants?

- In plants (including bacteria) a cell is always surrounded by a cell wall lined throughout with plasma lemma.
- The cell wall is found in plants and is absent in animals.
- This is built primarily by cellulose, a rod like polysaccharide of repeating glucose units linked with β(1-4) glycosidic linkage. In wood another compound, lignin, imparts added strength and rigidity to the cell wall.
- In case of animal cells, the outermost layer of cell is plasma lemma, which is also occasionally called 'cell membrane' or 'plasma membrane'.
- The cell wall is complex in nature and is differentiated into middle lamella, primary cell wall and secondary cell wall.

Middle lamella

- It is the outmost layer of plant cell wall and connects the two adjacent cells.
- It is composed of calcium and magnesium pectate and does not contain any cellulose.
- Middle lamella as intercellular substance or intercellular matrix.

Primary cell wall

- It is thin, elastic and lies between middle lamella and secondary cell wall.
- It is mainly composed of cellulose.
- It develops after middle lamella by deposition of hemicellulose, cellulose and pectin substances.

Secondary cell wall

- It is the inner most layer of cell wall and lies between primary cell wall and plasma membrane.
- It is relatively thick and is primarily composed of microfibrils of cellulose.
- In some tissues, besides cellulose, lignin and suberin are also found in the secondary cell wall.

The cell wall has minute apertures through which the cells of a tissue are interconnected. These apertures of cell wall are known as plasmadesmata. They are also referred to as canals of the cell wall.

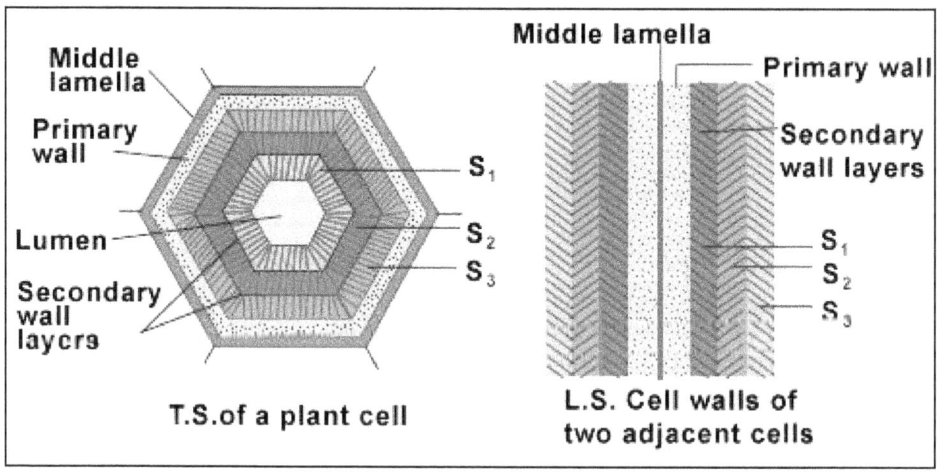

T.S.of a plant cell

L.S. Cell walls of two adjacent cells

The main functions of cell wall are

1. It determines the shape and size of a cell
2. It provides protection to the inner parts of a cell from the attack by pathogens.
3. It provides mechanical support to the tissues and act as a skeletal framework of plants.
4. It helps in transport of substances between two cells.

Explain in brief about the structure and functions of plasma lemma or plasma membrane?

- ☆ The term was coined by J.Q. Plower in 1931.
- ☆ This membrane is present just beneath the cell wall in plant cells, while it is the outer membrane in animal cell.
- ☆ In plants, it lies between the cytoplasm and the cell wall.
- ☆ It is a living, ultra thin, elastic, porous, semi-permeable membrane covering of cell.
- ☆ The plasma membrane is about 75-100 angstroms thick. In most of the cells, it is trilaminar (three layered) and made up of protein and lipids.
- ☆ The outer protein layer is 25 angstroms thick, the middle lipid layer is 25 to 30 angstroms thick and the inner protein layer is 25 to 30 angstroms thick.
- ☆ The three-layered protein-lipid-protein membrane is called a unit membrane. The outer and inner layers are made up of proteins and the middle layer is made up of lipids.
- ☆ Structure can be best explained by fluid mosaic theory.

- It is found to contain many pores through which exchange of molecules may occur.

The main functions of plasma membrane are

1. Primarily the plasma membrane provides mechanical support and external form to the protoplasm (cytoplasm and nucleus) and it also delimits the protoplasm from the exterior.
2. It checks the entry and exit of undesirable substances.
3. Due to its semipermeability, it transmits necessary materials to and from the cell (selective permeability) *i.e.* involved in the exocytosis and endocytosis of proteins and other macromolecules.
4. It permits only one way passage for molecules like minerals into the cell and restricts their outward movement.
5. The plasma membrane is also involved in communicating with other cells, in particular through the binding of ligands to receptor proteins on its surface.

Explain in brief about the structure and functions of nucleus in plant cell?

- Robert Brown first observed a cell nucleus in flowering plants in 1837.
- Generally a cell contains single nucleus.
- But there are a number of exceptions in which more than one nucleus is present. Plant cells with more than one nucleus are called coenocytes. *e.g.* Certain algae, fungi, vaucharia, rhizopus, where as animal cells with this character are called syncytia. *e.g.* Striated muscle cells of higher animals.
- The nucleus includes (a) Nuclear envelop/membrane, (b) Nucleoplasm or karyoplasms, (c) Nucleolus and (d) Chromatin.

Nuclear envelop/nuclear membrane

- It is a double membrane, semipermeable structure broken at numerous intervals by pores or openings.
- Under light microscope, it appears as a thin line between nucleus and cytoplasm.
- The space between the inner and outer membrane is known as the perinuclear space.
- Nuclear membrane joins the membrane of endoplasmic reticulum.
- The main function of nuclear membrane is to provide a pathway for the transport of materials between the nucleus and cytoplasm

Nucleoplasm/Karyolymph

- It is a fluid substance which escapes, if the nucleus is punctured.
- It fills the nuclear space around the chromosomes and the nucleolus.

- The karyolymph is composed primarily of protein materials and is rich in acidic proteins and RNA rich in bases, adenine and uracil.
- It is the site of certain enzymes in the nucleus.

Nucleolus

- Fontana first described the nucleolus in 1871.
- It is a relatively large, generally spherical body present within the nucleus.
- The number of nucleoli present in each nucleus depends upon the species and the number of the chromosomes or sets of chromosomes.
- In many plant and animal cells there is one nucleolus for each haploid set of chromosomes.
- Heterochromatic portions of specific chromosomes are found to be in contact with the nucleolus during interphase. These are called nucleolar organizing regions of the chromosomes and are responsible for producing much of nucleolar RNA.
- It is associated with sat chromosomes.
- Generally, the nucleolus disappears during cell division and reappears in daughter cells at the end of cell division in each daughter nucleus.
- However, a persistent nucleolus is found to present in Spirogyra and Euglena.
- The important functions of nucleolus are formation of ribosomes and synthesis of RNA.

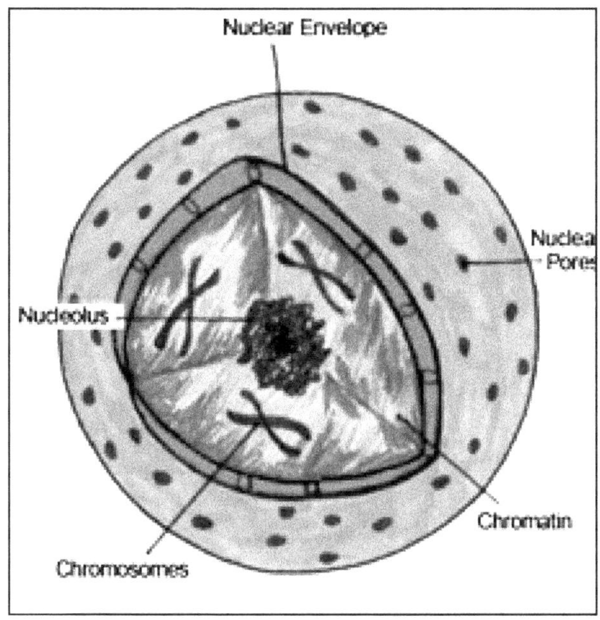

Chromatin

- ☆ The nucleus contain a darkly stained material called chromatin (Greek word, chromatin = colour), which is a combination of DNA, histone and other proteins that make up chromosomes.
- ☆ During interphase, the chromatin material is organized into a number of long, loosely coiled, irregular strands or threads called chromatin reticulum.
- ☆ When the cell begins to divide, the chromatin bodies condense to form shorter and thicker threads, which were termed chromosomes (Greek word, soma = body) by W. Waldeyer.

The main functions of chromatin are

1. It controls cellular activities and determine biological characteristics of organisms.
2. To package DNA into a smaller volume to fit in the cell,
3. To strengthen the DNA to allow mitosis and meiosis and to control gene expression and DNA replication.

What is meant by plastids? And explain its types?

- ☆ Plastids are the cytoplasmic organelles of the cells of plants and some protozoans such as Euglena.
- ☆ Whereas the cells of the bacteria, fungi and animals contain chromatophores instead of plastids.
- ☆ Plastids perform most important biological activities such as the synthesis of food and storage of carbohydrates, lipids and proteins.
- ☆ The term plastid is derived from the greek word "plastikas" means formed or moulded and was used by A.P.W. Schimper in 1885.
- ☆ He classified the plastids into the following types based on their structure, pigments and function. 1. Chromoplasts (coloured) 2. Leucoplasts (colourless).

Chromoplasts: (Greek words, chroma = colour; plast = living)

- ☆ These are the coloured plastids of plant cells.
- ☆ They contain a variety of pigments and synthesize the food through photosynthesis.
- ☆ Based on the type of pigment present in them, the chromoplasts of microorganisms and plant cells are as follows.

  a. **Chloroplasts:** (Greek words, chlor = green; plast = living)
     - ❐ These are most widely occurring chromoplasts of the plants.
     - ❐ They occur mostly in the green algae and higher plants.

- ❐ The chloroplasts contain the pigments chlorophyll A and chlorophyll B. They also contain DNA and RNA.

b. **Phaeoplasts:** (Greek words, phaeo = dark brown; plast = living)

- ❐ These contain the pigment "Fucoxanthin", which absorbs the light.
- ❐ They occur in the diatoms, dinoflagellates and brown algae.

c. **Rhodoplast:** (Greek words, rhodo = red; plast = living)

- ❐ The rhodoplast contains the pigment phycoerythrin which absorbs light.
- ❐ The rhodoplast occur in red algae.

Leucoplasts: (Greek words, leuco = white; plast = living)

- ☆ These are the colorless plastids which store the food material such as carbohydrates, lipids and proteins.
- ☆ The leucoplasts are rod like or spheroid in shape and occur in the embryonic cells, sex cells and meristematic cells.
- ☆ The most common leucoplasts of the plants cells are as follows

a. **Amyloplasts:** (Greek word, amyl = starch) these synthesize and store starch and occur in those cells which store starch.

b. **Elaioplasts:** These store lipids and occur in seeds of monocotyledons and dicotyledons.

c. **Proteinoplasts or proteoplasts:** These are the protein storing plastids which mostly occur in seed and contain few thylakoids.

Explain in brief about the structure and functions of the following cell organelles?

1. Chloroplasts
2. Mitochondria
3. Endoplasmic reticulum
4. Golgi complex
5. Ribosome

Chloroplasts: These are the most common plastids of many plant cells and perform the function of photosynthesis.

- ☆ Chloroplasts remain distributed homogeneously by in the cytoplasm of plant cells.
- ☆ Higher plant chloroplasts are generally biconvex or plano-convex.
- ☆ Generally 2-3 μ in thickness and 5-10 μ in diameter.
- ☆ Polyploid plant cells have larger chloroplasts than diploid plant cells.
- ☆ Number of chloroplasts varies from cell to cell and is related with the physiological state of the cell. But it usually remains constant for a

particular plant cell. The algae usually have a single huge chloroplast. The cells of higher plants have 20 to 40 chloroplasts.

- Chloroplast component consists of entirely glycosylglycerides rather than phospholipids. In glycosylglycerides the polar head of the group consists of galactose, digalactose without a phosphate group.
- Chloroplasts are bound by two unit membranes. Each membrane is trilaminar, lipoproteinaceous. Both unit membranes are separated from each other by a distinct space known as periplastidial space. The inner contents of the chloroplasts are heterogeneous and composed of (a) matrix or stroma and (b) grana.

Matrix or stroma

- The inner periplastidial space of the chloroplasts is filled with a watery, proteinaceous and transparent substance known as the matrix or stroma.
- The dark reaction of photosynthesis occurs in the matrix or stroma of chloroplasts and the stroma contains the multienzyme complex for the dark reactions.
- The grana and intergrana connecting membrane remain embedded in the matrix of chloroplasts.

Grana

- Chloroplast consists of many lamellar or membranous, granular and chlorophyll bearing bodies known as the grana, where the light reaction of photosynthesis takes place.
- The size of the grana may range from 0.3 to 2.7 µ.
- The chloroplasts may contain 40 to 60 grana in their matrix.
- Each granum of the chloroplast of a higher plant cell is composed of 10-100 disc like super imposed, membranous compartments known as thylakoids.

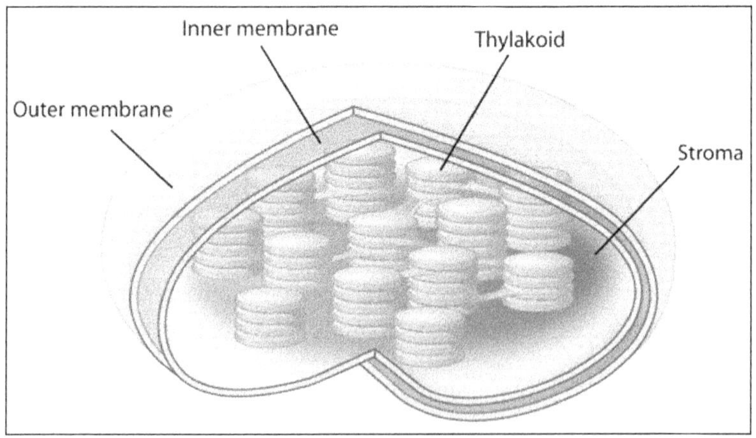

- In a granum, these thylakoids are arranged parallely to form a stack. Each thylakoid is separated from the stroma by its unit membrane. Within the thylakoid membrane, 4 sub units appear to be arranged as a functional entity called quantasomes.
- Mechanism of energy transfer known as photophosphorylation occurs within quantasomes.
- Grana are interconnected by network of membranous tubules called stroma lamella or Fret's channels.

The chloroplasts contain the ribosomes which are smaller than the cytoplasmic ribosomes. The ribosomes of the chloroplasts are 70s type and resemble the bacterial ribosomes. Woodcodk and Fernandez (1968) isolated segments of the DNA molecules from the chloroplasts. DNA of chloroplasts is present in stroma and plays an important role in cytoplasmic inheritance. The DNA of chloroplasts differs from the nuclear DNA in many aspects and it resembles closely the bacterial DNA.

- The most important and fundamental function of chloroplasts is the photosynthesis.
- The important processes of photosynthesis i.e, light and dark reactions occur within the chloroplast.
- The granum is the site of NADP reduction forming $NADPH^{+}H^{+}$ and photophosphorylation *i.e.*, formation of ATP in presence of light. Thus, light reaction of photosynthesis takes place in the granum region.
- The stroma is the main site for the dark reaction of photosynthesis.
- The chloroplast has its own genetic system and is self replicating. Thus, associated with cytoplasmic inheritance.

Mitochondria: (Greek word s, mitos = thread; chondrion = granule)

- These are first observed by Kolliker in 1880 who named them as granules. In 1882, Flemming named them as files, while in 1898, C. Benda gave the name mitochondria.
- In the cytoplasm of most cells occur many large-sized, round or rod-like structures called mitochondria.
- They are filamentous in shape and bound by two membranes composed of lipids and proteins.
- Each membrane trilamellar in nature with two protein layers sandwiching a bimolecular layer of lipid.
- The outer membrane forms a bag like structure around the inner membrane, and it contains porin proteins, which make it permeable to molecules of up to 10 KDa.
- And the inner membrane, which gives out many finger like folds into the lumen of the mitochondria. These folds are known as cristae.

☆ The space between the outer and inner mitochondrial membrane as well as the central space is filled up by a viscous mitochondrial matrix.

☆ The matrix, outer and inner membranes are found to contain many oxidative enzymes and co - enzymes.

☆ An average cell may have 200-800 mitochondria consists of proteins, lipids, RNA and DNA.

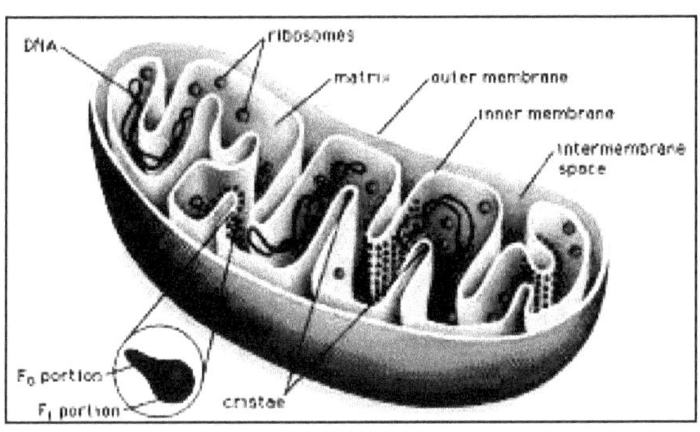

Functions

1. Mitochondria are the sites of cell respiration.
2. Oxidation of carbohydrates, lipids and proteins occurs in mitochondria.
3. The mitochondria also contain some amount of DNA within the mitochondrial matrix and are thus associated with cytoplasmic inheritance.
4. Mitochondria contain ribosomes and are capable of synthesis of certain proteins.
5. Oxidative decarboxylation and kreb's cycle takes place in the matrix of mitochondria, while respiratory chain and oxidative phosphorylation occurs in cristae of mitochondria.
6. ATP, the readily available form of energy is produced in mitochondria.
7. Krebs cycle takes place in the matrix of mitochondria.
8. Inner membrane is the site of oxidative phosphorylaion and electron transport involved in ATP production.
9. Heme synthesis occurs in mitochondria.
10. Controls the cytoplasmic Ca^{+2} concentration.

Since the major function of mitochondria is energy metabolism, during which ATP is synthesized, the mitochondria are also called power houses of the cell.

Endoplasmic Reticulum (ER)

☆ The endoplasmic reticulum was first observed by Porter in 1945 in liver cells of rats.

☆ Cytoplasmic matrix is transversed by vast reticulum or network of interconnecting tubules and vesicles known as endoplasmic reticulum.

☆ The endoplasmic reticulum is having a single vast interconnecting cavity which remains bound by a single unit membrane.

☆ The membrane of endoplasmic reticulum is supposed to have originated by impushing of plasma membrane in the matrix because it has an outer and inner layer of protein molecules sandwiching the middle layer of lipid molecules similar to plasma membrane.

☆ The membrane of endoplasmic reticulum may be either smooth when they do not have attached ribosomes or rough when they have ribosomes attached with it. Rough endoplasmic reticulum is present abundantly in pancreatic cells.

☆ One of the important functions of smooth endoplasmic reticulum is the synthesis of lipids and glycogen.

☆ Larger part of the endoplasmic reticulum resides in the outer layer of cytoplasm called the cell cortex.

☆ Rough endoplasmic reticulum is associated with the synthesis of proteins. The membrane of endoplasmic reticulum is found to be continuous with the nuclear membrane and plasma membrane.

The three principle forms of endoplasmic reticulum are:

1. **Cisternae:** Long, flattened, sac like, unbranched tubules arranged paralelly in bundles 40-50 mμ in diameter.
2. **Vesicles:** Oval, membrane bound, vacular structures, 25-230 mμ in diameter.
3. **Tubules:** Branched structures forming reticulum system along with cisternae and vesicles 50-190 mμ in diameter.

All three forms of endoplasmic reticulum are bound by a 50 A° thick single unit membrane of lipoproteinaceous nature.

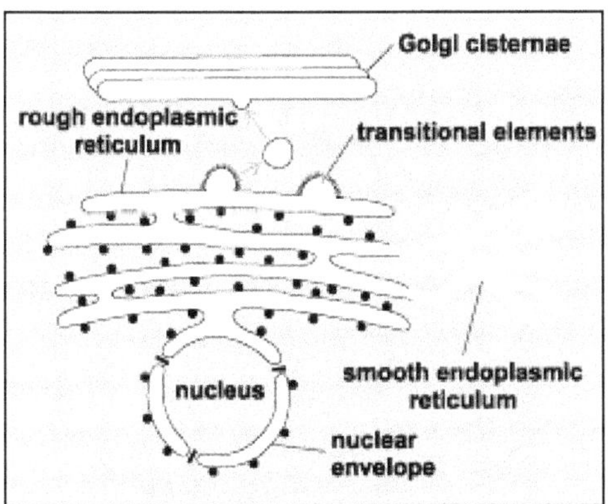

Functions

1. The endoplasmic reticulum forms the ultra structural skeletal frame work of cytoplasmic matrix and it provides mechanical support to it.

2. It also acts as an intracellular circulatory system and it circulates various substances into and out of the cell by the membrane flow mechanisms.
3. The endoplasmic reticulum acts as a storage and synthetic organ. *e.g.* It synthesizes lipids, glycogen, cholesterol, glyserides, hormones *etc.*
4. It acts as a source of nuclear membrane's material during cell division.
5. It protects cell from toxic effects by de -toxification.
6. In certain cases, it transmits impulses intracellularly. In such cases it is known as sarcoplasmic reticulum.
7. Endoplasmic reticulum which is of specialized nature and present in muscle cells is known as sarcoplasmic reticulum.
8. Rough endoplasmic reticulum is associated with the synthesis of proteins.
9. Smooth endoplasmic reticulum is associated with synthesis of lipids and glycogen.
10. It acts as an inter-cellular transport system for various substances.
11. It contains many enzymes which perform various synthetic and metabolic activities.

Golgi Complex

- ☆ It occurs in all cells except prokaryotic cells. In plant cells, they are called dictyosomes, which secrete necessary material for the formation of new cell wall during cell division.
- ☆ First reported by C. Golgi in 1898.
- ☆ It is a polymorphic structure having cisternae, vesicles and vacuoles.
- ☆ It is disc shaped and consists of central flattened plate - like compartments/ cisternae with a peripheral network of interconnecting tubules and peripherally occurring vesicles and golgian vacuoles.

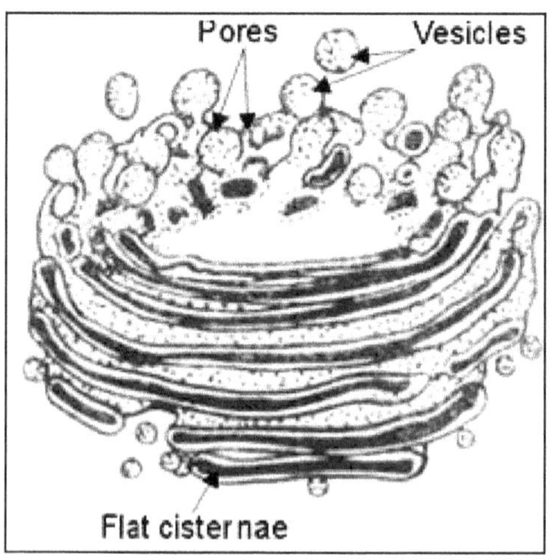

☆ The membranes of golgi complex are lipoproteinaceous and originate from membranes of endoplasmic reticulum.

Functions

1. Storage of proteins and enzymes which are secreted by ribosomes and transported by endoplasmic reticulum.
2. Secretory in function (exocytosis)
3. The dictyosome secretes necessary material for cell wall formation during cell division.
4. It has a role in the formation of plasma membrane.
5. It activates mitochondria to produce ATP, which is later utilized in respiratory cycle.
6. It helps in Packaging of proteins for exporting them.
7. It plays a role in sorting and processing of proteins for incorporation into organelles.
8. It is involved in the formation of the cell wall of plant cells, lysosomes and vacuoles.

Ribosomes

☆ Robinson and Brown in 1953 first observed ribosomes in plant cells, while Palade in 1955 first observed them in animal cells.

☆ They are small, dense, round and granular particles of about 200A° diameter and occurring either freely in mitochondrial matrix, cytoplasm, and chloroplasts or remain attached to membrane of endoplasmic reticulum forming the rough endoplasmic reticulum.

☆ They occur in all prokaryotic and eukaryotic cells and are hence called "universal components of all biological organisms".

☆ They originate in the nucleus and consist of mainly RNA and proteins. Each ribosome is composed of two structural sub units *viz.*, larger sub unit and smaller sub unit.

☆ The ribosomes are 70 S type in prokaryotes containing 50 S and 30 S subunits, while in eukryotes, they are 80 S type consisting of 60 S and 40 S subunits.

☆ The ribosome remains attached with the membranes of endoplasmic reticulum by larger subunit.

☆ The smaller subunit of ribosome is placed onto the larger subunit like a cap on the head.

☆ The ribosomes are essential for protein synthesis. They provide the platform for protein synthesis. They have the machinery for protein synthesis.

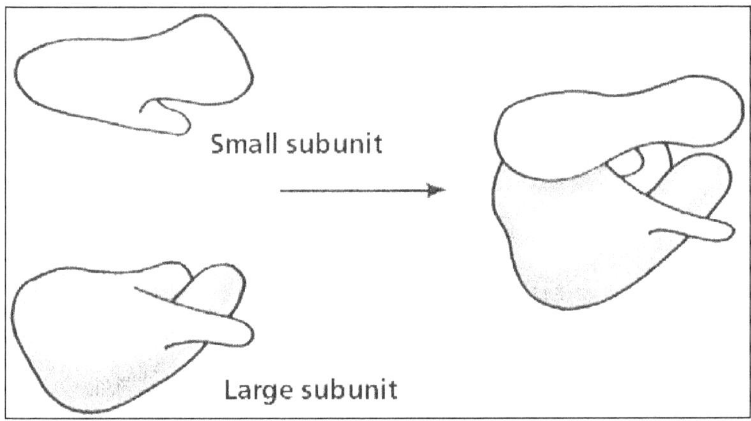

Write short notes on the following:

1. Lysosomes
2. Peroxisomes
3. Glyoxisomes
4. Spherosomes
5. Microtubules and microfilaments
6. Vacuole

Lysosomes

- ☆ Lysosomes are single membrane bound spheres present only in animal cells.
- ☆ These were first reported by Christian de Duve in 1955.
- ☆ The internal pH of these organelles is mildly acidic (pH: 4-5) and is maintained by integral membrane proteins which pumps H^+ ions into them.
- ☆ The lysosomes contain a range of hydrolyases that are optimally active at this acidic pH(acid hydrolyases) but which are inactive at the neutral pH of cytosol and extra cellular fluid.
- ☆ Lysosomes can be identified into three types:
  - ❒ Primary lysosomes: These are spherical vesicles filled with enzymatic fluid.
  - ❒ Secondary lysosomes: These are spherical vesicles which show the presence of food molecules or microorganisms.
  - ❒ Tertiary lysosomes: These are spherical vesicles which contain undigested food materials. These are also called residual bodies.

Functions

1. These are involved in digestion of food materials and phagocytosis of germs.
2. Enzymes present in lysosomes are involved in the degradation of host and foreign macromolecules in to their monomeric subunits *i.e.* proteases degrade proteins, lipases degrade lipids, phosphatises remove phosphate groups from nucleotides and phospholipids and nucleases degrade DNA and RNA.
3. Lysosomes are involved in the degradation of extracellular macromolecules that have been brought in to the cell by endocytosis as well as degradation and recycling of normal cellular components.
4. Lysosomes cause autolysis of cell contents when the cell suffers from starvation and are hence known as suicide bags of cells.

Peroxisomes

- ☆ These are spherical organelles bound by single lipoprotein membrane.
- ☆ These were first reported by Rhodin in 1954.
- ☆ In C_3 plants photorespiration takes place in peroxisomes, absent in C_4 plants.

Functions

1. During photorespiration, in these organelles, glycolic acid is oxidised to glyoxylic acid and finally converted in to amino acid, glycine.
2. They cause breakdown of hydrogen peroxide (H_2O_2) and protect the cells from toxic effects.
3. They cause oxidation of fatty acids and play an important role in the synthesis of phospholipids.

Glyoxysomes

- ☆ Glyoxisomes are spherical, single membrane bound organelles.
- ☆ These were discovered by Bridenbach in 1967.
- ☆ Glyoxysomes are temporary in that they occur during transient periods in the life cycle of a plant such as in certain beans and nuts which store fats in their seeds as energy reserves.
- ☆ Glyoxysomes appear in the first few days after seed germination in endosperm cells and associate closely with lipid bodies.
- ☆ They disappear after the storage fats are broken down and converted into carbohydrate.

Functions

Glyoxysomes are involved in the formation of sugars by the breakdown of fatty acids in germinating seeds *i.e.* glyoxalate cycle (conversion of fats into carbohydrates).

Spherosomes

- These are present in cytoplasm and rich in hydrolyzing enzymes, such as esterases, proteases, phosphatises and ribonucleases.

Functions

These are believed that these organelles are involved in the storage and translocation of lipids.

Microtubules and Microfilaments

- The microfilaments, microtubules and intermediate filaments constitute cytoskeleton, these elements consists of proteins.
- Microfilaments are rod like structures containing the protein actin.
- Microtubules are hollow cylindrical structures.

Function

1. Plays an important role in maintaining the structure and function of cells.
2. Microfilaments provide structural support and assist in cell movement.
3. Microtubules assist in movement of cilia, flagella and chromosomes; provides transport system.

Vacuole

- It is a membrane bound organelle found in plant cell and occupies most of the area in the plant cell.
- A vacuole is surrounded by a membrane called tonoplast.
- It is an enclosed compartment filled with water containing inorganic and organic molecules including enzymes in solution.
- Like lysosomes in animal cells, vacuoles have an acidic pH maintained by H^+ pumps in the membrane and contain a variety of degradative enzymes.
- Entry of water into the vacuole causes it to expand, creating hydrostatic pressure (turgor) inside the cell which is balanced by the mechanical resistance of the cell wall.

Functions

1. Isolating materials that might be harmful or a threat to the cell.
2. Stores waste products.
3. Maintains internal hydrostatic pressure or turgor within the cell as well as acidic internal pH

4. Exports unwanted substances from the cell
5. Allows plants to support structures such as leaves and flowers due to the pressure of the central vacuole.
6. Most plants stores chemicals in the vacuole that react with chemicals in the cytosol.
7. In seeds, stored proteins needed for germination are kept in protein bodies which are modified vacuole.

Explain in brief about the structure and functions of chromosomes?

- E. Strasburger in 1875 first discovered thread-like structures which appeared during cell division.
- These thread like structures were called chromosomes due to their affinity for basic dyes.
- The term chromosome is derived from two Greek words; chrom = colour, soma=body.
- This term was first used by Waldeyer in 1888.

Morphology

- The outer covering or sheath of a chromosome is known as pellicle, which encloses the matrix. Within the matrix lies the chromatin.
- Flemming introduced the term chromatin in 1879.
- The term chromatin refers to the Feulgen positive materials observed in interphase nucleus and later during nuclear division.

The chromosome morphology changes during cell division and mitotic metaphase is the most suitable stage for studies on chromosome morphology. In mitotic metaphase chromosomes, the following structural features can be seen under the light microscope.

1. Chromatid

- Each metaphase chromosome appears to be longitudinally divided into two identical parts each of which is called chromatid.
- Both the chromatids of a chromosome appear to be joined together at a point known as centromere.
- The two chromatids of chromosome separate from each other during mitotic anaphase (and during anaphase II of meiosis) and move towards opposite poles.
- Since the two chromatids making up a chromosome are produced through replication of a single chromatid during synthesis (S) phase of interphase, they are referred to as sister chromatids. In contrast, the chromatids of homologous chromosomes are known as non-sister chromatids.

2. *Centromere*

- ☆ Centromere and telomere are the most stable parts of chromosomes. The region where two sister chromatids appear to be joined during mitotic metaphase is known as centromere.
- ☆ It generally appears as constriction and hence called primary constriction.
- ☆ Centromere is a localized and easily detectable morphological region of the chromosomes which helps in the movement of the chromosomes to opposite poles during anaphase of cell division.
- ☆ The centromere divides the chromosomes into two transverse parts called arms.
- ☆ The centromere consists of two disk shaped bodies called kinetochores. The kinetochores do not form part of the chromatid but lie one on each side of the chromosome such that each chromatid is having its own kinetochore. One kinetochore is attached to the spindle fibres towards one pole and the other similarly towards the other pole.
- ☆ Depending on position of the centromeres, chromosomes can be grouped as:
  - **(a) Metacentric:** Centromere is located exactly at the centre of chromosome, *i.e.* both arms are equal in size. Such chromosomes assume 'V' shape at anaphase.
  - **(b) Submetacentric:** The centromere is located on one side of the centre point such that one arm is longer than the other. These chromosomes become 'J' or 'L' shaped at anaphase.
  - **(c) Acrocentric:** Centromere is located close to one end of the chromosome and thus giving a very short arm and a very long arm. These chromosomes acquire 'J' shape or rod shape during anaphase.
  - **(d) Telocentric:** Centromere is located at one end of the chromosome so that the chromosome has only one arm. These chromosomes are 'i" shaped or rod shaped.
- ☆ Normally chromosomes are monocentric having one centromere each.
- ☆ Acentric (without centromere) and dicentric (with two centromeres) chromosomes, if produced due to chromosomal aberrations, cannot orient properly on the equatorial plate and lag behind other chromosomes during anaphase movements.
- ☆ In certain organisms, centromere does not occupy a specific position, but is diffused throughout the body of chromosome. Such chromosomes, which do not have a localized centromere, are found in *Luzula* sps. and insects belonging to the order *Hemiptera*.

3. *Telomere*

- ☆ The two ends of chromosomes are known as telomeres.
- ☆ They are highly stable and do not fuse or unite with telomeres of other chromosomes due to polarity effect.

- Any broken end of a chromosome is unstable and can join with a piece of any other chromosome.
- But the telomeres impart stability to the chromosome, which retains its identity and individuality through cell cycle and for many cell generations.

4. *Secondary Constriction*

- The constricted or narrow region other than that of centromere is called secondary constriction.
- And the chromosomes having secondary constriction are known as satellite chromosomes or sat chromosomes.
- Chromosome may possess secondary constriction in one or both arms of it.
- Chromosomal end distal to the secondary constriction is known as satellite.
- Production of nucleolus is associated with secondary constriction and therefore it is also called nucleolus organizer region and satellite chromosomes are often referred to as nucleolus organizer chromosomes.

5. *Chromomere*

- In some species like maize, rye *etc.* chromosomes in pachytene stage of meiosis show small bead like structures called chromomeres.
- Chromomeres are visible during meiotic prophase (pachytene) and invisible in mitotic metaphase chromosomes.
- The distribution of chromomeres in chromosomes is highly characteristic and constant.
- The pattern of distribution being different for different chromosomes. They are clearly visible as dark staining bands in the giant salivary gland chromosomes.
- Chromomeres are regions of tightly folded DNA.
- Chromomeres of single chromosome show considerable variation in size.
- They may differ in size as in the case of maize or they may be of uniform size as in the case of rye.

6. *Chromonema*

- A chromosome consists of two chromatids and each chromatid consists of thread like coiled structures called chromonema (plural chromonemata).
- The term chromonema was coined by Vejdovsky in 1912.
- The chromonemata form the gene bearing portion of chromosomes.

7. *Matrix*

- The mass of acromatic material which surrounds the chromonemata is called matrix.

- ✩ The matrix is enclosed in a sheath which is known as pellicle.
- ✩ Both matrix and pellicle are non genetic materials and appear only at metaphase, when the nucleolus disappears.

Explain the different stages of mitosis along with the significance of mitosis?

- ✩ The term mitosis was coined by Flemming in 1882.
- ✩ Mitosis occurs in somatic organs like root tip, stem tip and leaf base *etc.* Hence it is also known as somatic cell division.
- ✩ The daughter cells are similar to the mother cell in shape, size and chromosome complement.
- ✩ Since the chromosome number is same in the daughter cells as compared to that of mother cell, this is also known as homotypic or equational division.

Mitotic cell cycle includes the following stages:

Interphase

- ✩ This is the period between two successive divisions.
- ✩ Cells in interphase are characterized by deeply stained nucleus that shows a definite number of nucleoli.
- ✩ The chromosomes are not individually distinguishable but appear as extremely thin coiled threads forming a faintly staining network.
- ✩ The cell is quite active metabolically during interphase.

Mitosis consist of four stage, *viz.* (a) Prophase, (b) Metaphase, (c) Anaphase and (d) Telophase.

Prophase

- ✩ The nucleus takes a dark colour with nuclear specific stains and also with acetocarmine/orcein.
- ✩ The size of the nucleus is comparatively big and the chromosomes that are thin in the initial stages slowly thicken and shorten by a specific process of coiling.
- ✩ The two chromatids of a chromosome are distinct with matrix coating and relational coiling.
- ✩ The disintegration of nuclear membrane denotes the end of prophase.

Metaphase

- ✩ After the disintegration of nuclear membrane, the shorter and thicker chromosomes will spread all over the cytoplasm.
- ✩ Later, the size of the chromosomes is further reduced and thickened.
- ✩ The distinct centromere of each chromosome is connected to the poles through spindle fibres.

- ☆ The chromosomes move towards equator and the centromere of each chromosome is arranged on the equator.
- ☆ This type of orientation of centromeres on the equator is known as autoorientation.
- ☆ The chromosomes at this stage are shortest and thickest.
- ☆ The chromatids of a chromosome are held together at the point of centromeres and the relational coils are at its minimum.

Anaphase

- ☆ The centromere of each chromosome separates first and moves to towards the poles.
- ☆ Depending on the position of the centromeres (metacentric, acrocentric and telocentric), the chromosomes show ' V ', ' L ' and ' i ' shapes respectively as the anaphase progresses.
- ☆ The sister chromatids move to the poles.
- ☆ The chromosome number is constant but the quantity of each chromosome is reduced to half.

Telophase:

- ☆ Chromosomes loose their identity and become a mass of chromatin.
- ☆ The nucleus will be re-organized from the chromatin.
- ☆ At late telophase stage, the cell plate will divide the cell into two daughter cells.

Cytokinesis

- ☆ The division of cytoplasm usually occurs between late anaphase and end of telophase.
- ☆ In plants, cytokinesis takes place through the formation of cell plate, which begins in the centre of the cell and moves towards the periphery in both sides dividing the cytoplasm into two daughter cells.
- ☆ In animal cells, cytokinesis occurs by a process known as cleavage, forming a cleavage furrow.

Significance of Mitosis

Mitosis plays an important role in the life of living organisms in various ways as:

1. Mitosis is responsible for development of a zygote into adult organism after the fusion of male and female gametes.
2. Mitosis is essential for normal growth and development of living organisms.
3. It gives a definite shape to a specific organism.

4. In plants, mitosis leads to formation of new organs like roots, leaves, stems and branches.
5. It also helps in repairing of damaged parts.
6. It acts as a repair mechanism by replacing the old, decayed and dead cells and thus it helps to overcome ageing of the cells.
7. It helps in asexual propagation of vegetatively propagated crops like sugarcane, banana, sweet potato, potato, *etc.* mitosis leads to production of identical progeny in such crops.
8. Mitosis is useful in maintaining the purity of types because it leads to production of identical daughter cells and does not allow segregation and recombination to occur.
9. In animals, it helps in continuous replacement of old tissue with new ones, such as gut epithelium and blood cells.

Explain the different stages of meiosis along with its significance?

The term meiosis was coined by J.B. Farmer in 1905.

- ☆ This type of division is found in organisms in which there is sexual reproduction.
- ☆ The term has been derived from Greek word; Meioum = diminish or reduce.
- ☆ The cells that undergo meiosis are called meiocytes.
- ☆ Three important processes that occur during meiosis are:
    1. Pairing of homologous chromosomes (synapsis)
    2. Formation of chiasmata and crossing over
    3. Segregation of homologous chromosomes

The first division of meiosis results in reduction of chromosome number to half and is called reduction division. The first meiotic division is also called heterotypic division. Two haploid cells are produced at the end of first meiotic division and in the second meiotic division, the haploid cells divide mitotically and results in the production of four daughter cells (tetrad), each with haploid number of chromosomes. In a tetrad, two daughter cells will be of parental types and the remaining two will be recombinant types. The second meiotic division is also known as homotypic division. Both the meiotic divisions occur continuously and each includes the usual stages *viz.*, prophase, metaphase, anaphase and telophase.

Meiotic cell cycle involves the following stages:

Interphase

Meiosis starts after an interphase which is not very different from that of an intermitotic interphase. During the premeiotic interphase DNA duplication occurs during the S phase

I. Meiosis-I

1. Prophase-I

It is of a very long duration and is also very complex. It has been divided into the following sub-stages:

a. Leptotene or Leptonema

- ✰ Chromosomes at this stage appear as long thread like structures that are loosely interwoven.
- ✰ In some species, on these chromosomes, bead-like structures called chromomeres are found all along the length of the chromosomes.

b. Zygotene or Zygonema

- ✰ It is characterized by pairing of homologous chromosomes (synapsis), which form bivalents.
- ✰ The paired homologous chromosomes are joined by a protein containing frame work known as synaptonemal complex. The bivalents have four chromatid strands

c. Pachytene or Pachynema

- ✰ The chromosomes appear as thickened thread-like structures.
- ✰ At this stage, exchange of segments between nonsister chromatids of homologous chromosomes known as crossing over occurs.
- ✰ During crossing over, only one chromatid from each of the two homologous chromosomes takes part. The nucleolus still persists.

d. Diplotene or Diplonema

- ✰ At this stage further thickening and shortening of chromosomes takes place. Homologous chromosomes start separating from one another.
- ✰ Separation starts at the centromere and travels towards the ends (terminalization).
- ✰ Homologous chromosomes are held together only at certain points along the length.
- ✰ Such points of contact are known as chiasmata and represent the places of crossing over. The process of terminalization is completed at this stage.

e. Diakinesis

- ✰ Chromosomes continue to undergo further contraction.
- ✰ The bivalents appear as round darkly stained bodies and they are evenly distributed throughout the cell.
- ✰ The nuclear membrane and nucleolus disappear.

2. Metaphase-I

- ✰ The chromosomes are most condensed and have smooth outlines.
- ✰ The centromeres of a bivalent are connected to the poles through the spindle fibres.

- ☆ The bivalents will migrate to the equator before they disperse to the poles.
- ☆ The centromeres of the bivalents are arranged on either side of the equator and this type of orientation is called co-orientation.

3. *Anaphase-I*

- ☆ The chromosomes in a bivalent move to opposite poles (disjunction).
- ☆ Each chromosome possess two chromatids.
- ☆ The centromere is the first to move to the pole. Each pole has a haploid number of chromosomes.

4. *Telophase-I*

- ☆ Nuclear membranes are formed around the groups of chromosomes at the two poles.
- ☆ The nucleus and nucleolus are reorganized.

II. Meiosis-II

The second meiotic division is similar to the mitotic division and it includes the following four stages:

1. *Prophase-II*

- ☆ The chromosomes condense again.
- ☆ The nucleolus and nuclear membrane disappear.
- ☆ The chromosomes with two chromatids each become short and thick.

2. *Metaphase -II*

- ☆ Spindle fibres appear and the chromosomes get arranged on the equatorial plane (auto-orientation).
- ☆ This plane is at right angle to the equatorial plane of the first meiotic division.

3. *Anaphase-II*

Each centromere divides and separates the two chromatids, which move towards the opposite poles.

4. *Telophase-II*

- ☆ The chromatids move to the opposite poles.
- ☆ The nuclear envelope and the nucleolus reappears. Thus at each pole, there is re -organization of haploid nucleus.

Cytokinesis

- ☆ The division of cytoplasm takes place by cell plate method in plants and by furrow method in animals.

- ☆ The cytokinesis may take place after meiosis I and meiosis II separately or sometimes may take place at the end of meiosis II only.

Significance of Meiosis

Meiosis plays a very important role in the biological populations in various ways as:

1. It helps in maintaining a definite and constant number of chromosomes in a species.
2. Meiosis results in production of gametes with haploid (half) chromosome number. Union of male and female gametes leads to formation of zygote which receives half chromosome number from male gamete and half from the female gamete and thus the original somatic chromosome number is restored.
3. Meiosis facilitates segregation and independent assortment of chromosomes and genes.
4. It provides an opportunity for the exchange of genes through the process of crossing over. Recombination of genes results in generation of variability in a biological population which is important from evolution points of view.
5. In sexually reproducing species, meiosis is essential for the continuity of generation. Because meiosis results in the formation of male and female gametes and union of such gametes leads to the development of zygotes and thereby new individual.

Explain the role of Cyclins, CDKs and CDKIs in cell cycle regulation?

- ☆ The cell cycle involves a series of carefully controlled events resulting in DNA duplication and cell division.
- ☆ Progression through each of the four distinct phases (G_1, S, G_2 and M) is carefully controlled by the sequential formation, activation and subsequent degradation or modification of a series of cyclins and their partners, the cyclin-dependent kinases (CDKs).
- ☆ In addition, a further group of proteins, the cyclin-dependent kinase inhibitors (CDKIs) are important for co-ordination of each stage.
- ☆ The transition from one stage to the next is regulated at a number of checkpoints which prevent premature entry into the next phase of the cycle.
- ☆ The degradation of various cyclins occurs at each checkpoint and it is this mechanism together with interaction of the CDKIs which allows the cell to enter the next phase.

Cyclins

Up to, 13 mammalian cyclins have now been identified, each one of which is required at a different stage of the cell cycle.

- ☆ They all contain a homologous region of about 100 amino acids called the cyclin box and it is this region of the protein which binds to the appropriate CDK.
- ☆ Each cyclin undergoes a characteristic pattern of synthesis and degradation dependent on the stage of the cycle at which it acts.
- ☆ The cyclins are broadly classified into G_1 and mitotic cyclins, according to the stage of the cycle during which they are produced.
- ☆ The G_1 cyclins are relatively short lived proteins and are degraded via their PEST sequences which lie C terminal to the cyclin box.
- ☆ The mitotic cyclins are longer lived and are degraded, prior to entry into mitosis, by proteinases via an ubiquitin-dependent pathway via the 'destruction' box located N terminal to the cyclin box.
- ☆ Cyclins are degraded in the cytosol by a large proteolytic complex called the 26s proteosome.
- ☆ Cyclin degradation results in CDK inactivation.

CDKs

The CDKS are key enzymes that control the transition between the different states of the cell cycle, and the entry of non dividing cells in to the cell cycle. There are at least 6 mammalian CDKs have so far been identified.

- ☆ The CDKs are activated by forming complexes with cyclin partners and by a pattern of phosphorylation and dephosphorylation at specific residues.
- ☆ CDKs are activated by phosphorylation of a conserved threonine residue at position 160 and by cyclin binding. *e.g.* Phosphorylation of the CDKs CDC2, CDK2 and CDK4 is carried out by the p40mol5 protein which in turn is activated by cyclin H.
- ☆ In addition, phosphorylation also may causes inhibition of cyclin/CDK complexes, *e.g.* The two CDKs, CDC2 and CDK2, are inactivated throughout interphase by phosphorylation on threonine 14 or tyrosine 15 by the wee1/mik1 protein kinase. At the end of G_1, the product of the CDC25 gene, activates the kinase by dephosphorylation of these residues.

CDKIs

There are at least 7 different CDKIs in mammalian cells which belong to two different classes.

- ☆ The first class comprises p21, p27 and p57 which preferentially bind to the G_1/S class of CDKs.
- ☆ The second class of CDKs, referred to as the INK4 (Inhibitor of CDK4) family, is comprised of ankyrin repeat proteins and includes p15, p16, p18 and p19.
- ☆ These inhibitors act on cyclin D complexed either to CDK4 or CDK6.

Explain in detail about regulation/control of mitotic cell cycle?

Each of the cyclin-CDK complexes, together with the CDKIs, are responsible for controlling different stages of the cell cycle by preventing progression through checkpoints in the presence of DNA damage.

Late G1 Checkpoint

- ☆ The D type cyclins are linked to the regulation of the first checkpoint at G_1/S.
- ☆ They are synthesised in response to growth factors and are very short lived.
- ☆ The D type cyclins are found in partnership with four kinases, CDK2, -4 -5 and -6 although CDK4 appears to be the main partner in most cell types.
- ☆ Activation of CDK4 by complexing with cyclin D and phosphorylation on threonine 160 drives cells through this checkpoint by phosphorylation of the Rb protein which releases the E2F transcription factor allowing it to activate genes necessary for DNA synthesis.

G1/S phase

- ☆ The E type cyclins are believed to act after the D type and to be important for the initiation of DNA replication.
- ☆ Cyclin E is expressed towards the end of G_1, and complexes with CDK2 to activate it.
- ☆ As with cyclin D-CDK4, phosphorylation of the threonine residue (160) is necessary for activation.
- ☆ After cells have entered S phase, cyclin E is rapidly degraded and CDK2 is released to be complexed by cyclin A at the next stage.
- ☆ Normally, cells which have suffered DNA damage are prevented from entering S phase and are blocked at G_1, a p53-dependent process through its it transcriptional regulation of the cyclin-dependent kinase inhibitor, p21.
- ☆ Activation of p53 by DNA damage results in increased p21 levels.
- ☆ It can then bind to a number of cyclin-CDK complexes including cyclin D-CDK4, cyclin E-CDK2 and cyclin A-CDK2 thereby preventing phosphorylation of pRb causing the cell cycle to arrest in G_1.

S phase

- ☆ Once cells enter S phase, a further set of cyclins and CDKs are required for continued DNA replication.
- ☆ In mammalian cells, cyclin A-CDK2 performs this function.
- ☆ Cyclin A is expressed from S phase through G_2 and M. Cyclin A binds to two different CDKs.

☆ Initially during S phase it is found complexed to CDK2 and during G_2 and M it is complexed to CDC2.

Mitosis

☆ Entry into the final phase of the cell cycle, mitosis, is signalled by the activation of the cyclin B-CDC2 complex.

☆ This complex accumulates during S and G_2 but is kept in the inactive state by phosphorylation of tyrosine 15 and threonine 14 residues, a process regulated by the weel/mikl kinases.

☆ At the end of G_2, the CDC25c phosphatase is stimulated to dephosphorylate these residues thereby activating CDC2.

☆ Cyclin B is located in the cytoplasm during interphase but is translocated to the nucleus at the beginning of mitosis.

☆ Cyclin B-CDC2 plays a major role in controlling the rearrangement of the microtubules in mitosis.

☆ In addition, the complex plays a role in disassembling the nucleus and allowing the cell to round up and divide.

There is one final checkpoint which occurs at the end of metaphase and at this point the correct assembly of the mitotic apparatus and the alignment of chromosomes on the metaphase plate are monitored. Normal cells arrest at this point if there are any defects, whereas in tumour cells abnormalities of spindle formation are found, suggesting that the checkpoint control is lost. Cyclin B is degraded as cells enter into anaphase. Re-establishment of interphase can then be initiated.

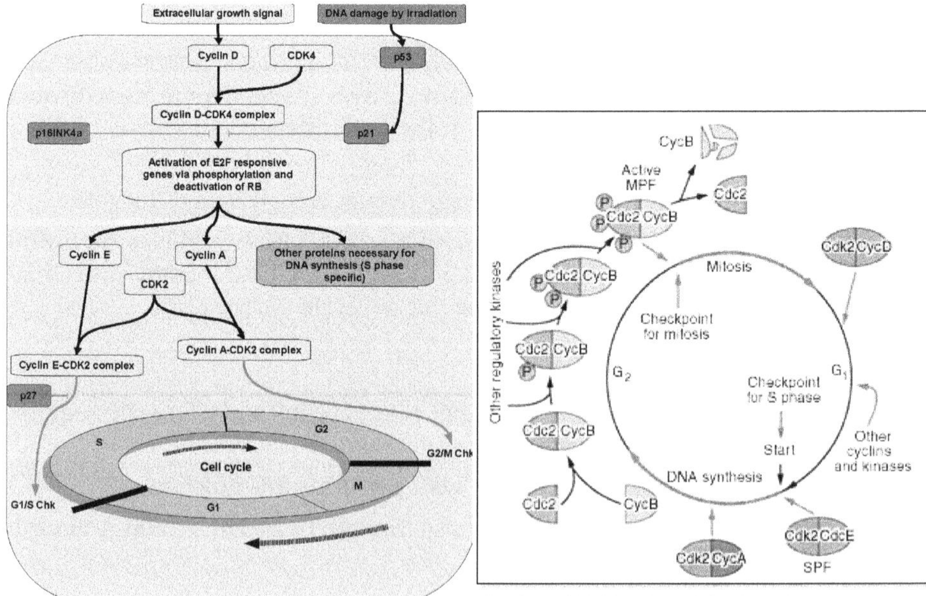

Explain in brief about the cell cycle genes causing disorders?

Cyclin D1

- Cyclin DI is encoded by the CCNDI gene on chromosome11q13 and is identical to the PRAD1 proto-oncogene. Its overexpression is associated with a number of tumours such as esophageal, breast and gastric cancers.
- Chromosome rearrangements involving 11q13 are found in several tumors such as parathyroid adenomas and B-cell lymphomas.

Cyclin D2

- Cyclin D2 is encoded by the CCND2 gene on chromosome 12p13 and has been identified as the VIN1 proto-oncogene and has been shown to be amplified in human colorectal cancer.

Cyclin A

- Cyclin A was one of the first cyclins to be implicated in tumour development.
- The cyclin A gene has been found to be the site of integration of HBV in a case of hepatocellular carcinoma resulting in the production of a chimeric protein absent of the 'destruction' box necessary for the degradation of the cyclin in mitosis.
- However, gene rearrangements involving cyclin A have rarely been observed in other liver tumours and there have been few reports of other abnormalities of cyclin A in tumors from other sites.

CDK4

- CDK4 has also been shown to have a potential role in tumorigenesis as the gene encoding this protein, located on 12q13, has been shown to be amplified in sarcomas and gliomas.

CDKIs

- The CDKIs have also been shown to be involved in tumour development.
- The CDK4/6 inhibitor, p16, acts by competing directly with cyclin D, thereby preventing these kinases from phosphorylating the Rb protein.
- p16 gene aberrations are frequently found in a wide range of cancers. p16 mutations occur at high frequencies in biliary tract and esophageal cancers.
- Homozygous deletions at the p16 locus occur commonly in gliomas, nasopharyngeal carcinomas, acute lymphocytic leukaemias, sarcomas and bladder and ovarian tumours.
- Pancreatic, head and neck and non-small-cell lung carcinomas sustain both p16 deletions and mutations.

2

Molecular Genetics and Gene Expression

Define gene? And explain in brief about evolutionary concept of gene?

Classical definition of gene: Gene is the unit of function (one gene specifies one character), recombination and mutation.

Modern definition of gene: Gene may be defined as the unit of function, *i.e.* the sequence of DNA, that specifies one polypeptide or one RNA molecule; this unit includes the 5′- (upstream) and 3′- (downstream) noncoding regulatory sequences as well as introns of the gene. Such a gene can be identified by complementation test.

Summary of the Evolutionary Concept of Gene

Functional Definition of Gene

- ☆ G.J. Mendel (1866) - A 'unit factor' that controls one specific phenotypic trait (one gene-one character)
- ☆ Sir A.E. Garrod (1902) - One mutant gene-one mutant block
- ☆ G.W. Beadle and E.L. Tatum (1940) - One gene specifies one enzyme (one gene-one enzyme hypothesis)
- ☆ U.M. Ingram (1957) - One gene encodes one polypeptide (one gene-one polypeptide hypothesis)
- ☆ C. Yanofski, and co-workers, and A. Sarabhai and coworkers (1960) - Gene is a continuous sequence of nucleotides specifying a collinear sequence of amino acids in a polypeptide. (Difficulty: Overlapping genes, gene-within gene, interrupted genes, antibody gene segemnts)

Recombination Test

- Before 1940 - Recombination occurs between, but not within genes (gene is the unit of recombination).
- C. Oliver (1940) - Intragenic recombination shown to occur in the *lozenge* gene of *Drosophila.*

Complementation Test (Operational Definition of Gene)

- E.B. Lewis (Early 1940s) - Complementation test for functional allelism in *Drosophila:* later used by S. Benzer to analyse the r^{II} locus of T_4 bacteriophage.

Answer the following questions.

1. **Define the genome?**
2. **Explain in brief about the organization of prokaryotic and eukaryotic genome?**
3. **Eukaryotic DNA is more complex than prokaryotic DNA. Explain with suitable example?**

Genome

H. Winkler (1920) coined the term genome, defined as "the complete set of chromosomal and extra chromosomal genes of an organism" or the total genetic content contained in a haploid set of chromosomes in eukaryotes, in a single chromosome in bacteria or in the DNA or RNA of viruses.

Every organism possess a genome that consists the biological information needed to construct and maintain a living example of that organism. The organization of genome varies from prokaryotes to eukaryotes.

Prokaryotic Genome

A typical prokaryote genome is contained a single, circular DNA molecule in the cytosol called the nucleoid, which lacks a nuclear envelope, in contrast to the multiple linear, compact, highly organized chromosomes found in eukaryotic cells.

- In addition to this, many important genes are stored in separate circular DNA structures called plasmids.
- Majority of the prokaryote genome is a single, circular DNA molecule but *Streptomyces coelicolor* and *Agrobacterium tumefaciens* DNA is linear molecules have free ends.
- Another variation in some prokaryotes is the presence of multipartite genomes *i.e.* genomes that are divided in to two or more DNA molecules. It leads to problem in distinguishing a genuine part of the genome from a plasmid.
- In general, prokaryotic genes are shorter than their eukaryotic counterparts, the average length of a bacterial gene being about two-thirds of a eukaryotic gene, even after the introns have been removed.

- ☆ The noncoding DNA in the genome accounts for only 11 per cent of the total and it is distributed around the genome in small segments.
- ☆ There are no introns in prokaryotic genome.
- ☆ Discontinuous genes are virtually absent in prokaryotes, the few exceptions occurring normally among the archea.
- ☆ The infrequency of repetitive sequences *i.e.* most of the prokaryotic genome do not have anything equalent to the high copy number genome of wide repeat families found in eukaryotic genomes.
- ☆ One characteristic feature of prokaryotic genome is presence of operons. An operon is group of genes that are located adjacent to one another in the genome with perhaps just one or two nucleotide between the end of one gene and the start unit. *e.g.* Lactose operon in *E. coli*

Organization of DNA in Prokaryotes

The organization of DNA in the nucleoid is supercoiled. Supercoiling is an ideal way to package a circular molecule in to a small space.

- ☆ The supercoiling is generated and controlled by two enzymes, DNA gyrase and DNA topoisomerase I.
- ☆ DNA attached to protein core from which 40-50 supercoiled loops radiate out in to the cell.
- ☆ Each loop contains approximately 100 kb of supercoiled DNA, the amount of DNA that becomes unwound after a single break.
- ☆ The protein component of the nucleoid includes DNA gyrase and DNA topoisomerase I, the two enzymes that are primarily responsible for maintaining the supercoiled state.
- ☆ And a set of at least four proteins have a more specific role in packaging the bacterial DNA. *e.g.* HU proteins (The HU protein is a small, basic, heat-stable DNA-binding protein that is well-conserved in prokaryotes and is associated with the bacterial nucleoid).
- ☆ Bacteria lack the histone proteins that are found bound to DNA in eukaryotic genome.

Eukaryotic Genome

Eukaryotic genome is rather complex and is enclosed by a typical nuclear membrane which forms a true nucleus.

- ☆ Eukaryotes have 2-3 genomes like nuclear genome, mitochondrial genome and plastid genome.
- ☆ The nuclear genome is split into a set of linear DNA molecules each contained in a chromosome. The only variability at this level of eukaryotic genome structure lies with chromosome number.

- ☆ The eukaryotic genome is distributed in several linear chromosomes and is made up of DNA and proteins are complexed together to form nucleoprotein fibres called chromatin. The chromatin fibres coil and fold to form the chromosome.
- ☆ Chromatin is composed of DNA (30-40 per cent), RNA (1-10 per cent) and proteins (50-60 per cent).
- ☆ These constituents vary in different organisms and even in the different tissues of the same species, even in the same cell.
- ☆ The proportion of DNA, RNA and proteins vary with the stage of cell cycle.

DNA

Three broad classes of DNA *i.e.* single copy protein coding genes and repetitive sequences are scattered throughout the chromosome, chromosomal banding patterns reflect the levels of compartmentalization of the DNA.

- ☆ Single copy protein encoding genes (unique DNA) are the sequences which are present in a single copy, in each genome.
- ☆ Repetitive DNA elements can be divided in to two major groups, distinguished by their genome organization and localization on the chromosomes, tandem repeats and interspersed repeats.

Tandem Repeats

These repeats are also called as satellite DNA because fragments containing tendemly repeated sequence form satellite bands when genomic DNA is fractioned by density gradient centrifugation.

- ☆ Tandem repeats are very large arrays of tandem repeating, non-coding DNA and the length of repeat ranges from 5-20 bp depending on species.
- ☆ DNA elements arranged in tandem arrays include different types of satellite DNAs, the telomeric repeat and the rDNA.
- ☆ It is the main structural constituent of heterochromatin.
- ☆ It is broadly categorized into either minisatellites or microsatellites, which serve as useful DNA markers such as VNTRs/SSRs/STRs.

Interspersed Repeats

It refers to the distribution of individual repeat unit around the genome in random fashion. Repeat units are 100-1000 base pairs long.

- ☆ Copies are similar but not identical to each other.
- ☆ Dispersed repetitive DNA elements are scattered throughout the genome, interspersed with other sequences and distributed along the chromosome.
- ☆ Mobile elements in interspersed repeats are DNA transposable elements and retro elements and other dispersed repeats.

RNA

RNA constitutes a very small proportion of chromatin; it is 3.5 per cent in human, 0.3 per cent in cow and 10 per cent in pea.

- ☆ But apart from these, a special class of RNAs called "chromosomal RNA" is associated with the chromosomes *i.e.* small nucleolar RNA (snoRNA), small nuclear RNA (snRNA) and micro RNA (miRNA).
- ☆ Chromosomal RNA constitutes about 5 per cent of the total chromosome weight.
- ☆ These RNAs are small molecules containing 40 to 60 nucleotides.
- ☆ These RNAs mainly involved in the structural organization of chromatin fibres and gene regulation.

Protein

Proteins constitute more than half of the total mass of chromosomes.

- ☆ They belong to two classes: Histone or basic proteins and non histone proteins.
- ☆ Another class of protein called "protamines" are found associated with chromosomes in sperms of certain animals.

Histone Proteins

Histones are basic proteins or acid soluble proteins and have a not positive charge.

- ☆ These proteins are of 5 types namely H1, H2A, H2B, H3 and H4.

Non-histone Proteins

Non histone proteins occur in much lower proportion than histones, and their proportion in the total chromosome mass varies considerably in the different organisms.

- ☆ Nonhistone protein consists of various enzymes involved in different functions. *e.g.* DNA polymerase, RNA polymerase, nucleases, polynucleotide ligase, DNA methylase, proteases, histone methylases, histone acetylases, histone deacetylases and histone kinases *etc.*
- ☆ Apart from these enzymes, certain nonhistone proteins are found that have high electrophoretic mobility. These are called high mobility group (HMG) proteins.

Organization of DNA in Eukaryotes

In eukaryotes, very long DNA molecules are packaged in to chromosomes of much smaller size. This organization of DNA was explained by two methods *i.e.* Folded fibre model and Nucleosome – Solenoid model. However, nucleosome – solenoid model of DNA organization is most widely accepted model and it was proposed by Korenberg and Thomas in 1974.

- ✰ Due to solenoid coiling of nucleosome containing fibre, three levels of condensation are required to package 103 – 105 μm of DNA in eukaryotic chromosomes in to metaphase structure of few microns long.
- ✰ The three levels of organization of DNA are
  - i. The first level of condensation involves packaging DNA as a negative super coil in to nucleosome, to produce the 10 nm diameter interphase chromatin fibre. This clearly involves an octamer of histone molecules, two each of histones H_2a, H_2b, H_3 and H_4. Each nucleosome has 6 nm height and 11 nm diameter.
  - ii. This second level of condensation involves the additional folding of this 10 nm nucleosome fibre to produce 30 nm chromatin fibre characteristic mitotic and meiotic chromosomes. Histone H_1 is involved in this super coiling of 10 nm nucleosome fibre to produce 30 nm chromatin fibre. This process is called as solenoid formation.
  - iii. Finally, non histone chromosomal proteins forms scaffold that is involved in condensing the 30 nm chromatin fibre in to highly packed metaphase chromosome of 84 nm diameter.

Eukaryotic DNA is more Complex than Prokaryotic DNA

The chromosomes of eukaryotes are more complex than prokaryotes due to the presence of repetitive base sequences in the haploid chromosome complement rather than unique (non repeated) base sequence.

This genome complexity can be studied by

- i. DNA density gradient analysis
- ii. DNA renaturation kinetics

DNA Density Gradient Analysis

- ✰ When DNA of a prokaryote (*e.g. E. coli*) is isolated, fragmented, and centrifuged at high speeds for long periods of time in a 6M Cesium chloride (CsCl) solution, the DNA will form a single band in the centrifugation tube at the position where the CsCl density is equal to the density of CsCl solution.
- ✰ While the centrifugation of DNAs from eukaryotes to equilibrium conditions in such CsCl solutions usually reveals the presence of one large main band of DNA and one to several small bands.
- ✰ These small bands of DNA are called satellite band and the DNA in these bands referred to as satellite DNA.
- ✰ *e.g.* The genome of *Drosophila virilise*, contains three distinct satellite DNAs, each composed of a repeating sequence of seven base pairs.
- ✰ The bands mainly form at a position where the CsCl density is equal to the density of DNA containing about 50 percent A:T and 50 percent G:C base pairs.

- The extra hydrogen bond in a G:C base pair results in tighter association between the bases and thus a higher density than for A:T base pairs.
- Thus DNA density increases with increasing G:C content, this intern represents the repetitive DNA.

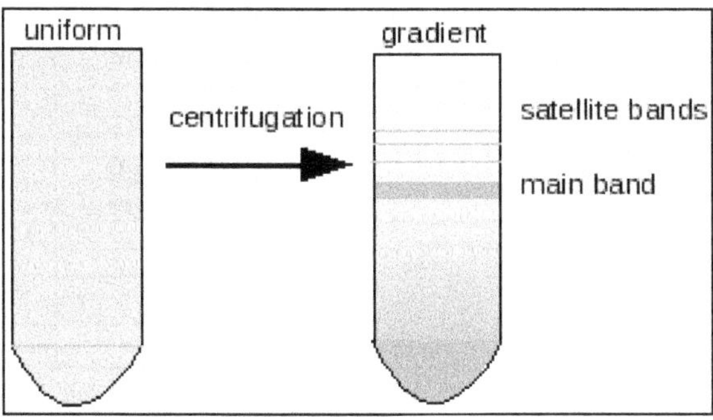

DNA Renaturation Kinetics

In this, DNA complexity can be studied by C_0t analysis technique. It is a biochemical technique that measures how much repetitive DNA is in a genome. It mainly involves.

i. Heating a sample of genomic DNA until it denatures into the single standard form *i.e.* Denaturation.

ii. Various concentrations of the DNA are incubated for varying times at a temperature allowing reassociation *i.e.* Renaturation.

iii. And then measuring and plotting of the fraction of the DNA which is not reassociated in each sample.

- Renaturation rate in a denatured sample of DNA is mainly a function of DNA concentration and time allowed for renaturation.
- These two parameters are expressed as C_0t (initial concentration of single stranded DNA (C_0) × time (t)) which is estimated as moles per litre × seconds as follows.

$$C/C_0 = 1/1 + K.\, C_0t$$

Where C_0 and C are the concentrations in moles of nucleotides per litre (Concentration of single stranded DNA) at the initial time t_0 and after reassociation has progressed for time t, and K is a reassociation rate constant.

- The rate of renaturation of a given DNA segment is proportional to the number of copies of that segment present in a unit DNA solution *i.e.* the more the number of copies, the fast would be renaturation.
- A useful parameter is obtained as $C_0t1/2$, *i.e.*, the value required for half reassociation *i.e.* $C_0t1/2$ is the product of the concentration and time required for reassociation to proceed half way.

☆ A greater value of C_ot implies a slower reaction. The value of $C_ot1/2$ is directly related to the amount of DNA in the genome *i.e.*, the greater the amount of DNA in a genome, the greater will be the value of $C_ot1/2$.

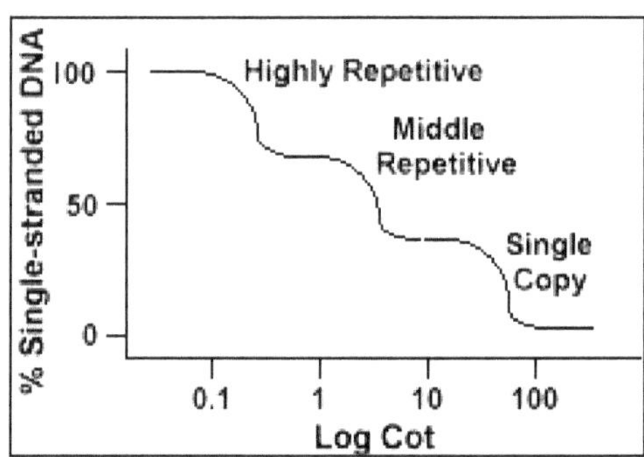

☆ Highly repetitive DNA generally has a $C_ot1/2 < 10^{-2}$. It accounts for 1 per cent to >15 per cent of the genome, with most genome in the 10-15 per cent range. It is mainly located near centromeres, near telomeres and is associated with he7terochromatin. In some organisms it is may also be interspersed with gene- containing regions.

☆ Moderately repetitive DNA consists of families of related sequences, not necessarily identical to one another. The stringency of reannealing allows some mismatched base pairs to form. Thus, retained moderately repetitive DNA melts with a lower T_m than renaturated unique DNA or native DNA.

☆ And also the proportion of double stranded DNA in the sample which is undergoing renaturation can be measured by any of the following methods.

i. Hydroxylapatite (HAP) is used to distinguish single stranded from double stranded DNA. All DNA is bound at low salt concentrations. At intermediate salt concentrations, partially or completely double stranded DNA remains bound to HAP, while single stranded DNA elutes.

ii. Single stranded DNA has a higher optical density (OD) at a specific wavelength than does double strand DNA. Therefore, OD of a renaturating DNA sample would decline with time. And the change in OD may be used to estimate the amount of double stranded DNA.

iii. The sample may be treated with specific nuclease (*e.g.* S1 nuclease), which digests only single stranded DNA. The DNA left in the sample after digestion will be double stranded, and its quantity can be estimated.

Write the differences between single copy genes and satellite DNA? And explain the role of repetitive DNA in genome function?

Sl.No.	Single Copy Genes	Satellite DNA (Highly Repetitive Sequences)
1.	A single copy gene has one locatable region on a DNA molecule	Satellite DNA consists of highly repetitive sequences that can repeat up to 100,000 times in various places on a DNA molecule.
2.	These genes can make up to 1-2 per cent of the human genome	These sequences constitutes more than 5 per cent of the human genome
3.	These genes corresponds to a unit of inheritance *i.e.*, a protein	These sequences are not involved with inheritanoo
4.	These genes transcribed to make RNA, which intern is translated to make a protein.	These sequences are not transcribed
5.	These genes are usually thousands of base pairs in length	These sequences are typically between 5 and 300 base pairs per repeat
6.	These genes are less and useful for DNA profiling	These sequences have high rate of mutation making it useful for DNA profiling

Role of Repetitive DNA in Genome Function

1. DNA is data storage medium, and genomes function as computational information organelles.
2. Generic repeated signals in the DNA are necessary to format expression of unique coding sequence files and to organize additional functions essential for genome replication and accurate transmission to progeny cells.
3. Repetitive DNA elements such as retrotransposons provide the physical basis for integrating different regions of the genome and for coordinating interdependent aspects of genome function like packaging, expression and transmission.
4. Cooperative interactions between repeated DNA elements and integrated protein domains are essential to the formation of nucleoprotein complexes that carryout basic genome operations.
5. The distribution of repetitive signals confers characteristic genome system architecture independent of coding sequence content. Different genome system architectures can have distinct transmission and expression properties even with the same coding sequences.
6. Recognition by repeat derived small interfering siRNA molecules provides a mechanistic basis for analogous chromatin formatting by repetitive arrays with different sequence contents.
7. Repetitive DNA (Microsatellites) has very high mutation rates (where a 'mutation' means a change in repeat number). Thus they are often variable within a population and useful for population genetics. This property also makes them useful for "DNA fingerprinting".

What is meant by C- value? And explain in brief about the C- value paradox?

C- value

C- value means "constant" (or "characteristic") value of haploid DNA content per nucleus, typically measured in pictograms. The C- value of an organism is the amount of DNA in the organism's genome.

Thomas coined the term C-value paradox to denote the unexpected lack of relationship between the presumed complexity of an organism and its C- value *i.e.* lack of correlation between genome size and genetic complexity.

There are two types of paradox were present, which are

i. N- Value paradox: Complexity does not correlate with gene number.

ii. K- Value paradox: Complexity does not correlate with chromosome number.

- ☆ 'C' values range from < 106 bp (mycoplasma) to > 1011 bp (some plants and amphibians).
- ☆ *e.g.* Toad xenopus and man have genomes of essentially the same size, though we assume that man is more complex in terms of genetic development.
- ☆ It is not understood why natural selection allows this variation and whether it has evolutionary significance.
- ☆ There are two main reasons for 'C' value variations.

  a. Organization of eukaryotic genes: In eukaryotes the actual coding sequence of a gene is much smaller than the average size of genes.
    - ☆ Due to the presence of 'introns'-intervenal sequences interrupting coding sequences 'exons'.
    - ☆ Such genes are called 'split' genes of interrupted genes.
    - ☆ The size of eukaryotic in RNA is about 1/5th of the size of the gene coding for it.

  b. Repetitive DNA:
    - ☆ The actual number of genes per genome is much lower than the number estimated on the basis of haploid DNA content.
    - ☆ This is mainly due to the presence of 'repetitive DNA'.

Answer the following questions

1. **What are transposable elements? And give a brief description on it?**
2. **What potential role transposable elements have played in genome evolution?**
3. **How transposable elements have helped in functional genomics. Explain?**

Transposable Element

A genetic element that can move from one position to another in a DNA molecule. And the process is called transposition.

- ☆ The movement of transposable elements to a new site in the genome is sponsored by the enzyme transposase.
- ☆ Transposition may due to
  - i. Conservative process, which involves the excision of the sequence from its original position followed by its reinsertion resulting in the changing its position in the genome with increasing its copy number. *e.g.* IS elements, Composite elements, Ac-Ds and P- elements.
  - ii. Replicative process, which results in an increase in copy number because during this process the original element remains in place while a copy is inserted at the new position. *e.g.* Tn3 elements.

Types of Transposons

I. RNA transposons (Transposition via RNA intermediate i.e. Retrotransposition)

a. RNA transposons with long terminal sequences (LTRs)

e.g. Ty1/Copia retroelements or Ty3/gypsy retroelements in yeast.

b. RNA transposons that lack LTRs

e.g. LINES (Long interspersed nuclear elements) and SINES (Short interspersed nuclear elements).

II. DNA transposons (Transposons donot require RNA intermediate)

a. DNA transposons in prokaryotes

e.g. Insertion sequences (IS1, IS186), Composite transposons (Tn10, Tn5 and Tn903), Tn3 type transposons and transposable phages.

b. DNA transposons in eukaryotes

e.g. Ac/Ds elements in maize and P- elements in drosophila

Transposable Elements in Genome Function

The transposable elements have a number of effects on evolution of genome.

- i. The most significant of this is the ability of transposons to initiate recombinant events that lead to genome rearrangements.
- ii. In many cases, the resulting rearrangement will be harmful because important genes will be deleted, but some instances where the result has been beneficial. *e.g.* Recombination between pair of LINE-1 elements causes of the β-globin gene duplication that results in the G and A members of the gene family.
- iii. Movement of transposons from one site to another can also have impact on genome evolution.

e.g. Movement of exons and other gene segments might also be brought about by DNA transposons called Mutator like transposable elements (MULEs), which are found common in plants *e.g.* Rice.

iv. Transposition has also been associated with altered pattern of gene expression.

e.g. The efficiency with which DNA- binding proteins that are attached to upstream regulatory sequences can activate transcription of a gene might be affected if a transposon moves into a new site immediately upstream of the gene.

v. The presence of promoters and/or enhancers within the transposon might also affect transcription by subjecting the adjacent genes to an entirely new regulatory regime.

e.g. Transposon – directed gene expression occurs with the mouse gene *Slp*, which codes for a protein involved in the immune response.

vi. Transposable elements induced mutations are somatically unstable because they revert to wild type and hence reconstitute the expression of mutated gene. The frequent somatic expression of the transposable elements results in a variegated phenotype.

vii. Transposable elements have the hall marks of selfish DNA, there are numerous cases where transposable element components have been co-opted by the host to create new genes or modify gene regulation.

e.g. Epigenetic regulation has been transformed from a process to silence invading transposable elements and viruses in to a key strategy for regulating plant genes.

Transposable Elements in Functional Genomics

Transposable elements are powerful mutagens for functional genomics in plants.

i. Multicopy transposons can be used effectively in plants with large genomes to heavily mutagenize and recover insertions in to genes of interest.

ii. Single copy transposons, on the other hand, allow the use of enhancer and gene trap reporter gene to monitor patterns of gene expression as well as gene disruption.

iii. Libraries of single copy insertions can be sequenced systematically and screened for mutant phenotypes. These libraries represent the most economical method for systematic function search in plant genomes.

iv. In *A. thaliana*, the comparison of gene trap tag sequences with genomic and expressed sequence tag data bases will allow the location of every insertion to be determined with nucleotide precision, allowing their use as tools in positional cloning as well as in wide scale gene disruption.

v. By examining reporter gene expression patterns in viable heterozygotes, the role of essential genes can be assessed in later development even if insertions are homozygous or haploid lethal.

vi. By combining insertions in homologous genes, the function of redundant genes can also be assessed.

Answer the following questions

1. **Define and explain in brief about the exonic and intronic sequence of complexity?**
2. **Define exon shuffling? And explain in brief about the mechanisms of exon shuffling?**
3. **Explain in brief about the functions of introns?**

The sequence of DNA comprises an interrupted gene are divided into two catogeries:

Exons: These are the sequences represented in the mature RNA. By definition, a gene starts and ends with exons that correspond to the 5′ and 3′ ends of the RNA.

Introns: These are the intervening sequences that are removed when the primary transcript is processed to give the mature RNA *i.e.* introns usually do not code for proteins.

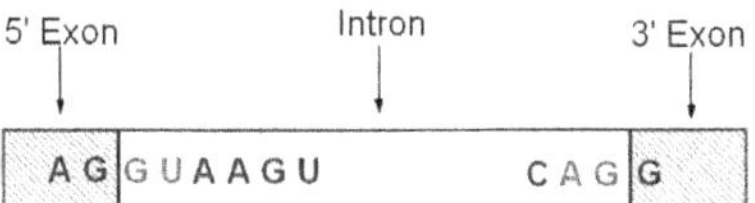

Some of the important illustrations regarding the introns and exons:

1. The exon sequences are in the same order in the gene and in the RNA, but an interrupted gene is longer than its final RNA product.
2. because of the presence of the introns.
3. Introns are removed by the process of RNA splicing, which occur only in cis on an individual RNA molecule.
4. Only mutations in exons can affect protein sequence; however mutations in introns can affect processing of the RNA and therefore prevent production of protein.
5. Mutations that affect the splicing are usually deleterious. The majority are single base substitutions at the junctions between introns and exons. They may cause an exon to be left out of the product, cause an intron to be included to make splicing occur at an aberrant site. The most common result is to introduce a termination codon that result in truncation of the protein sequence.
6. In yeast most genes are uninterrupted and in higher eukaryotes most genes are interrupted and the introns are usually much longer than exons.

7. When a gene is uninterrupted, the restriction map of its DNA corresponds exactly with the map of its mRNA. When a gene possesses an intron, the map at each end of the gene corresponds with the map at each end of the message sequence.
8. Introns can be detected by the presence of additional regions when genes are compared with their RNA products by restriction mapping or electron microscopy.
9. The position of introns is usually conserved when homologous genes are compared between different organisms. The lengths of the corresponding introns may vary greatly.
10. Comparisons of related genes in different species show that the sequences of the corresponding exons are usually conserved but the sequences of the introns are much less well related.
11. Introns evolve much more rapidly than exons because of the lack of selective pressure to produce a protein with a useful sequence.
12. Exons are usually short, typically coding for <100 amino acids. Introns are short in lower eukaryotes, but range up to several 10s of kb in length in higher eukaryotes. The overall length of a gene is determined largely by its introns.

Exon Shuffling

Exon shuffling is a process through which two or more exons from different genes can be brought together ectopically, or the same exon can be duplicated, to create a new exon-intron structure.

The mechanisms of exon shuffling involves crossing over during sexual recombination of parental genomes, transposon mediated shuffling and illegitimate recombination.

Crossing over during sexual recombination of parental genomes:

- ✰ It is the process which behind the evolution of eukaryotes.
- ✰ Since introns are longer than exons most of the cross overs occur in noncoding regions.
- ✰ Introns with large numbers of transposable elements and repeated sequences which promote recombination of non-homologous genes.
- ✰ In addition, mosaic proteins are composed of mobile domains which have spread to different genes during evolution and which are capable of folding themselves. *e.g.* Hemostatic proteins.

Transposon Mediated Shuffling

LINEs, helitron, and LTRs are three types of transposons that can facilitate evolution of new genes.

Long Interspersed Element (LINE)-1

It is a potential mechanism for exon shuffling is the long interspersed element (LINE) -1 mediated 3′ transduction.

- It is transcribed by RNA polymerase II to give an mRNA that code for two proteins: ORF1 and ORF2, which are necessary for transposition.
- Upon transposition, L1 associates with 3′ flanking DNA and carries the non-L1 sequence to a new genomic location.

Helitron

It is nother mechanism through which exon shuffling occurs is by the usage of helitrons.

- Helitron transposons were first discovered during studies of repetitive DNA segments of rice, worm and the thale crest genomes.
- Helitron encoded proteins are composed of a rolling-circle (RC) replication initiator (Rep) and a DNA helicase (Hel) domain.
- The Rep domain is involved in the catalytic reactions for endonuclelytic cleavage, DNA transfer and ligation.
- In addition this domain contains three motifs. The first motif is necessary for DNA binding. The second motif has two histidines and is involved in metal ion binding. Lastly the third motif has two tyrosines and catalyzes DNA cleavage and ligation.
- There are three models of gene capture by Helitrons: the ′read-through″ model 1 (RTM1), the ′read-through″ model 2 (RTM2) and a filler DNA model (FDNA).
- According to the RTM1 model an accidental "malfunction" of the replication terminator at the 3′ end of the Helitron leads to transposition of genomic DNA. It is composed of the read-through Helitron element and its downstream genomic regions, flanked by a random DNA site, serving as a "de novo" RC terminator.
- According to the RTM2 model the 3′ terminus of another Helitron serves as an RC terminator of transposition. This occurs after a malfunction of the RC terminator.
- Lastly in the FDNA model portions of genes or non-coding regions can accidentally serve as templates during repair of dsDNA breaks occurring in helitrons.
- *e.g.* Helitrons in maize cause a constant change of genic and nongenic regions by using transposable elements, leading to diversity among different maize lines.

Long-terminal Repeat (LTR) Retrotransposons

Long-terminal repeat (LTR) retrotransposons are part of another mechanism through which exon shuffling takes place.

- They usually encode two open reading frames (ORF). The first ORF named gag is related to viral structural proteins. The second ORF named pol is a polyprotein composed of an aspartic protease (AP) which cleaves the polyprotein, an Rnase H (RH) which splits the DNR-RNA hybrid, a reverse transcriptase (RT) which produces a cDNA copy of the transposons RNA and a DDE integrase which inserts cDNA into the host's genome.
- Additionally LTR retrotransponsons are classified into five subfamilies: Ty1/copia, Ty3/gypsy, Bel/Pao, retroviruses and endogenous retroviruses.
- Retrotransponsons synthesize a cDNA copy based on the RNA strand using a reverse transcriptase related to retroviral RT.
- The cDNA copy is then inserted into new genomic positions to form a retrogene.
- *e.g.* this mechanism has been proven to be important in gene evolution of rice and other grass species through exon shuffling.

Illegitimate Recombination

Illegitimate recombination (IR) is the recombination between short homologous sequences or nonhomologous sequences. There are two classes of IR:

- The first corresponds to errors of enzymes which cut and join DNA (*i.e.*, DNases.)
- The second class of IR corresponds to the recombination of short homologous sequences which are not recognized by the previously mentioned enzymes.

Functions of Introns

The functions of introns are mainly associated with the life span of introns, which can be divided into five phases.

- The first phase is the genomic intron, which is the DNA sequence of the introns.
- The second phase is the transcribed intron, which is the phase in which the intron is under active transcription.
- The third phase is the intron, in which the spliceosome is assembled on the intron and is actively excising it.
- The fourth phase is the excised intron, which is the intronic RNA sequence released upon the completion of the splicing reaction.
- The final phase is the exon junction complex (EJC) - harbouring transcript, which is the mature mRNA in which the location of exon-exon junctions is marked by the EJC.

Functions Associated with the Genomic Intron

At this level, introns have a potential to serve as repositories of cis elements, participating in the regulation of transcription, and genome organization.

- Transcription initiation – the effect on the expression is especially strong for specific intron, implying that it is its sequence, rather than splicing per se, that underlies the function.
- *e.g.* Intron of the shrunken-1 (*sh1*) locus in maize increased expression at least 10 times more efficiently than other maize introns.
- Transcription termination – intronic sequence elements regulate 3′ end processing in a splicing independent manner *i.e.* removal of intron or its replacement by other introns substantially reduces the efficiency of the 3′ end formation. *e.g.* Intron of the human β- globin gene.
- Genome organization – the ability of nucleosomes to form in some genes is severely perturbed when their introns are deleterious.

Functions Associated with Transcribed Introns

- This unique function of introns mainly associated with the fact that they are transcribed, regardless of their sequence content, or their position, or of the fact that they are later excised from the pre-mRNA.
- Many introns, therefore requires minute, hours, and even days to transcribe.
- This raises the intriguing possibility that introns may serve as tools to orchestrate time delays between activation of a gene, and the appearance of its protein product.

Functions Associated with Spliced Introns

- Transcription regulation – splicing of most of the introns occurs concomitantly with transcription, and that these two cellular processes are strongly coupled mainly through the carboxyl terminal domain (CTD) of RNA polymerase II.
- Alternative splicing – Introns not only passively allow for alternative splicing because of their more existence, but also actively regulate splicing by hosting splicing regulatory elements (SREs).

Functions Associated with Excised Introns

- Once an intron had been excised, it typically becomes a part of post-splicing complexes that lead to efficient debranching and degradation.
- But when an RNA gene is embedded within the intron, it is expressed upon intron removal, and outlives its intronic host.
- Many families of non-coding RNAs (ncRNAs) have been characterised, such as microRNAs (miRNAs), small nucleolar RNAs (snoRNAs), small-interfering RNAs (siRNAs) and various long noncoding RNAs (lncRNAs).

Functions Associated with EJC – Harbouring

EJC participates in a range of mRNA- related cellular processes.

- ✰ Nonsense – mediated decay (NMD) is a eukaryotic surveillance mechanism that selectively degrades mRNAs harbouring premature termination codons (PTCs).
- ✰ The link between splicing and export of mature mRNA from nucleus to the cytoplasm is presumably caused by the fact that spliceosome assembly on the pre-mRNA facilitates the recruitment of export factors. It can be done directly by the spliceosome, or by the EJC that is deposited near the exon - exon junction.
- ✰ mRNA localization is achieved with the help of a diverse family of shuttle proteins, so EJC plays an important role in recruiting shuttle proteins to the mRNA.

Intron-Positional Conservation

Intron position refers as the point of intron insertion along the mRNA.

- ✰ The level of conservation of intron position may be correlated with the functional importance of the intron.
- ✰ If an intron becomes associated with a function of any type, its chances to be lost will decrease.
- ✰ Analysis of intron – exon structure by comparing of their respective intron position will directly provide the gene architecture of orthologous genes.

Answer the following questions.

1. **Define the split gene?**
2. **How and when the split genes were discovered?**
3. **Explain in brief about the evolutionary pattern of Interrupted genes?**
4. **Discuss in brief about the mechanism of production of mRNA from a split gene?**
5. **Explain in brief about the over lapping genes and pseudo genes?**

Split gene is a eukaryotic gene in which the coding sequence is divided in to two or more exons that are interrupted by a number of noncoding intervening sequences (introns). It is also known as interrupted gene.

- ✰ The split or interrupted gene may include nuclear genes for proteins, nuclear genes coding for rRNA, nuclear genes coding for tRNA, mitochondrial genes in yeast, chloroplast genes in a wide variety of plants, genes in archaebacteria and genes in bacteriophages of *E.coli*. However, they are entirely absent in eubacterial genomes.

Discovery of Split Genes

The discovery of split genes was made in 1977 by several groups in a variety of materials which includes:

i. Two research groups separately neaded by Phillip A. Sharp and Richard J. Roberts studied genes of Adenovirus-2

ii. Research groups of D.S. Hogness, I.B. David and N. Dvidson studied genes for 28S rRNA in drosophila.

iii. Research groups of P. Chambon, P. Leder and R. A. Flavell studied β globin genes, ovalbumin genes and tRNA genes.

- In all these cases the genes were found to be interrupted by intervening sequences.
- However, the credit for discovery of split genes goes to Phillip Sharp and Richard Roberts, who won the Nobel Prize for medicine for their work on split genes in 1993.

Sharp and Roberts Experiment Summary

- They wanted to know the site of DNA sequence, where exactly the making of mRNA starts.
- So they need to locate the corresponding DNA sequence, and identify the DNA sequence that precedes the promoter sequence.
- To accomplish this, they developed a technique that allowed them to catch short sequences from the very start of the Adenovirus-2 mRNAs.
- They were expecting to find 15-20 different promoter sequences which code for the different mRNAs.
- But they found that there is only one sequence for all the mRNAs and that the main parts of the mRNA were encoded a long way apart from its very start *i.e.* a single mRNA molecule correspond to no less than four well separated regions in the DNA molecule.
- Then they concluded that genes present in multiple, well separated strands in the DNA molecule *i.e.* split genes.
- They also analysed the hybrids of late mRNA of Adenovirus-2 with the adenovirus genomic DNA.
- When these mRNA- DNA hybrids were examined under electron microscope, the adjoining sequences of mRNA were found to hybridize with discontinuous stretches of genomic DNA of Adenovirus-2.
- The intervening DNA sequences were observed as loops and the phenomenon was later described as R- looping.

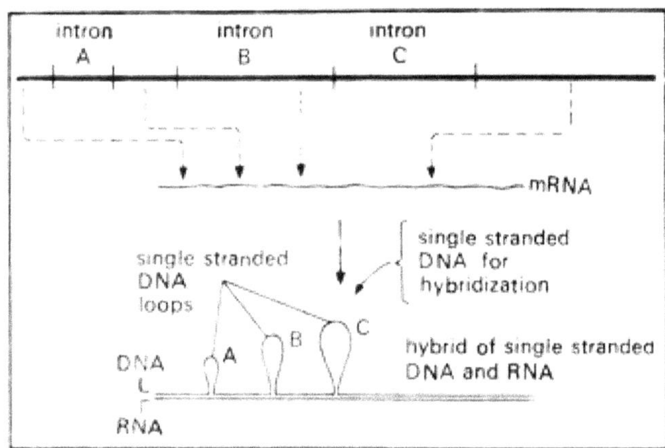

Evolution of Interrupted Genes

The manner in which interrupted genes evolved is not clear.

1. **Introns early model:** The current proteins have evolved by combining over a period of time, together ancestral proteins that were originally separate and encoded by different genes.
   - ✰ These genes were brought together in a sequence, in a manner that the sequences flanking the genes were retained as introns, while the sequences encoding the modules formed to the exons.
   - ✰ Primitive state and the source of non-interrupted genes.
2. **Introns late model:** The non-interrupted state is primitive and introns were subsequently inserted to create the split gene.
   - ✰ The significance of split organization of eukaryotic genes is not clear.
   - ✰ Introns are relics of evolutionary processes that brought together different ancestral genes to form new larger genes.
   - ✰ It is also possible that at least some introns have been introduced within certain exons during the evolution.
   - ✰ Introns may also provide for increased recombination rates between exons of a gene and thus be of some significance in the creation of genetic variation.
   - ✰ Some introns may code for enzymes involved in the processing of hnRNA.
   - ✰ Thus different introns may play different role; but some introns may have no known biological function.

Mechanism of Production of mRNA from Split Genes

- ✰ In split genes, the transcription of DNA leads to production of mRNA having only exonic sequences and the sequence representing introns being entirely absent.

- ☆ The mechanism of production of mRNA from DNA sequences having intervening sequences called introns was a matter of discussion during 1978 and 1979.
- ☆ Later it was discovered that both exons and introns are first transcribed and that this primary transcript is then modified.
- ☆ The sequences corresponding to introns are removed and the sequences corresponding to exons are joined together in correct order to give rise to mRNA.
- ☆ And also the order of exons on DNA is the same as the order in which they are found in processed mRNA.
- ☆ Diagrammatic representation of a split gene having three exons and two introns and its relationship with mRNA was given below (Redrawn from Science. Vol. 204, 1979).

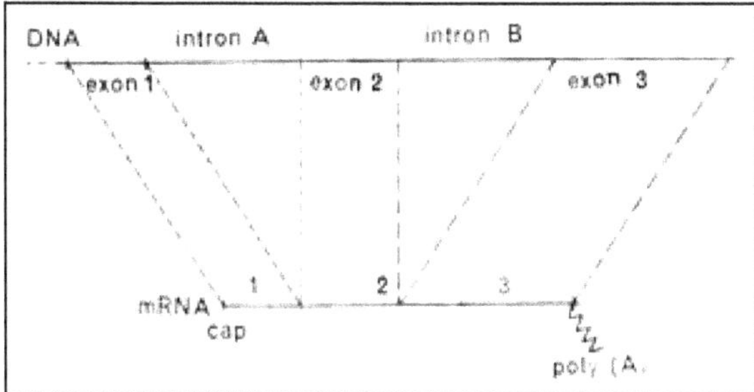

Overlapping Genes

An overlapping gene is a gene whose expressible nucleotide sequence partially overlaps with the expressible nucleotide sequence of another gene.

PROTEIN A:	– – Gly – Lys – STOP
DNA:	. . . G G A A A A T G A G A A A A . . .
PROTEIN K:	– – – Glu – Asn – Glu – Lys – – – –
PROTEIN C:	Met – Arg – Lys – – – –

- ☆ Overlapping genes uses the two different reading frames of a single DNA sequence encode two different proteins.
- ☆ Key feature of these overlapping genes is presence of initiation of triplet TAC (specifying AUG codon in mRNA) within coding sequence of another gene.

  e.g. Bacteriophage/x174, SV40 and MS2 phage.

Pseudo Genes

Pseudo genes are related, often very similar to functional genes but their sequences cannot be translated in to proteins.

- ☆ Pseudo genes are non functional gene copies due to mutations.
- ☆ Pseudo genes represent the inactive versions of currently active genes.
- ☆ They have lost their protein coding ability and have stop codons in middle of gene.
- ☆ Pseudo genes may arise by tandem duplications or accumulating mutations, and are usually recognizable by a lack of an open reading frame.
- ☆ Pseudo genes prevent the transcription, RNA splicing and translation due to frequent occurrence of termination codon. *e.g.* Rabbit pseudo gene dikkhaβ_2.
- ☆ There are two types of pseudo genes
  a. Conventional pseudo genes: The gene has been inactivated because of its nucleotide sequence has been changed due to mutations.
  b. Processed pseudo genes: These are produced by abnormal adjunct to gene expression. It is a copy of an mRNA molecule, it does not contain any introns that were present in the parent gene. *e.g.* Mouse dikkhaα_3 globin genes.

Answer the following questions.

1. **Define RNA splicing? And explain in brief about the different types of intron excision from RNA transcripts?**
2. **Explain in brief about the alternative splicing?**
3. **Define RNA editing? And explain how information content of mRNA molecule was altered by editing?**
4. **Explain in brief about the 5′ capping and 3′ poly-adenylation of mRNA?**
5. **Explain in brief about the RNA interference (RNAi) mechanism?**

RNA Splicing

It refers to removal of introns and joining of the exons in their natural order to generate mature mRNAs.

- ☆ The two boundaries between the two exons and the intron lying between them are known as splicing junctions.
- ☆ During splicing the phosphodiester bonds at the two splicing junctions are cleaved and a new phosphodiester bond is formed between the two exons and introns removed from the precursor RNA.

Three distinct patterns of intron splicing:

1. The introns of tRNA precursor are excised by precise endonucleolytic cleavage and ligation reactions catalyzed by splicing endonuclease and ligase activities.

- ✩ A yeast tRNA splicing endonuclease heterotetramer composed of TSEN54, TSEN2, TSEN34 and TSEN15, cleaves pre-tRNA at two sites in the acceptor loop to form a 5′-half tRNA, terminating at a 2′,3′- cyclic phosphodister group, and a 3′- half tRNA, terminating at a 5′ – hydroxyl group, along with a discarded intron.
- ✩ Yest tRNA kinase then phosphorylates the 5′- hydroxyl group using adenosine triphosphate.
- ✩ Yeaset tRNA cyclic phosphodiesterase cleaves the phosphodiester group to form a 2′- phosphorylated 3′ end.
- ✩ Yeast tRNA ligase adds adenosine monophosphate group to the 5′ end of the 3′ – half and joins the two halves together.

2. Introns of rRNA precursors are removed autocatalytically in a unique reaction mediated by the RNA molecule itself *i.e.* no protein and enzymatic activity is involved here.
   - ✩ There are three kinds of self splicing introns, Group I, II and III. Group I and II introns perform splicing similar to the spliceosome without requiring any protein.
   - ✩ The mechanism in which group I introns are spliced is as follows.
     a. 3′ OH of a free guanine nucleoside or a nucleotide cofactor attacks phosphate at the 5′ splice site.
     b. 3′ OH of the 5′ exon becomes a nuclephile and the transesterification results in the joining of the two exons.
   - ✩ The mechanism in which group II introns are spliced is as follows.
     a. The 2′ OH of a specific adenosine in the intron attacks the 5′ splice site, thereby forming a lariat.
     b. The 3′ OH of the 5′ exon triggers the transesterification at the 3′ splice site thereby joining the exons together.
   - ✩ *e.g.* In *Tetranema thermophila*, introns of rRNA removes by without protein catalytic activity, but here splicing mechanism involves a series of phosphodiester bond transfers, with no bond lost or gained in the process.
3. The introns of nuclear pre-mRNA (hnRNA) transcripts are spliced out in two step reactions carried out by complex ribonucleoprotein particle called spliceosomes. These two steps are:
   a. The complete spliceosome participates in the cleavage of phosphodiester linkage at the 5′ intron splice site and the formation of phosphodiester linkage between 5′ guanine of the intron and the conserved adenine near the 3′ end of the intron.
   b. The 3′ intron splice site is cleaved and the two exons are joined by phosphodiester linkage.
      - ✩ Two types of spliceosomes have been identified which contain different SnRNPs (Small nuclear ribonucleoproteins).

☆ The major spliceosome splices introns containing GU at the 5′ splice site and AG at the 3′ splice site. It is composed of the U1, U2, U4, U5 and U6 SnRNPs and is active in the nucleus. Whereas U3 localized in nucleolus and probably involved in formation of ribosomes.

☆ The minor spliceosome is very similar to the major spliceosome, and it has different but functionally analogous SnRNPs for U1, U2, U4 and U6, which are respectively called U11, U12, U4atac. Where U5 SnRNP is same for both major and minor spliceosomes.

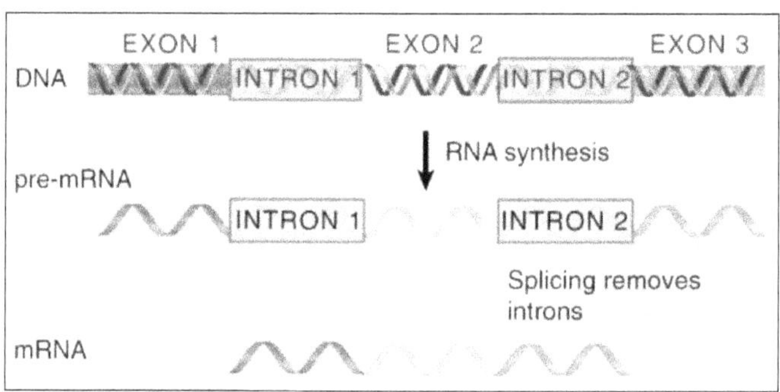

Alternative Splicing

It is a process which creates a range of unique proteins by varying the exon composition of the same mRNA. It happens more than regular splicing.

☆ It makes same gene at same locus codes for different product. So genes of this type do not fit the one gene – one polypeptide hypothesis. *e.g.* Mammalian tropomyosin genes – family of protein isoforms

☆ In this exons can be extended or spliced, and introns can be retained.

☆ The pattern of alternative splicing may differ in following aspects:

i. Variable 5′ splicing site may be connected to a constant 3′ site.

ii. A constant 5′ splicing site may be connected to alternative downstream sites.

iii. Splicing sites are used to add or substitute exons or introns to generate different types of mRNAs.

☆ Alternative splicing of pre-mRNA transcripts is regulated by a system of trans-acting proteins (activators and repressors) that bind to cis-acting sites or elements (enhancers and silencers) on the pre-mRNA transcript itself.

☆ And the effects of regulatory factors are many times position-dependent. *e.g.* A splicing factor that serves as a splicing activator when bound to an intronic enhancer element may serve as a repressor when bound to its splicing element in the content of an exon, and vice versa.

- Alternative splicing will also have implication on sex determination *i.e.* different types of splicing of genes will lead to different sexes discovered in drosophila.

RNA Editing

It is a process through which base sequence of mRNA is changed after transcription at the level of RNA.

- RNA editing is confined to a single base, or it may affect the entire mRNA.
- RNA editing is relatively rare, and common forms of RNA processing like splicing, 5′-capping and 3′-polyadenylation are not usually included as editing.
- RNA editing mainly occurs in two ways
  - i. By changing the structure of individual bases. *e.g.* Apolipoprotein – B (apo-B) genes in intestine of mammals involves C to U conversion. It results in formation of UAA codon, leads to premature chain termination *i.e.* truncation of apolipoprotein.
  - ii. By inserting or deleting uridine monophosphate residues. *e.g.* In mitochondria of trypanosomes, uredine monophosphates are inserted into gene transcripts.
- RNA editing process is mediated by guide RNA (gRNA).
- These guide RNAs serve as templates for editing. *e.g.* 'U's are inserted in the gaps in pre-mRNA molecules opposite the 'A's in the guide RNAs.
- These guide RNAs not only acts as template but also donates uridine monophosphates that are inserted in to the pre-mRNAs by two step mechanism.
  - i. Cleavage of an ester bond in pre-mRNA and formation of an ester bond between the guide RNA and the 3′ portion of the pre-mRNA.
  - ii. Cleavage of the ester bond between the two terminal 'U's of the guide RNA, reforming the pre-mRNA, but with a 'U' inserted at the site of initial cleavage step1.

5' capping and 3' poly-adenylation of mRNA

Eukaryotic gene transcripts usually undergo three major modifications *i.e.* the addition of 7-methyl guanosine caps to 5′ termini, the addition of poly (A) tails to 3′ ends and the excision of non coding intron sequences.

5' capping of mRNA

- After transcription, the 5′ ends of eukaryotic pre-mRNAs are modified by the addition of 7-methyl guanosine (7-MG) caps.
- These 7-MG caps are added when the growing RNA chains are only about 30 nucleotides long.

- ☆ The 7-MG cap contains an unusual 5′-5′ triphosphate linkage and two or more methyl groups.
- ☆ The 7-MG caps are recognized by protein factors involved in the initiation of translation and also help to protect the growing RNA chains from degradation by nucleases.

3'-Poly adenylation of mRNA

- ☆ The 3′ ends of RNA transcripts synthesized by RNA polymeraseII are produced by endonucleolytic cleavage of the primary transcripts rather than by the termination of transcription.
- ☆ After cleavage, the enzyme poly (A) polymerase adds poly (A) tails, tracts of adenosine monophosphate residues about 200 nucleotides long, to the 3′ ends of the transcripts.
- ☆ The addition of poly (A) tails to eukaryotic mRNA is called polyadenylation.
- ☆ The formation of poly (A) tails on transcripts requires a specificity component that recognizes and binds to the AAUAAA sequence, a stimulatory factor that binds to the G-U- rich sequence, an endonuclease, and the poly (A) polymerase.
- ☆ The poly (A) tails of eukaryotic mRNAs enhances their stability and plays an important role in their transport from the nucleus to the cytoplasm.

RNA Interference (RNAi) Mechanism

RNA molecules involved here are two types of 1. Short interfering RNAs (Si RNAs) 2. Micro RNAs (mi RNAs). Steps involved in RNA interference mechanism are

- ☆ A large double stranded RNA molecule is diced in to small, double stranded interfering 21-28 bp long fragments by dicer enzyme.
- ☆ The small interfering RNAs and proteins assemble in to ribonucleoprotein particles
- ☆ The small interfering RNA in ribonucleoprotein particle is unwound to produce an RNA- induced silencing complex (RISC)
- ☆ RISC targets a sequence in a messenger RNA that is complementary to the interfering RNA
- ☆ The RISC's interfering RNA basepairs with its target in the mRNA
- ☆ If perfectly base paired, mRNA is cleaved. It leads to degradation of mRNA. The RISC associated with this RNA interference is called short interfering RNAs.
- ☆ If imperfectly base paired, translation of the mRNA is arrested *i.e.* polypeptide synthesis from the mRNA is repressed. The RISC associated with this RNA interference is called Micro RNAs.

Answer the following questions.

1. **What is promoter? What is role of promoter in gene regulation? And explain in brief about the types of promoters?**
2. **Write the basic differences between structure of eukaryotic and prokaryotic promoters? And explain in brief about the CaMV 35S promoter?**
3. **Explain in brief about the prokaryotic gene regulation system?**
4. **Explain in brief about the eukaryotic gene regulation system?**
5. **Write the basic difference between prokaryotic and eukaryotic gene control system?**

Promoter

Promoter is the DNA segment of an operon or a transcription unit, to which RNA polymerase binds before it initiates transcription.

Role of Promoters in Gene Regulation

- ✰ The promoter region is usually assumed to be the key cis – acting regulatory region that controls the transcription of adjacent coding regions in to mRNA, which is then directly translated in to proteins.
- ✰ DNA sequences within promoters can be identified as binding sites for trans – acting factors, "transcription factors", which may cause activation or repression of transcription.

Types of Promoters

Promoters are of different types according to the intended type of control of gene expression.

i. Constitutive promoters: These promoters direct expression in virtually all tissues and are largely, if not entirely, independent of environmental and developmental factors. Constitutive promoters are usually active across species and even across kingdom.

ii. Synthetic promoters: These promoters comprise consensus DNA sequences of common elements of natural promoter regions.

iii. Inducible promoters: These promoters expressed only under the presence of factors/compounds. Because their expression is normally restricted to certain tissues, so they can also be considered as tissue-specific.

- ✰ Based on nature of the factors that trigger their expression, these are divided in to two groups.
    i. Chemically regulated, where chemical compounds usually not naturally found within plants, switch on promoter activity. *e.g.* Promoters that respond to antibiotics, copper, alcohol and herbicides *etc.*

ii. Physically regulated, where abiotic and external factors such as light, heat, mechanical injury induce promoter activity.

iv. Tissue specific promoters: These promoters operate in particular tissues and at certain developmental stages of plant.

☆ These promoters may be induced by endogenous and exogenous factors, so they may be classified as inducible.

Structure of Eukaryotic and Prokaryotic Promoters

Promoter is the main determinant for the initiation of transcription and modulation of levels and timing of gene expression.

Eukaryotic Promoters

☆ Promoters in eukaryotic organisms like plants, animals comprise multiple elements, some of which are found in nearly all promoters. These include:

a. CAAT box: A consensus sequence close to -80 bp from the start point (+1). It plays an important role in promoter efficiency, by increasing its strength. And it seems to function in either orientation. This box is replaced in plants by a consensus sequence called AGGA box.

b. TATA box: A sequence usually located around 25 bp upstream of the start point. The TATA box tends to be surrounded by GC rich sequences. The TATA box binds RNA polymerase II and a series of transcription factors (TFII x, being 'x' a letter that identifies an individual transcription factor to form an initiation complex).

c. GC box: A sequence rich in guanidine (G) and cytidine (C) nucleotides is usually found in multiple copies in the promoter region, normally surrounding the TATA box.

d. CAP site: A transcription initiation sequence or start point defined as +1, at which the transcription process actually starts.

Conserved Eukaryotic Promoter Elements	*Consensus Sequence*
CAAT box	GGCCAATCT
TATA box	TATAA
GC box	GGGCGG
CAP site	TAC

☆ The transcription process only takes place in the downstream direction, *i.e.* from left (5′) to right (3′).

☆ RNA polymerase II enzyme works in conjunction with other transcription factors that recognize signals embodied in the promoter region.

☆ RNA polymerase II starts its journey at the TATA region where it binds and travels along the DNA until it reaches the CAP site where the actual synthesis of RNA starts.

Prokaryotic Promoters

Promoters of prokaryotic organisms like bacteria have similar elements as the eukaryotic promoters with a few basic differences.

- Prokaryotic promoters contain at least three conserved features defining the region where the RNA polymerase binds:
  a. The start point, defined as +1
  b. The TATA box is located at -10 position to the start point in contrast to the -35 bp in eukaryotic promoters.
  c. The TTGACA sequence, also called the -35 sequence located around 35 bp upstream of the start point.
- The one more common feature of prokaryotic organism is that a promoter serves to initiate the transcription of multiple structural genes that are immediately adjacent to it. This arrangement is called an operon.
- A single transcribed mRNA is translated in to several proteins whose functions are interrelated.
- In operons, promoters have adjacent, juxtraposed, or interspersed regulatory sites to which regulatory proteins bind.
- But in eukaryotic promoters, the regulatory sites are spread out over a longer distance.

CaMV 35S Promoter

It is one of the most widely constitutive promoter for driving the expression of genes in transgenic plants.

- At the beginning of the 1980's, Chua and collaborators at the Rockfeller university isolated the promoter responsible for the transcription of the whole genome of a Cauliflower mosaic virus (CaMV) infecting turnips.
- The promoter was named CaMV 35S promoter (35S promoter) because the coefficient of sedimentation driven by this promoter is 35S.
- The 35S promoter in particular is referred to as the promoter for the full length mRNA of the CaMV genome.
- This is very strong promoter causing higher levels of gene expression in dicot plants. But it is less effective in monocots, especially in cereals.
- The differences in behaviour are probably due to differences in quality and/or quantity of regulatory factors.
- The promoter responsible for the transcription of another part of the genome of CaMV, the CaMV 19S promoter, is also used as constitutive promoter, but is not as widely used as the 35S promoter.

Regulation of Prokaryotic Gene Expression

- Gene expression in prokaryotes is regulated at several different levels: transcription, mRNA processing, mRNA turnover, translation and post translation. However, the regulatory mechanisms with the largest effects on phenotype at the level of transcription.

- ✰ Based on regulation of transcription, the various regulatory mechanisms seem to fit into two categories:
  1. Mechanisms that involve the rapid turn-on and turn-off of gene expression in response to environmental change.
  2. Mechanisms referred to as programmed circuits or cascades of gene expression. In these cases, the sequential expression of genes is genetically pre-programmed, and the genes cannot usually be turned on out of sequence. *e.g.* When a lytic bacteriophage infects a bacterium, the viral genes are expressed in a predetermined sequence, and this sequence is directly correlated with the temporal sequence of gene-product involvement in the reproduction and morphogenesis of the virus.
- ✰ Constitutive, inducible and repressible gene expression:
  1. In prokaryotes, genes that specify housekeeping functions such as rRNAs, tRNAs, and ribosomal proteins are expressed constitutively. Other genes usually are expressed only when their products are needed.
  2. Genes that encode enzymes involved in catabolic pathways often expressed only in the presence of the substrate of the enzymes *i.e.* their expression is inducible.
  3. Genes that encode enzymes involved in anabolic pathways are turned off in the presence of the end product of the pathway *i.e.* their expression is repressible.
- ✰ Positive and negative control of gene expression:
  1. In positive control mechanism, the product of regulatory gene, an activator, is required to turn on the expression of structural genes.
  2. In negative control mechanisms, the product of a regulator gene, a repressor, is necessary to turn off the expression of the structural genes.
  3. Activators and repressors regulate gene expression by binding to sites (operator) adjacent to the promoter of structural genes.
  4. Whether or not the regulator proteins can bind to their binding sites depends on the presence or absence of small effector molecules that form complexes with the regulator proteins.
  5. The effector molecules are called inducers in inducible systems and co-repressors in repressible system.
- ✰ Operons: co-ordinately regulated units of gene expression
  1. In bacteria, genes with related functions frequently occur in co-ordinately regulated units called operons.
  2. Each operon contains a set of contiguous structural genes, a promoter (the binding site for RNA polymerase) and an operator (the binding site for a regulatory protein called a repressor).

3. When a repressor is bound to the operator, RNA polymerase cannot transcribe the structural genes in the operon. When the operator is free of repressor, RNA polymerase can transcribe the operon.

☆ Lactose operon in E. *coli*: Induction and catabolic repression.

1. The *E. coli* lac operon is a negative inducible and catabolic repressible system.
2. The three structural genes in the lac operon are transcribed at high levels only in the presence of lactose and the absence of glucose.
3. In the absence of lactose, the lac repressor binds to the lac operator and prevents RNA polymerase from initiating transcription of the operon.
4. Catabolic repression keeps operons such as lac encoding enzymes involve in catabolism from being induced in the presence of glucose, the preferred energy source.

☆ Tryptophan operon in E. *coli*: Repression and attenuation.

1. The *E. coli* trp operon is a negative repressible system, transcription of five structural genes in the *trp* operon is repressed in the presence of significant concentrations of tryptophan.
2. Operons such as *trp* that encode enzymes involved in amino acid biosynthetic pathways often are controlled by a second regulatory mechanism called attenuation.
3. Attenuation occurs by the premature termination of transcription at a site in the mRNA leader sequence (the sequence 5′ to the coding region) when tryptophan is prevalent in the environment in which the bacteria are growing.

Regulation of Eukaryotic Gene Expression

Gene expression in eukaryotes can be regulated at the several levels like transcriptional, processing, translational and post translation.

☆ However, the genome of eukaryotes normally expressed in controlled fashion *i.e.* spatial and temporal regulation of genes.

i. Spatial regulation of tubulin genes in plants. *e.g.* α and β tubulin genes in *Arabidopsis* are expressed in tissue specific manner.
ii. Temporal regulation of globin genes in animals. *e.g.* α and β globin genes in vertebrates are expressed at different times during development.

☆ Regulation in eukaryotes can occur in the nucleus at either the DNA or RNA level or in the cytoplasm at either the RNA or polypeptide level.

At Transcription Level

☆ Transcription occurs in prokaryotes when negative regulation molecules such as the 'lac' repressor protein have been removed from the vicinity of a gene and positive regulator molecules such as the catabolic activator protein (CAP)/cyclic AMP complex have been bound to it.

- The eukaryotic gene transcription mainly induced by environmental factors such as heat and light and by biological factors like hormones and growth factors.
  i. Temperature. *e.g.* In drosophila, heat shock protein called HSP70 is mediated by heat-shock transcription factor (HSTF).
  ii. Light. *e.g.* RUBISCO activity in plants is mainly induced when plants exposed to light only.
  iii. Hormones. *e.g.* Hormones like estrogen and progesterone play important role in female reproductive cycles while testosterone is a hormone of male differentiation.
- As in prokaryotes, eukaryotic transcriptional regulation is mediated by protein-DNA interactions. Positive and negative regulator proteins bind to specific regions of the DNA and stimulate or inhibit transcription. These proteins are called "transcription factors".
- These factors bind to response elements or more generally to sequences called enhancers located in the vicinity of a gene.
- Enhancer is a cis-acting element that increases the efficiency or activity of any promoter located in its vicinity in either orientation.
- Many eukaryotic transcription factors have characteristic structural motifs that result from association between amino acids within their polypeptide chains. These motifs mainly are:
  i. Zinc finger motif, is a short peptide loop that forms when two cycteines in one part of the polypeptide and two histidines in another part nearby jointly bind a zinc ion and is described as Cys_2/His_2 finger.
  ii. Helix turn helix is a stretch of three short helices of amino acids separated from each other by turns. And in transcription factors, this motif mainly coincides with a highly conserved region of 60 amino acids called the homeodomain. Homeodomain proteins can be either an activator or repressor of transcription.
  iii. Leucine zipper is a stretch of amino acids with a leucine at every seventh position. The region adjacent to the leucine zipper domain is highly basic and it may be involved in DNA binding.
  iv. Helix loop helix, is a stretch of two helical regions of amino acids separated by a non helical loop. It mainly involves in protein dimerization and DNA binding. DNA binding is mainly accomplished by a basic region near this motif.
- After transcription, RNA is processed by capping at its 5′ end, polyadenylation at its 3′ end and intron splicing. However, prokaryotic RNAs do not undergo these terminal and internal modifications.

- This 3′ polyadenylation mainly influence the stability of mRNA. The sequence of the 3′ untranslated region (3′UTR) preceding a poly (A) tail also affect the mRNA stability.

At Translation Level

- Regulation of eukaryotic genome at translation may involves the components of translation like ribosomes, tRNA, mRNA, enzymes aminoacyl transferases and amino acyl – tRNA synthetases.
  - i. Ribosomes from metaphase cells are inactive in translation due to their association with a protein.
  - ii. If the activity of aminoacyl transferase modified, it leads to the alteration of the rate of addition of aminoacids to the growing polypeptide chain.
  - iii. Degeneracy of tRNA *i.e.* single aminoacid is coded by more than one codon, it leads to variation in the type of tRNA for that amino acid.
  - iv. The regulation of aminoacyl - tRNA synthetases enzyme is by tissue specific expression.
- After translation, a single gene may give rise to different, functional polypeptides in different tissues due to a variation in protein processing.
- In addition to above regulation mechanisms at transcription and translation levels, the phenomenon like position effects, controlling elements like Ac-Ds system, paramutation, dosage compensation, genetic imprinting and epigenetic inheritance will influence the gene action in eukaryotes.

Sl.No.	*Prokaryotes*	*Eukaryotes*
1.	As the life span of prokaryotes is very short so it cannot have multiple stages/levels of gene regulation	Eukaryotes have multiple stages/levels of gene regulation
2.	Prokaryotes exhibit lower levels of genome organisation	Eukaryotes exhibit higher level of genome organisation
3.	Cellular and organ level diversity. It is a product of genetic and epigenetic control	Spatial and temporal regulation of genes
4.	Abiotic stresses like nutrients are lacking for survival of prokaryotes *i.e.* nutrients are limiting factors for prokaryotes helps in induction and repressing of expression of genes	Nutrients and temperature are driving forces behind the gene reguation
5.	Operon – group of structurally/functionally related genes that are under common genetic control and transcribed as a single unit (Polycistronic unit)	Operons are not present in eukaryotes but operon like pathways is present, but the transcript is not a polycistronic unit/monocistronic unit.
6.	In prokaryotes, group of genes have same promoter	In eukaryotes, these clusters of genes will have separate promoter sites of themselves.

Explain in brief about the role of following phenomenon in concern with gene expression in eukaryotes?

1. **Paramutation**
2. **Dosage compensation**
3. **Position effects**
4. **Genetic imprinting**
5. **Epigenetic inheritance**

Paramutation

Paramutation is an interaction between two alleles at a single locus, whereby one allele induces a heritable change in the other allele.

- It mainly refers to a change in the function of allele *i.e.* paramutable allele of a structural gene that may last for several generations, and is induced by its presence in the same nucleus with another allele *i.e.* paramutagenic allele, of the same gene.
- Paramutation was first discovered and studied in maize (*Zea mays*) by R.A. Brink in the 1950s.
- Paramutation may share common mechanisms with other epigenetic phenomena, such as gene silencing and genomic imprinting.
- In maize, paramutation mainly due to RNA-directed DNA-methylation. So that *r1* loci in maize, shows weaker expression state adopted by a paramutant allele can range from completely colorless to nearly fully colored kernels.
- The main mechanisms behind the paramutation includes,
  a. The paramutagenic allele causes an increase in the number of copies of heterochromatic sequence called metamere, which is associated with the paramutable allele.
  b. The sequence of paramutable allele changes by a process of gene conversion.
  c. The paramutable allele undergoes methylation under the influence of paramutagenic allele.
  d. The chromatin structure at the paramutable allele may change due to an interaction with the paramutagenic allele.

Dosage Compensation

Dosage Compensation is the equalization of gene expression between the males and females of a species. Because sex chromosomes contain different numbers of genes, different species of organisms have developed different mechanisms to cope with this inequality.

Dosage Compensation Mechanisms

1. Hyperactivation of X – chromosome present in male. *e.g.* Drosophila.
   - ☆ In drosophila, hyperactivation results due to complex proteins, these are male specific lethal proteins.
   - ☆ Five genes code for these proteins are Msl_1, Msl_2, Msl_3, Mle (maleless) and Mof (male absent).
   - ☆ Here RNA also involved and coded by genes rox1 and rox2
2. Inactivation of one X – chromosome present in female. *e.g.* Cat and mice.
   - ☆ It was first observed by Mary Lyon in embryonic cells of cat and mice.
   - ☆ It affects the fur colour of cat and mice and it is controlled by two alleles *i.e.* B (black pigmentation) and b (orange pigmentation).
   - ☆ Here inactivation is random process, so equal chance of inactivation of two chromosomes. So it shows mosaicism. This type of phenotype is tortoise shell phenotype.
   - ☆ Murraya bar also observed inactive body called bar body in females only not in males.
   - ☆ Molecular analysis of bar body shows that methyl groups attached to this inactivated X- chromosome.
   - ☆ Inactivation of X – chromosome will started at any point in chromosome is called X – chromosome inactivation centre (XIC). It is due to binding of X – chromosome inactivation specific transcript (XIST) to that site.
3. Hyperactivation *i.e.* partial repression of X – linked genes in the somatic cells of hermaphrodites. *e.g.* Nematode - *Caenorabdities elegens*

Position Effects

Position effect is the effect on the expression of a gene when its location in a chromosome is changed, often by translocation.

Position effects mainly classified in to three catogeries.

1. Variegated position effects: It mainly involves heterochromatinization *i.e.* when euchromatin segment placed close to heterochromatin, it changes in to heterochromatin.

   e.g. in drosophila, the gene *w+* is located in 3C2 band of the X chromosome, and produces dull red eye colour. When paracentric inversion places the 3C2 band near or in the heterochromatic region close to centromere, it produces mottled eye colour.
2. Cis-trans position effects:
3. Duplication position effects: It is mainly due to unequal crossing over. *e.g.* Bar eye gene of drosophila.

Genetic Imprinting

It is the epigenetic phenomenon by which certain genes are expressed in a parent-of-origin-specific manner.

- ☆ It mainly refers to difference in the behaviour or activity of the alleles of a gene contributed by the two parents of the individual.
- ☆ It mainly involves DNA methylation and histone methylation without altering the genetic sequence.
- ☆ The grouping of imprinted genes within clusters allows them to share common regulatory elements, such as non-codingRNAs and differentially methylated regions (DMRs).
- ☆ When these regulatory elements control the imprinting of one or more genes, they are known as imprinting control regions (ICR).
- ☆ Example for genetic imprinting is gene encoding IGF-II (insulin like growth factor II) in mice. The expression of an allele of IGF-II gene in embryos depends upon the sex of the parent from which it was inherited. This allele inherited from father is expressed while that from mother was not expressed.
- ☆ This differential expression is due to methylation of the IGF-II allele present in eggs, and nonmethylation of that present in the sperms.

Answer the following questions.

1. **Define translation?**
2. **Explain in detail about the different steps involved in protein synthesis?**
3. **Write the basic differences in translation process of prokaryotes and eukaryotes?**

Translation

The process of protein synthesis where amino acids are brought by the tRNA and sequentially arranged based on the codons of the mRNA is known as translation (*i.e.* the nucleotide sequence is translated into the amino acid sequence).

- ☆ The translation process requires mRNA, rRNA, ribosomes, 20 kinds of aminoacids and their specific tRNAs and many translation factors.

The process of translation (protein synthesis) consists of five major steps *viz.*,

1. Activation of aminoacids
2. Transfer of aminoacids to tRNA
3. Chain initiation
4. Chain elongation and
5. Chain termination. Each step is governed by specific enzymes and cofactors.

Differences in Protein Synthesis between Prokaryotes and Eukaryotes

Sl.No.	*Feature*	*Prokaryotes*	*Eukaryotes*
1.	Initiation	☆ Translation initiation involves base-pairing between 16S rRNA and mRNA in the shine-dalgarno consensus sequence ☆ Formyl methionine is incorporated by the initiation codon AUG ☆ Initiation site of mRNA first binds with the 30S subunit of ribosome, which then accepts formyl-methionyl-$tRNA_f^{met}$ complex ☆ The formation of initiation complex requires a number of initiation factors *viz.*, GTP and the factors IF1, IF2 and IF3.	☆ Translation initiation is based on the recognition of the 5'-cap of mRNA by the 40S subunit of ribosomes and by some proteins. ☆ Methionine is incorporated at the initiation codon AUG ☆ Methionyl-$tRNA_i^{met}$ first binds to the 40S subunit, to which mRNA molecule is then attached to form the initation complex. ☆ The formation of initiation complex requires a number of initiation factors *viz.*, GTP and the factors eIF2 – eIF6.
2.	Elongation	☆ In prokaryotes, 15 amino acids are added in one second ☆ Here elongation factor EF-Tu is mainly responsible for binding of aminoacyl-tRNA to the A site of the ribosome by forming EF-Tu-GTP complex. ☆ Inactive form of EF-Tu-GDP can be reactivated by EF-Ts factor by forming EF-Tu-EF-Ts complex. ☆ Translocation of the ribosome along the mRNA is mediated by the factor EF-G.	☆ In eukaryotes, 2 amino acids are added in one second. ☆ The amino-acyl-tRNA is brought to the A site is mediated by the factor eEF1α, which is homologous to EF-Tu ☆ The function of EF-Ts is replaced by Eef-1β in eukaryotes. ☆ Translocation is mediated by the factor eEF2 in eukaryotes.
3.	Termination	In prokaryotes, two classes of releasing factors like RF1 recognize UAA and UAG, while RF2 recognizes UGA and UAA and RF3 provides GTP.	In eukaryotes, a single releasing factor eRF1 functions in termination.

Steps

1. In the cytoplasm, the amino acids are activated in the presence of ATP and linked to their respective tRNAs by a process called charging of tRNA in the presence of an enzyme aminoacyl synthetase.

$$\text{Aminoacid} + \text{ATP} \xrightarrow[\text{Synthetase}]{\text{Aminoacyl} - \text{tRNA}} \text{Aminoacyl - AMP} + \text{Pyrophosphate}$$

2. Thus a number of tRNA molecules, pick up aminoacids freely floating in the cytoplasm and forms aminoacyl-tRNAs.

$$\text{Aminoacyl - AMP} + \text{tRNA} \xrightarrow[\text{Synthetase}]{\text{Aminoacyl} - \text{tRNA}} \text{Aminoacyl - tRNA} + \text{AMP}$$

3. The processed mRNA enters the cytoplasm and binds to ribosomes, which serve as work benches for protein synthesis.
4. The ribosome consists of rRNAs and different proteins. Ribosome contains two subunits; the large subunit and the small subunit.
5. The process of translation starts when an initiating aminoacylated tRNA base pairs with an initation codon of an mRNA molecule that has been located by the small subunit of ribosome.
6. Then the larger subunit joins. Two separate and distinct sites are available in the ribosome to which the tRNAs can bind; A (acceptor or aminoacyl attachment) site and P (peptidyl) site.
7. An aminoacyl-tRNA first attaches to site A (acceptor site or aminoacyl attachment site) the kind of aminoacyl t-RNA being determined by the sequence of mRNA (codon) attached to site A.
8. The peptide bonds are formed between the aminoacids which is catalysed by the enzyme peptidyl transferase.
9. The peptidyl tRNA along with the mRNA codon moves to the P (peptidyl) site making the A site available for the attachment of a new aminoacyl-tRNA.
10. Thus the translation proceeds and at the end a releasing factor binds to the stop codon terminating the translation.
11. The ribosome releases the polypeptide and mRNA and subsequently dissociates into two subunits.
12. Further processing of polypeptide chain into proteins and enzymes is done in the cytoplasm itself and depends upon the bonding properties of the amino acids joined in them.

Answer the following questions.

1. **Define Genetic code?**
2. **Write the properties of genetic code?**
3. **Is genetic code is universal. Discuss it with suitable examples?**

Genetic Code

The rules that determine which triplet of nucleotide codes for which amino acid during protein synthesis.

- ☆ It specifies how the nucleotide sequence of an mRNA is translated in to the amino acid sequence of a protein.

Properties of Genetic Code

There are several important features or properties of a genetic code.

1. **The code is triplet:** The triplet code was first suggested by Gammow in 1954. In a triplet code, three RNA bases code for one amino acid.
   - ☆ Codon comprises of three nucleotides and is required to count all 20 aminoacids found in proteins.
   - ☆ A two letter code would have $4^2 = 16$ codons, which is not enough to account for all 20 amino acids, whereas a three letter code would give $4^3 = 64$ codons.
2. **The code is universal:** The same genetic code is applicable to all forms of organisms from microbes to human beings.
3. **The code is comma less:** The genetic code is without a comma or break. The codons are continuous and there are no breaks between the codons.
   - ☆ A change or deletion of a single base in the code will alter the entire sequence of aminoacid to be synthesized.
4. **The code is non-overlapping:** Three nucleotides or bases code for one aminoacid and six bases will code for two amino acids. In a non – overlapping code, one base or letter is read only once.
5. **The code is non-ambiguous:** Out of the 64 codons, 61 code for 20 different aminoacid, while 3 are nonsense codons. None of the codons code for more than one aminoacid.
6. **The code is degenerate:** In most cases, several codons code for the same aminoacid.
   - ☆ Only two amino acids, *viz.*, tryptophan and methionine are coded by one codon each.
   - ☆ Nine amino acids are coded by two codons each, one amino acid (isoleucine) by three codons each, five amino acids by four codons each and three amino acids by six codons each.
   - ☆ This type of redundancy of genetic code is called degeneracy of genetic code.

- ✰ Such a system provides protection to the organisms against many harmful mutations. If one base of codon is mutated, there are other codons, which will code for the same amino acid and thus there will be no alteration in polypeptide chain.

7. **The code has polarity:** The code has a definite direction for reading of message, which is referred to as polarity.
   - ✰ Reading of code in opposite direction will naturally specify for another amino acid example: GUC codes for valine, if reversed, CUG codes for Leucine.

The Genetic Code is not Universal

This was first discovered in 1979 by Frederick Sanger's group in Cambridge, U.K. They found that several human mitochondrial mRNA contain the sequence 5′-UGA-3′, which normally codes for termination, but it is a tryptophan codon in human mitochondria.

- ✰ Modifications are less common among prokaryotes, but one example is known in *Mycoplasma* species *i.e.* Codon- dependent codon reassignment.
- ✰ Codon- dependent codon reassignment occurs when the protein to be synthesized contains either selenocysteine or pyrolysine. But selenocysteine is coded by 5′-UGA-3′ and pyrolysine by 5′-UAG-3′, but these codons have a dual mening because thay are still used as termination codons in the organisms.

3

Molecular Biology Techniques

Expalin in brief about the overview of nucleic acid isolation and purification?

Isolation and purification of nucleic acids is the first in molecular biology studies and all recombinant DNA techniques.

Extraction Methods

The extraction of nucleic acids from biological material requires cell lysis, inactivation of cellular nucleases and seperation of the dersired nucleic acid from cellular debris.

- ☆ Cell membrane disruption and inactivation of intracellular nucleases may be combined. *e.g.* A single solution may contain detergents to solubilize cell membranes and strong chaotropic salts to inactivate intracellular enzymes.
- ☆ After cell lysis and nuclease inactivation, cellular debris may easily be removed by filtration or precipitation.

Purification Methods

Methods for purifying nucleic acids from cell extracts are combination of extraction/precipitation, chromatography and cetrifugation are frequently used to remove proteins.

i. Extraction/Precipitation

- ☆ Solvent extraction is used to eliminate contaminants from nucleic acids *e.g.* A combination of phenol and choloform are frequently used to remove proteins.

- ☆ Selective precipitation can also purify nucleic acids. *e.g.* High concentration of salt (salting out) or changes in pH can be used to precipitate proteins.
- ☆ Precipitation may also be used to concentrate nucleic acids. *e.g.* The target nucleic acids are precipitated with isopropanol or ethanol. If the amount of target nucleic acid is low, an inert carrier (*e.g.* glycogen) can be added to the mixture to increase precipitation efficiency.

ii. Chromatography

Chromatography methods may utilize gel filtration, ion exchange, selective adsorption or affinity binding.

i. **Gel filtration** exploits the molecular seiving properties of porous gel particles.
   - ☆ A matrix with defined pore size allows smaller molecules to enter the pores by diffusion whereas bigger molecules are excluded from the pores.
   - ☆ Thus, molecules are eluted in order of decreasing molecular size.

ii. **Ion exchange chromatography** depends on an electrostatic interaction between a target molecule and a functional group on the coulmn matrix.
   - ☆ The technique allows concentration and seperation of molecules from a large volume in a short time.
   - ☆ Nucleic acids – highly negetively charged, linear polyanions – can be eluted from ion exchange coulmns with simple salt buffers.

iii. **Adsorption chromatography,** nucleic acids adsorb selectively onto silica or glass in the presence of chaotropic salts, while other biological molecules do not.
   - ☆ A low salt buffer or water then elutes the nucleic acids, thereby producing a sample that could be used directly in most downstream application.

iv. **Affinity chromatography** is a highly specific adaptation of adsorption chromatography.
   - ☆ An immobilized ligand recognizes and binds a particular structure on a bimolecule. Washes then remove unbound components (with different structures).
   - ☆ Finally, a "competitor molecule" (which also recognizes the immobilized ligand) floods the binding sites on the affinity matrix, releasing the bound molecules.

iii. **Centrifugation:** Selective cetrifugation is a powerful purification method. *e.g.* Ultracetrifugation in self-forming CsCl gradients at high g-forces has long been used for plasmid purification
   - ☆ Frequenly, centrifugation is combined with another method. *e.g.* Spin column chromatography combines gel filtration and centrifugation

to purify DNA or RNA from smaller contaminants (salts, nucleotides *etc.*), for buffer exchange, or for size selection.

- Some procedures combine selective adsorption on a chromatographic matix with centrifugal elution to selctively purify one type of nucleic acids.

iv. **Electrophoresis:** Nucleic acids may be seperated electrophoretically according to their size. This seperation is most commonly done on a agarose gels. In the presence of ethidium bromide, the seperated nucleic acids may be seen under UV light.

- Electrophoresis seperation is also frequcnlty used to determine size and purify of DNA. *e.g.* After PCR, electrophoresis is used to quickly check product length and purity (absence of biproducts).

v. **Affinity purification:**

- In this, more and more purification methods have combined affinity immobilization of nucleic acids with magnetic seperation. *e.g.* Poly $(A)^+$ mRNA may be bound to streptavidin-coated magnetic particles by biotin-labeled oilgo (dT) and the particle complex removed from the solution (and unbound contaminants) with a magnet.
- This soild phase technique simplifies nucleic acid purification, since it can replace several centrifugation, organic extraction and phase seperation steps with a single, rapid magnetic seperation step.

Write the working principles for the following techniques?

i. **UV – visible spectrophotometer**

ii. **Agarose gel electrophoresis unit**

iii. **Polyacrylamide gel electrophoresis unit**

UV-visible Spectrophotometer

It is used to measure the chemical constituents of the sample and to get spectrometric analysis of the given samples.

Principle

If beam of light of intensity falls on a cell containing the sample, the emergent radiation intensity I is less than the incident radiation, since a portion of it is aborbed by the sample. The aborption is different at different wavelenghts and is characterisitc of the sample. This characterisitc quantity is called the aborbance and may be calculated from the Beer-Lambert principle.

- The Beer-Lambert law states that in a sample, each successive portion along the path of the incident radiation, containing an equal number of aborbing molecules aborbes an equal fraction of the radiation that transverse it.
- A spectrometre meassures the relative amount of light energy passed through a substance that is absorbed or transmitted.

- This instrument is mainly used to determine how much light of (a) certain wavelength (s) is adsorbed by (or transmitted through) a solution.
- Transmittance (T) is the ratio of transmitted light to incident light.

  Aborbance (A) = - log T. Absorbance is usually the most useful measure, because there is a linear relationship between absorbance and concentration of a substance. This relationship is shown by the Beer-Lambert law: A = ebc; Where e = extinction coefficient (a proportionality constant that depends on the absorbing species), b = path length of the cuvette. Most standard cuvettes have a 1 cm path and thus, this can be ignored; c = concentration.

Agarose Gel Electrophoresis Unit

Agarose gel electrophoresis is the horizontal gel electrophoresis which is the easiest and most popular way of seperating and analyzing the DNA. It seperates the DNA fragments based on their molecular weight.

Principle

In this, DNA molecules are seperated on the basis of charge by applying an electric field to the electrophoretic apparatus. Shorter molecules migrate more easily and move faster than longer molecules through the pores of the gel and this process is called sieving. The gel might be used to look at the DNA inorder to quantify it or to isolate a particular band. The DNA can be visualized in the gel by the addition of Ethidium bromide through gel documentation unit.

- Agarose is polysaccharide obtained from the red algae *Porphyra umbilicalis.*
- Its systematic name is (1 4)-3, 6- anhydro-a-L-galatopyranosyl- (1 3)-β-D-galactopyran.
- Agarose makes an inert matrix.
- Most agarose gels are made between 0.7 per cent and 2 per cent of agarose.
- A 0.7 per cent gel show good seperation of large DNA fragments (5-10 kb) and a 2 per cent gel will show good resolution for small fragemts with size range of 0.2-1 kb.
- Low percentage gels are very weak (leads to break when lifting) but high percentage gels are usually brittle and donot set evenly.
- Agarose gel electrophoresis is a powerful seperation method frequently used to analyze DNA fragments generated by restriction enzymes, and it is a convenient analytical method for determining the size of DNA molecule in the range of 500 to 30,000 base pairs.

Polyacrylamide Gel Electrophoresis Unit (PAGE)

Polyacrylamide Gel Electrophoresis, is an analytical method used to separate components of a protein mixture based on their size.

- The gels typically consist of acrylamide, bisacrylamide, the optional denaturant (SDS or urea), and a buffer with an adjusted pH.

- The solution may be degassed under a vacuum to prevent the formation of air bubbles during polymerization.
- Alternatively, butanol may be added to the resolving gel (for proteins) after it is poured, as butanol removes bubbles and makes the surface smooth.
- A source of free radicals and a stabilizer, such as ammonium persulfate and TEMED are added to initiate polymerization.
- The polymerization reaction creates a gel because of the added bisacrylamide, which can form cross-links between two acrylamide molecules.
- The ratio of bisacrylamide to acrylamide can be varied for special purposes, but is generally about 1 part in 35.
- The acrylamide concentration of the gel can also be varied, generally in the range from 5 per cent to 25 per cent.
- Lower percentage gels are better for resolving very high molecular weight molecules, while much higher percentages are needed to resolve smaller proteins.

Principle

The technique is based upon the principle that a charged molecule will migrate in an electric field towards an electrode with opposite sign.

- The general electrophoresis techniques cannot be used to determine the molecular weight of biological molecules because the mobility of a substance in the gel depends on both charge and size.
- To overcome this, the biological samples needs to be treated so that they acquire uniform charge, then the electrophoretic mobility depends primarily on size.
- For this different protein molecules with different shapes and sizes, needs to be denatured (done with the aid of SDS) so that the proteins lost their secondary, tertiary or quaternary structure.
- The proteins being covered by SDS are negatively charged and when loaded onto a gel and placed in an electric field, it will migrate towards the anode (positively charged electrode) are separated by a molecular sieving effect based on size.
- After the visualization by a staining (protein-specific) technique, the size of a protein can be calculated by comparing its migration distance with that of a known molecular weight ladder (marker).
- The gel used is divided into an upper "stacking" gel of low percentage (with large pore size) and low pH (6.8), where the protein bands get squeezed down as a thin layer migrating toward the anode and a resolving gel (pH 8.8) with smaller pores. Cl^- is the only mobile anion present in both gels.

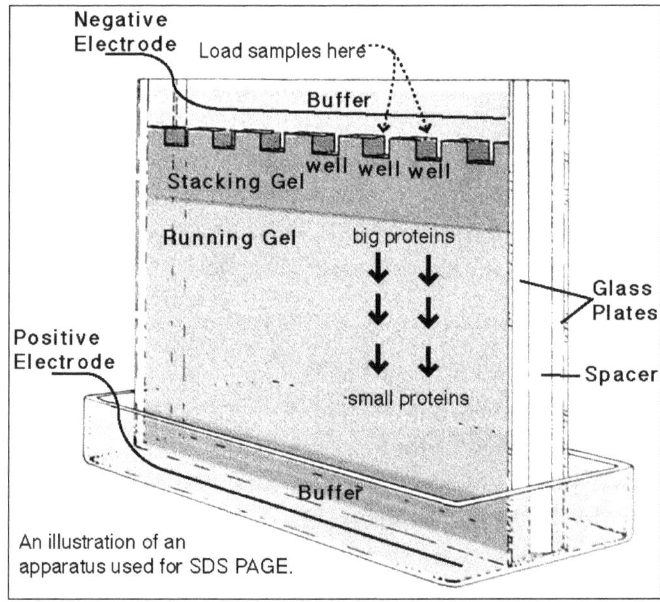

An illustration of an apparatus used for SDS PAGE.

- ☆ When electrophoresis begins, glycine present in the electrophoresis buffer, enters the stacking gel, where the equilibrium favors zwitterionic form with zero net charge.
- ☆ The glycine front moves through the stacking gel slowly, lagging behind the strongly charged, Cl^- ions. Since these two current carrying species separate, a region of low conductivity, with high voltage drop, is formed between them.
- ☆ This zone sweeps the proteins through the large pores of the stacking gel, and depositing it at the top of the resolving gel as a narrow band.
- ☆ When glycine reaches resolving gel it becomes negatively charged and migrates much faster than protein due to higher charge/mass ratio.
- ☆ Now proteins are the main carrier of current and separate according to their molecular mass by the sieving effect of pores in gel.

What are molecular probes? How can these obtained and labeled? And discuss their utility?

A molecular probe is a small (15-30 bp long) nucleotide sequence used in molecular biology to study the properties of other molecules or structures. Radioactive DNA or RNA sequences are used as molecular probes in molecular genetics.

These molecular probes can be obtained in several ways like:

i. DNA segments isolated from the genome of the organism or cDNA molecules prepared by using mRNA can be cloned in *E. coli* and used as probes.

ii. Single stranded DNA probes can be prepared by cloning the concerned DNA segment in a vector like phage M13 of *E. coli* and used as probes.

iii. Single stranded cDNA probes can be prepared by limiting the copying of mRNA by reverse transcriptase to only one strand.

iv. Highly purified mRNA can be used as probe. Single stranded RNA probes can be readily prepared by cloning the corresponding DNA sequence inserted in to a special vector like pGEM.

v. PCR can be used to generate single stranded copies of a DNA sequence used as probes. This is called asymmetric PCR.

Labeling of Probes

The probes can be labeled with radioactivity or with non-radioactive labels.

- ☆ The technique of radioactive labeling includes, nick translation, primer extension, end labeling and direct labeling methods.
- ☆ Of all the methods nick translation is widely used to label the double stranded DNA molecules.

Nick Translation

It mainly involves the

1. DNA sample to be labeled is treated with DNaseI to induce the nicks in the DNA molecules and the positions of the nicks being random.
2. Thus each nick has a free 3′-OH group due to the breakage of phosphodiester bond on one strand of a double stranded DNA molecule.
3. And the existing copy of the strand of duplex is digested by the 5′-3′ exonuclease activity of the enzyme.
4. In this reaction, deoxynucleoside triphosphates are radiolabelled with P^{32}.
5. The labeled nucleotides are progressively incorporated in to newly synthesized segment and it has the same base sequence as the strand it replaces.
6. This process is called 'nick translation' because there is a movement of the nick along the DNA duplex due to the activities of *E. coli* DNA polymerase I.
7. In non-radiactive labeling, the Biotin-labelled probes are prepared by nick translation in which biotin- conjugation nucleotides are used.

Applications

- ☆ In DNA cloning, to identify the recombinant clones carrying the desired DNA insert.
- ☆ DNA finger printing for the unequivocal identification of plant varieties, criminals and potential relationships *etc.*

- *In situ* hybridization for determining the locations of specific sequences in specific chromosomes.
- Confirmation of the integration of DNA insert in the host genome by southern hybridization and its expression in transformed cells by northern hybridization.
- Development of genome maps of eukaryotes. *e.g.* RFLP maps.

Write the principle, procedure and applications of the following techniques?

1. **Southern blotting technique**
2. **Northern blotting technique**
3. **Western blotting technique**
4. **Dot blot technique**
5. **Colony blot hybridization technique**
6. **DNA foot printing**
7. **Fluorescent *insitu* hybridization technique**
8. **Enzyme linked immunosorbent assay**

Southern Blotting Technique

A Southern blot is a method used in molecular biology for detection of a specific DNA sequence in DNA samples. Southern blotting combines transfer of electrophoresis-separated DNA fragments to a filter membrane and subsequent fragment detection by probe hybridization.

- The technique was developed by E.M. Southern in 1975.

Procedure

1. Extract and purify DNA from cells: Isolate the DNA from the rest of the cellular material in the nucleus.
2. Restriction digestion: Cut the DNA into different sized fragments using restriction endonucleases.
3. Gel electrophoresis for separation of DNA molecules based on the charge and size.
4. Denaturation and blotting:
   - DNA is then denatured with alkaline solution such as NAOH. This causes the double stranded becomes single stranded.
   - The process of transferring the DNA from the gel to membrane is called as blotting.
   - The blot is usually done on a sheet of nitrocellulose paper or nylon by either electro blotting or capillary blotting.
   - DNA is then neutralized with NaCl to prevent re-hybridization before adding the probe.

5. Hybridization: The labeled probe is added to the membrane in buffer and incubated for several hours to allow the probe molecules to find their targets.
6. Autoradiography: The sheet is removed from the bag, washed thoroughly to remove unhybridized probes and viewed using autoradiography or ultraviolet light depending on the labels used (radioactive of fluorescent).

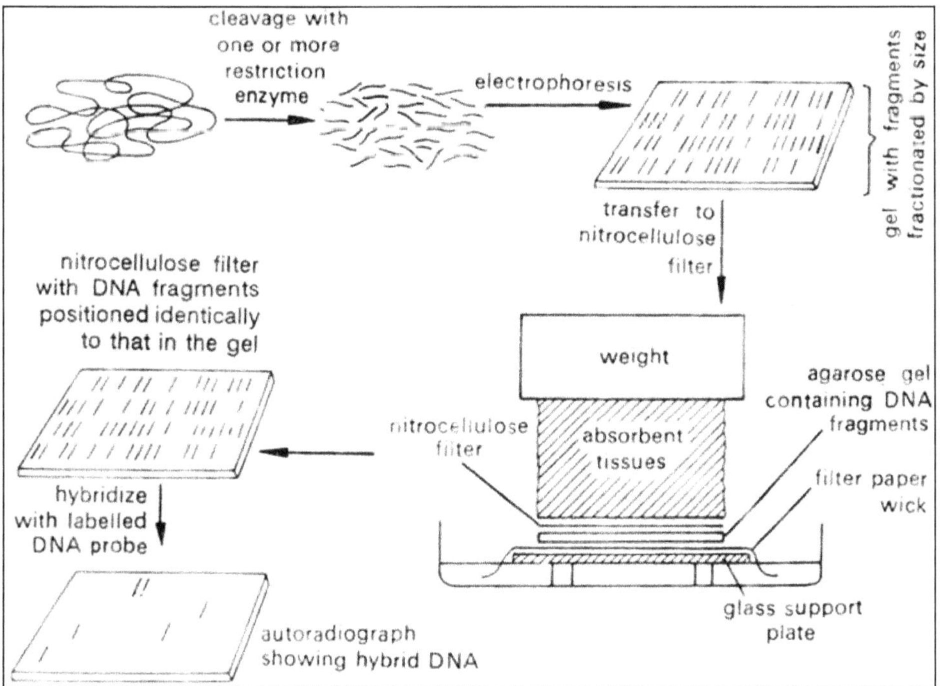

Applications

- To identify the specific DNA in DNA sample *i.e.* identification and cloning of specific gene.
- To isolate desired DNA for construction of rDNA.
- Used in genome mapping through construction of RFLP mapping.
- Southern transfer may also be used for homology-based cloning on the basis of amino acid sequence of the protein product of the target gene.
- To identify mutations, deletions and gene rearrangements.
- Used in prognosis of cancer and in diagnosis of genetic diseases *e.g.* Diagnosis of HIV-1 and infectious diseases.
- Used in phylogenetic analysis.
- In DNA finger printing: *e.g.* Paternity and maternity testing, Criminal identification and forensics *etc.*

☆ Southern blotting can also be used to identify methylated sites in particular genes. *e.g.* Particularly useful are the restriction nucleases *MspI* and *HpaII*, both of which recognize and cleave within the same sequence. However, *HpaII* requires that a C (cytosine) within that site be methylated, whereas *MspI* cleaves only DNA unmethylated at that site.

Northern Blotting Technique

Northern blot is a laboratory technique used to detect a specific RNA sequence in a blood or tissue sample. The sample RNA molecules are separated by size using gel electrophoresis. The RNA fragments are transferred out of the gel to the surface of a membrane.

☆ An adaptation of Southern blotting is Northern blotting, in which RNA molecules are electrophoresed through the gel instead of DNA.

☆ The technique was developed by Alwine and his colleagues in 1979.

☆ The term "Northern" has no specific significance just a misnomer.

Procedure

1. Extract and purify mRNA from cells.
2. Gel electrophoresis for separation of RNA molecules based on the charge and size.
3. Gel is immersed in depurination buffer for 5-10 minutes and washed with water.
4. Transfer to amynobenzyloxymethyl filter paper through blotting technique.
5. After transfer membrane is baked at 80°C.
6. Add labeled DNA probe for hybridization to take place.
7. Wash off unbound probe and autoradiograph to detect mRNA – DNA hybrid.

Applications

☆ Detecting a specific mRNA in a sample.

☆ Used to observe a particular gene expression pattern between tissues, organs, developmental stages, environmental stress levels, pathogen infection and over the course of treatment.

☆ Used in the screening of recombinants by detecting the mRNA produced by the transgene.

☆ In disease diagnosis.

☆ Used to show overexpression of oncogenes and downregulation of tumor-suppressor genes in cancerous cells when compared to normal tissue.

Western Blotting Technique

Western blot (Protein immunoblot) is a analytical technique used to detect specific proteins in a sample of tissue homogenate or extract.

- It is an adaptation of the Southern blot procedure, used to identify specific amino-acid sequences in proteins.
- It was devised by Towbin in 1979.
- The term western has no scientific significance just a misnomer.
- It is mainly based on antigen-antibody interaction *i.e.* immune detection method.

Procedure

1. Extract and purify proteins from cells.
2. Separated by SDS-PAGE electrophoresis.
3. Transfer proteins from gel to nitrocellulose paper.
4. Blocking nonspecific antibody sites on the nitrocellulose paper with Bovine Serum Albumin (BSA) or milk powder.
5. Probing electro blotted proteins with primary antibody.
6. Washing away nonspecifically bound primary antibody.
7. Detecting bound antibody by horseradish peroxidase-anti-lg-congugate or formation of a diaminobenzidine (DAB) precipitate, radiolabelling or use of fluorescently labeled secondary antibody.
8. Autoradiography or photographing or fluorescent detection.

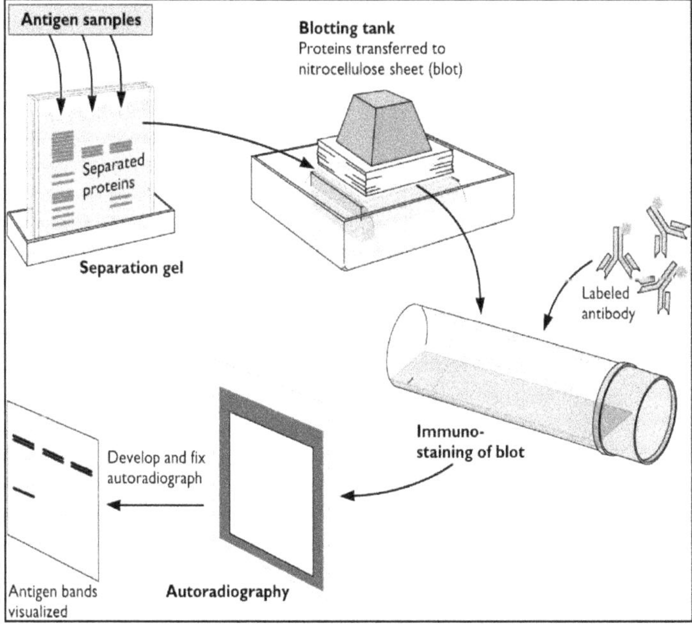

Applications

- ✰ Highly sensitive method to detect a specific protein even in very low quantity.
- ✰ Used in clinical diagnosis. *e.g.* The confirmatory HIV test employs a western blot to detect anti-HIV antibody in a human serum sample.
- ✰ A western blot is also used as the definitive test for bovine spongiform encephalopathy (BSE, commonly referred to as 'mad cow disease').
- ✰ A western blot can also be used as a confirmatory test for Hepatitis B infection.
- ✰ Quantifying a gene product (gene expression studies).

Dot Blot Technique

A dot blot (or slot blot) is a technique in molecular biology used to detect biomolecules, and for detecting, analyzing, and identifying proteins. It represents a simplification of the northern blot, Southern blot, or western blot methods.

- ✰ In this non-fractionated or non-electrophoresed samples are directly blotted and immobilized on a nitrocellulose or nylon membrane as dots or spots for hybridization.
- ✰ Compared to southern blotting, in this, sample from different sources can be tested in a single run.
- ✰ In dot blotting the nucleic acids are blotted as circular blots, whereas in slot blotting they are blotted into rectangular slots.

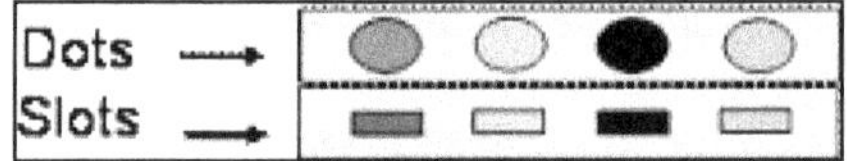

- ✰ In this black dots represent the samples where target DNA is present or probe has bound.

Procedure

1. Extract and purify DNA or RNA from different sources
2. Apply directly as small dots on nitrocellulose or nylon membrane.
3. If DNA, denature it with mild alkali to form single strands.
4. Immobilize by baking at 80°C for 2-3 hrs.
5. Add labeled probe for hybridization to take place
6. Wash off unbound probe and autoradiography for detection of biomolecules or proteins.

Applications

- ✰ Detection of specific DNA or RNA in a sample without the step of electrophoresis.

- ☆ Many samples can be screened in a single run.
- ☆ Used to detect the *Chlamydia trachomatis* infection and other sexually transmitted diseases.
- ☆ Dot blot is used to detect Antidiacyltrehalose Antibodies in Tuberculous patients and Typhoid Fever.

Colony Blot Hybridization Technique

Colony blot hybridization is applied to DNA or RNA released from blotted microbial colonies.

1. The microbial colonies are transferred (blotted) to a membrane. The cells are lysed in place to release the nucleic acids.
2. The RNA or DNA (after denaturation) is fixed to the filter and hybridized with a labeled probe.
3. Blocking reagent may be added prior to the probe to prevent unspecific binding.
4. Excess probe is washed away and the membrane is visualized by UV or autoradiography.
5. Colony blot hybridization can be used for screening clones or bacterial isolates.

DNA Foot Printing

DNA foot printing is a method of investigating the sequence specificity of DNA – binding proteins *in vitro*. It can be used to study protein – DNA interactions both outside and within cells.

Procedure

1. Isolate and clone the DNA fragment known to bind to a given protein.
2. Preparation and labeling of sample of the DNA fragment at one end of on strand only.
3. Add protein of interest to a portion of the labeled template DNA.
4. Add a cleavage agent to both portions of DNA fragment.
5. The cleavage agent is a chemical or enzyme that will cut at random locations in a DNA sequence. The reaction should occur just long enough to cut each DNA molecule in only one location.
6. A protein that specifically binds a region within the DNA fragment will protect the DNA from cleavage agent.
7. Run both samples side by side on a gel electrophoresis.
8. The portion of DNA template without protein will be cut at random locations, and thus it will produce a ladder like distribution on a gel.

9. The DNA fragment with the protein will result in ladder distribution with a break in it, the 'foot print', where the DNA has been protected from the cleavage agent.

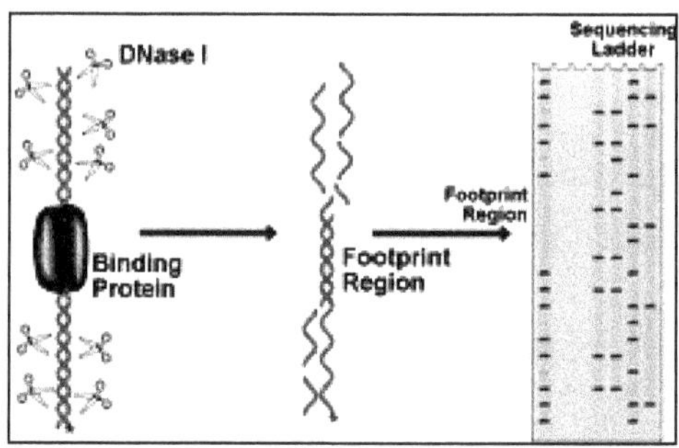

Applications

- ☆ It is mainly used to determine the locations and lengths of binding sites of various proteins that bind to DNA.
- ☆ It can also apply to determine the base sequence of binding sites.
- ☆ *In vivo* foot printing combined with immune precipitation can be used to assess protein specificity at many locations throughout the genome.
- ☆ It can be used to assess the binding strength of a protein to a region of DNA.
- ☆ It can also used to identify the binding sites of transcription factors. But it is possible by analysis of DNA fragment produced by PCR coupled with fluorescent molecule such as carboxy fluorescein (FAM) to the primers.

Fluorescent In situ Hybridization Technique

Fluorescence in situ hybridization (FISH) is a laboratory technique for detecting and locating a specific DNA sequence on a chromosome. The technique relies on exposing chromosomes to a small DNA sequence called a probe that has a fluorescent molecule attached to it.

The FISH technique involves:

1. Denaturation of DNA (nuclear, probe),
2. Labeling of probe with reporter molecule (biotin, dioxigenin),
3. Hybridization with the probe and finally incubation in immuno-fluorescent reagent before taking photographs with fluorescent microscope.
   - ☆ If there is affinity between reporter molecule and fluorescent molecule, the fluorescent molecule will be deposited at the site of *in-situ* hybridization.
   - ☆ The probes used in FISH can be total genomic DNA or chromosome specific DNA or single copy sequences. For, this reason FISH can be used for gene mapping and study of structural and numerical aberration of chromosome.

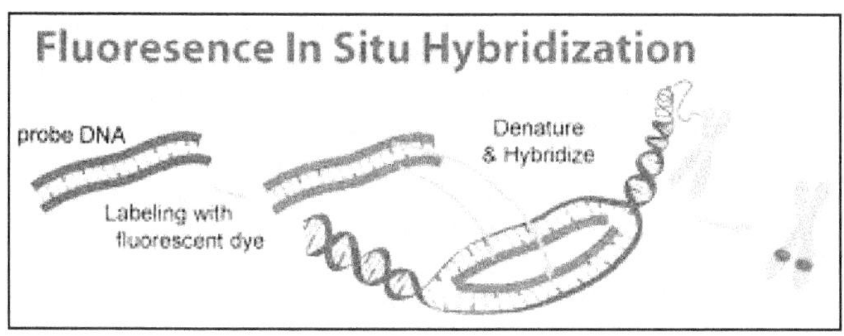

Applications

1. Unique sequences, chromosomal subregions, or entire genomes can be specifically highlighted in metaphase or interphase cells by fluorescence in situ hybridization (FISH).
2. This technique can be used to identify chromosomes, detect chromosomal abnormalities or determine the chromosomal location of specific sequences.
3. FISH can be used in cytogenetic analysis of diseases *e.g.* Karyotype of 32 Armenian patients with chronic myeloid leukemia have been studied for the presence of Philadelphia (Ph) chromosome resulting from translocation t(9;22)(q34;q11) by means of conventional karyotype analysis (CK) and fluorescence in situ hybridization (FISH) in the bone marrow and/or peripheral blood specimens.

Enzyme Liked Immune Sorbent Assay (ELISA)

Enzyme-linked immunosorbent assay is one of immunoassay method using antibodies to capture an antigen and an enzyme labeled antibody to estimate the amount of antigen.

- This technique was introduced by Engvall and Pearlmann in 1971.
- The main basic principle involved in ELISA is Lock and Key concept *i.e.* Antigen (key), substance when introduced into the body produces antibodies and Antibody (lock), protein in the body that is used by immune system to identify and neutralize foreign targets (referred to as antigens) – Key fits into the lock.
- ELISA can be used to detect the presence of antigens that are recognized by the antibody or it can be used to test for antibodies that recognize the antigen.

ELISA Methods

Direct ELISA

- It is suitable for the detection of proteinaceous antigens and may require pre purification of the sample.
- Direct ELISA can be performed when desired antibody is available in a pre conjugated state.

Indirect ELISA

☆ If the primary antibody is not conjugated, then indirect ELISA required in which a conjugated secondary antibody is targeted to the isotope (*e.g.* mouse IgG1, goat IgM, chicken IgY *etc.*) of the primary antibody.

Sandwich ELISA

☆ The sandwich ELISA measures the amount of antigen between two layers of antibodies *i.e.* capture and detection antibody.

☆ The antigen to be measured must contain at least two antigenic sites capable of binding to the antibody, since at least two antibodies act in the sandwich.

Competitive ELISA

It involves the competition between two antibodies for one antigen, or two antigens for one antibody.

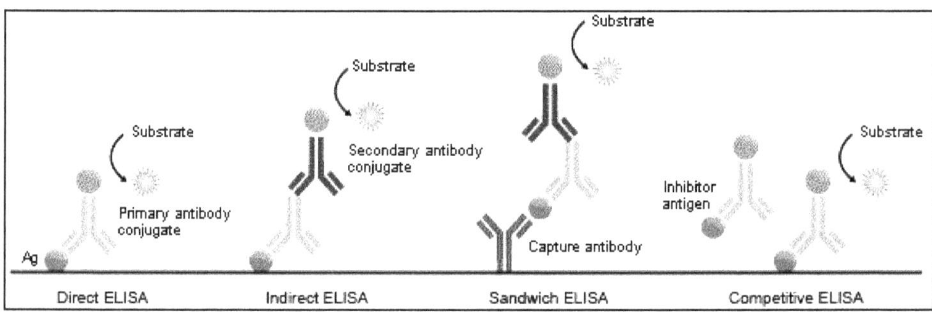

Procedure

1. Standard solutions or assay samples are added to the antibody-coated wells, and incubated for several hours so as to the antigen molecules are captured by "capture antibody".
2. After this binding reaction, the reaction mixture is discarded, and wells are washed to remove excessive materials.
3. The second antibody which recognizes another epitope in antigen is added. This second antibody has been labeled with an enzyme such as horseradish peroxidase (HRP).
4. The enzyme-labeled second antibody will bind to the antigen which is bound to the capture 10 antibody on the bottom area of wells. This means that the enzyme (HRP) is also fixed on the bottom of wells. The amount of the antigen captured is proportional to fixed enzyme.
5. Enzyme activity is measured by adding a chromogenic substrate of this enzyme. In the case of HRP, tetramethylbenzidine (TMB) is often used.

6. After incubation for some period, the chromogenic substrate is changed to a colored product.
7. The reaction is stopped by adding a reaction stopper, *e.g.* Diluted sulfuric acid, and absorbance is measured using a plate reader.
8. The standard curve is prepared from the concentration of standard solutions and their absorbance. And the sample assay values are obtained from the absorbance using the standard curve (calibration curve).

Applications

- Useful for determining serum antibody concentrations.
- Detecting potential food allergens. *e.g.* In milk, peanuts, walnuts, almonds and eggs.
- Useful in disease outbreaks *i.e.* for tracking the spread of diseases. *e.g.* HIV, bird flu, colds and cholera *etc.*
- Detection of antigens. *e.g.* Pregnancy hormones, drug allergen, GMO and mad cow disease *etc.*
- Detection of antibodies in blood sample for past exposure to disease. *e.g.* Lyme disease, trichinosis, HIV and bird flu *etc.*

Define mono clonal antibodies? And explain the method of production of mono clonal antibodies through hybridoma technology?

Mono Clonal Anti Bodies

Monoclonal antibodies (mAb or moAb) are monospecific antibodies that are made by identical immune cells that are all clones of a unique parent cell. Monoclonal antibodies have monovalent affinity, in that they bind to the same epitope.

Hybridoma Technology

- A hybridoma, which can be considered as a harry cell, is produced by the injection of a specific antigen into a mouse,
- Procuring the antigen-specific plasma cells (antibody-producing cell) from the mouse's spleen and the subsequent fusion of this cell with a cancerous immune cell called a myeloma cell.
- The hybrid cell, which is thus produced, can be cloned to produce many identical daughter clones.
- These daughter clones then secrete the immune cell product.
- Since these antibodies come from only one type of cell (the hybridoma cell) they are called monoclonal antibodies.
- The advantage of this process is that it can combine the qualities of the two different types of cells; the ability to grow continually, and to produce large amounts of pure antibody.

- ☆ HAT medium (Hypoxanthine Aminopetrin Thymidine) is used for preparation of monoclonal antibodies.
- ☆ Laboratory animals (*e.g.* mice) are first exposed to an antigen to which we are interested in isolating an antibody against.
- ☆ Once splenocytes are isolated from the mammal, the B cells are fused with immortalized myeloma cells - which lack the HGPRT (hypoxanthine-guanine phosphoribosyl transferase) gene - using polyethylene glycol or the Sendai virus.
- ☆ Fused cells are incubated in the HAT (Hypoxanthine Aminopetrin Thymidine) medium.
- ☆ Aminopterin in the myeloma cells die, as they cannot produce nucleotides by the de novo or salvage medium blocks the pathway that allows for nucleotide synthesis. Hence, unfused D cell die.
- ☆ And also unfused B cells die as they have a short life span.
- ☆ Only the B cell-myeloma hybrids survive, since the HGPRT gene coming from the B cells is functional. These cells produce antibodies (a property of B cells) and are immortal (a property of myeloma cells).

Define polyclonal antibodies? And write the differences between polyclonal and monoclonal anti bodies?

Polyclonal antibodies (pAbs) are antibodies that are secreted by different B cell lineages within the body (whereas monoclonal antibodies come from a single cell lineage). They are a collection of immunoglobulin molecules that react against a specific antigen, each identifying a different epitope. Or

The immune response to an antigen generally involves the activation of multiple B-cells all of which target a specific epitope on that antigen. As a result a large number of antibodies are produced with different specificities and epitope affinities these are known as polyclonal antibodies (See table on next page).

How somatic cell hybridization technique is used to assign genes on human chromosomes? Explain.

Techniques mainly used to assign genes on human chromosomes are

1. Chromosomal banding
2. Somatic cell hybridization

Somatic Cell Hybridization

In this one cell is hybridized with other cell.

e.g. Human cell is hybridized with mouse cell.

- ☆ Littlefield explained the technique, how we can identify the hybridized cells, so that in mammalian cells one protein called aminopterin, prevents the enzymes required for DNA replication.

Characters	Polyclonal Anti-bodies	Monoclonal Anti-bodies
Basic feature	☆ Recognize multiple epitopes on any one antigen. ☆ High specificity. Detect only one epitope on the antigen.	☆ Polyclonals are made up mainly of IgG subclass. ☆ They will consist of only one antibody subtype *e.g.* IgG1, IgG2, IgG3.
Antibody production	☆ Technology and skills required for production less complex ☆ Production time scale is short and inexpensive to produce compared to monoclonal antibodies.	☆ High technology required. ☆ Training is required for the technology used. ☆ Timescale is long for hybridomas.
Advantages	Polyclonals will recognize multiple epitopes on any one antigen which has the following advantages: ☆ High affinity. Polyclonals can help amplify signal from target protein with low expression level, as the target protein will bind more than one antibody molecule on the multiple epitopes. ☆ Due to recognition of multiple epitopes, polyclonals can give better results in IP/ChIP. ☆ Polyclonals are more tolerant of minor changes in the antigen (*e.g.* polymorphism, heterogeneity of glycosylation or slight denaturation) than monoclonal (homogenous) antibodies. ☆ Polyclonals will identify proteins of high homology to the immunogen protein or can be used to screen for the target protein in tissue samples from species other than that of the immunogen. This also makes it important to check immunogen sequence for any cross reactivity. ☆ Polyclonal antibodies are often the preferred choice for detection of denatured proteins. ☆ Multiple epitopes generally provide more robust detection.	Monoclonals detect one epitope only on any one antigen which has the following advantages: ☆ Monoclonals usually have less background from staining of sections and cells. As they are more specifically detecting one target epitope, they are less likely to cross-react with other proteins. ☆ Due to their high specificity, monoclonal antibodies are excellent as the primary antibody in an assay, or for detecting antigens in tissues, and will often give significantly less background staining than polyclonal antibodies ☆ Compared to polyclonal antibodies, homogeneity of monoclonal antibodies is very high. If experimental conditions are kept constant, results from monoclonal antibodies will be highly reproducible, between experiments (*e.g.* very low lot-to-lot variation). ☆ Specificity of monoclonal antibodies makes them extremely efficient for binding of antigen within a mixture of related molecules, such as in the case of affinity purification.
Disadvantages	☆ Prone to batch-to-batch variability. ☆ They produce large amounts of non-specific antibodies which can sometimes give background signal in some applications. ☆ Multiple epitopes make it important to check immunogen sequence for any cross-reactivity. ☆ Polyclonal antibodies not useful for probing specific domains of antigen, because antiserum will usually recognize many domains.	☆ Monoclonal antibodies may be too specific (*e.g.* Less likely to detect in across a range of species). ☆ More vulnerable to the loss of epitope through chemical treatment of the antigen than polyclonal antibodies are. This can be offset by pooling two or more monoclonal antibodies to the same antigen (*e.g.* cocktail antibodies).

These two enzymes are HPRT (Hypoxanthin phosphoribosyl transferase) and TK (Thymidine kinase)

$HPRT^-/TK^-$ - are unable to synthesize DNA.

$HPRT^+/TK^+$ - are able to synthesize DNA.

- ☆ Here, he mixed the mouse cell with $HPRT^+/TK^-$ and human cell with $HPRT^-/TK^+$. And then these cells kept in HAT medium (Hypoxanthine, Aminopterin and Thymidine).
- ☆ So by the presence of aminopterine, $HPRT^-/TK^-$ cells were not survive and only $HPRT^+/TK^+$ will survive.
- ☆ This is not only useful for identification of hybrid cells but also for mapping of TK gene on chromosome 17.

Synteny test: In this, two loci present on one chromosome was identified.

e.g. B – locus link with lactate dehydrogenase also link with B- locus of peptidase.

Assignment test: In this, loci present on particular chromosome are identified.

e.g. Blood clotting, glycoprotein tissue factor - III, present on chromosome-1 was identified.

Explain the different methods of DNA sequencing.

DNA Sequencing

The basic objective of DNA sequencing is to determine the order of bases in a DNA molecule. There are two basic techniques of DNA sequencing.

1. *Sanger-Coulson method*

- ☆ Chain termination DNA sequencing is based on the principle that single stranded DNA molecules differ in length by just a single nucleotide can be separated from one another by polycrylamide gel electrophoresis.
- ☆ This technique is also called as dideoxy-mediated chain termination method as it involves the use of chain terminating dideoxy nucleoside triphosphate (ddNTPs).
- ☆ In this use single stand of the DNA to be sequenced as a template for DNA synthesis catalyzed by a DNA polymerase.
- ☆ In this, in addition to the usual mixture of four dNTPs, 2′3′ – dideoxy nucleotide triphosphates are added to the reaction mixture.
- ☆ When these base analogues are incorporated in to the growing chain of DNA, the polymerization reaction will stop because the 3′ ends of the analogues lack a hydroxyl group.
- ☆ These chain terminations are included at relatively low concentrations so that they will not be included at every possible position.
- ☆ These nested set of DNA fragments produced can be resolved by gel electrophoresis and also the use of radioactive labels enables visualization of the fragments on an autoradiography of the gel.

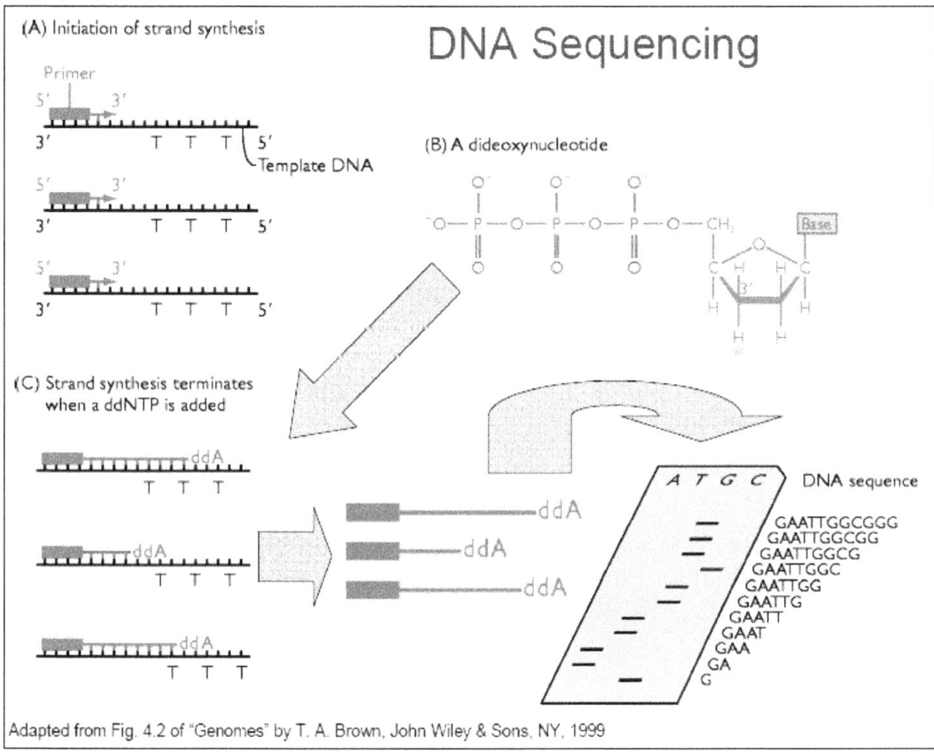

Adapted from Fig. 4.2 of "Genomes" by T. A. Brown, John Wiley & Sons, NY, 1999

2. *Maxam-Gilbert Method*

This method is not commonly used and involves chemical degeneration of original DNA rather than synthesis.

- This method utilizes chemicals that break DNA after certain nucleotide.
- The DNA to be sequenced must be labeled so that after electrophoresis the fragments of DNA containing the label can be identified by autoradiography. Usually P^{32} or S^{35} is used to label the 5′ end of the DNA.
- The DNA to be sequenced is made single stranded and is devided in to four samples.
- Each of the four samples is treated with a chemical that produces breakes in the DNA at specific bases.
- To achieve the desired cleavage specificity the base must first be chemically modified. *e.g.* Dimethyl sulphate methylates guanine.
- The fragments generated in four samples, upon treatments with the various chemicals, are resolved on a polyacrylamide gel.
- Each of the four samples is loaded in a separate lane in the gel.

☆ The gel usually contains urea to produce a gel with very high resolution and also to prevent secondary structures formation.

☆ After electrophoresis, the complete fragments of DNA containing the radioactive label can be visualized by autoradiography. The autoradiography image will provides the results of sequencing.

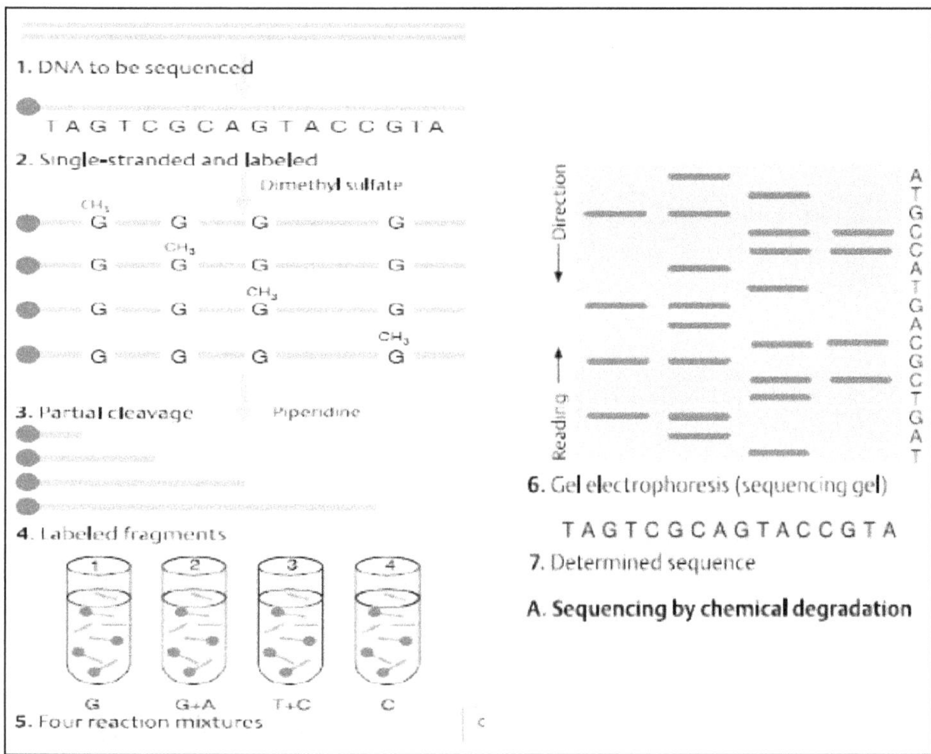

A. Sequencing by chemical degradation

Pyrosequencing is used for rapid determination of very short sequences. Pyrsequencing does not require electrophoresis or any other fragment separation procedure and so is more rapid than either chain termination or chemical degradation method. It can only generate a few tens of base paires per experiment, but it is becoming an important technique in situations where many short sequences must be generated as quickly as possible, for example SNP typing.

Next Generation Sequencing Technologies

Next generation sequencing (NGS) is a type of DNA sequencing technology that uses parallel sequencing of multiple small fragments of DNA to determine sequence.

☆ This 'high- throughput' technology has allows a dramatic increase in the speed and reduces the cost of which an individual's genome can be sequenced.

Comparing Features of Next Generation Sequencers

Features	*Roche (454)*	*Illumina*	*SOLiD*
Chemistry	Pyrosequencing	Polymerase-based	Ligation-based
Amplification	Emulsion PCR	Bridge Amp	Emulsion PCR
Paired ends/sep	Yes/3kb	Yes/200 bp	Yes/3 kb
Mb/run	100 Mb	1300 Mb	3000 Mb
Time/run	7 h	4 days	5 days
Read length	250 bp	32-40 bp	35 bp
Cost per run (total)	$8439	$8950	$17447
Cost per Mb	$84.39	$5.97	$5.81

Answer the following questions.

1. **Define complete genome sequencing?**
2. **Discuss in brief about the shot gun and hierarchical methods of DNA sequencing?**
3. **Write the features of rice genome project?**

Complete Genome Sequencing is a laboratory process that determines the complete DNA sequence of an organism's genome at a single time. This includes sequencing all of an organism chromosomal DNA as well as DNA present in the mitochondria and, for plant in chloroplast.

- ☆ Sequence of longer DNA molecules have one major limitation *i.e.* even with the most sophisticated technology it is rarely possible to obtain a sequence of more than 750 base pairs in a single experiment.
- ☆ But this can be achieved by employing the shot gun method of sequence assembly.

Shot Gun Method

In this method, the DNA molecule is broken in to small fragments, each of which is sequenced and the master sequence is assembled by individual fragments.

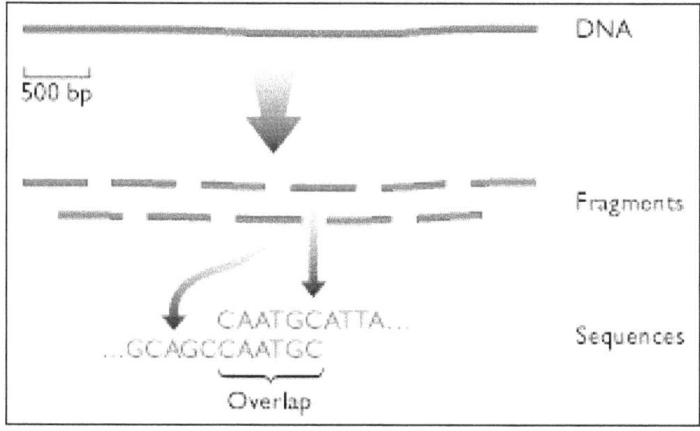

- ✰ But this is most preferable for sequencing small prokaryotic genomes, but much more difficult with larger genomes because the required data analysis becomes disproportionately more complex as the number of fragments increases *i.e.* for 'n' fragments, the number of possible overlaps is given by $2n^2$-2n.
- ✰ One more problem with shot gun method is that it can lead to errors when repetitive regions of a genome are analyzed, when repetitive sequences are broken in to fragments, many of the resulting pieces contain the same or very similar sequence motifs.
- ✰ So it may leads to the error *i.e.* internal region of tandem repeat or the segment of DNA between the two repeats being left out of the master sequence.
- ✰ But when a genome map is available, it provides a guide for the sequencing experiments by showing the positions of genes and other distinctive features.
- ✰ Based on this, the sequencing phase of the project can proceed in either of two ways:

  i. **Whole genome shot gun method:** This method utilizes the distinctive features on the genome map as land marks to aid assembly of the master sequence from the huge numbers of short sequences.

  - ✰ This method mainly involves:
    - a. Random sequencing of the entire genome
    - b. This results in pieces of contiguous sequence, possibly hundreds of kilobases in length
    - c. If a contiguous sequence contains a marker, than it can be positioned on the genome map
  - ✰ Thus presence of genome map ensures the regions containing repetitive DNA are assembled correctly.
  - ✰ The whole genome shot gun method is a rapid way of obtaining a draft of eukaryotic genome sequence.

  ii. **Clone contig method:**

  - ✰ This method involves the
    - a. Brokening of genome in to manageable segments,
    - b. Sequencing of segments by shot gun method
    - c. And placing of master sequence at its known position on the map.
  - ✰ This step by step approach takes longer than whole genome shot gun sequencing, but produces a more accurate and error free sequences.

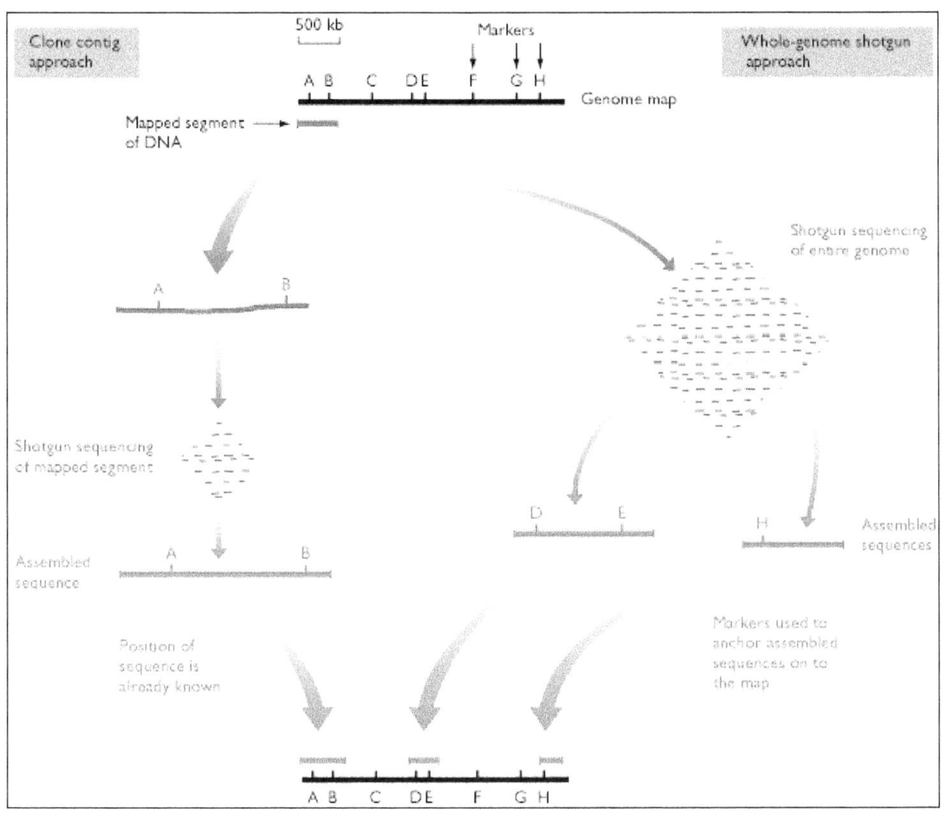

Features of Rice Genome Project

1. The rice genome is well mapped and well characterized, and it is the smallest of the major cereal crop genome at an estimated 400 to 430 Mb and consists of 32 to 50 thousand genes.
2. The International Rice Genome Sequencing Project (IRGSP) began in September 1997, at a workshop held in conjuction with the International symposium on Plant Molecular Biology in Singapore.
3. India, previously an unfunded member of the IRGSP, has a now Rice Genome Programme (represented by Akhilesh Tyagi of the University of Delhi and Nagendra Singh of the IARI) and will begin work on chromosome 11.
4. Sequencing of the rice genome is being performed mainly from genomic BAC or PAC libraries created from the *Nipponbare* (*Japonica*) variety, which was chosen as the common template throughout the IRGSP; China, working on the sequencing of chromosome 4, is the only member to use a different variety, *Indica Guang Lu Ai4.*

5. Rice genome composed of repetitive DNA is estimated to constitute at least 50 per cent of the rice genome and as much as 70 per cent of the maize genome.
6. The Rice Genome Research Programme uses a shot gun approach to sequence PAC or BAC clones.
7. Finger printing and physical mapping are used to create minimal tiling paths and to anchor BAC clones to physical positions along the length of chromosome.
8. After sequencing, softwares such as PHRED and PHRAP is used to order the subclone sequences and reassemble the entire BAC sequences.
9. Monsanto announced on April 4, 2000, that the company had completed a working draft of the rice genome, which would be made available to the IRGSP and helped to complete the sequencing in the year 2002.

4

PCR and Gene Cloning

What is meant by PCR? And discuss the steps involved in PCR along with applications and limitations?

Polymerase Chain Reaction

PCR technique is developed by Kary mullis 1985. It is one of the most powerful molecular biology technique used to multiply minute (or) trace amounts (μg microgram quantities) of DNA copies of the desired DNA by multiple cycles of cooling and heating in a reaction catalysed by a heat stable DNA polymerase enzyme. (The PCR is carried out *in vitro*).

- ☆ PCR is based on the features of semi conservative DNA replication carried out by DNA polymerase in prokaryotic and eukaryotic cells.

Procedure of PCR

Amplification of DNA is achieved by a repetitive series of cycles involving 3 steps.

1. **Denaturation:** The DNA double helix separated into two complimentary single strands by heating reaction mixture to temperature between 90-98°C that ensures DNA denaturation. The duration of this step in the first cycle of PCR is usually, 2 min at 94°C, but in subsequent cycles it is of only 1min duration.
2. **Annealing:** The mixture is now cooled to a temperature of generally 40-60°C that permits annealing of primer to the complimentary sequences in the DNA. The duration of annealing step is usually 1 min during the first as well as the subsequent cycles of PCR. The annealed primers are then extended (*i.e.* synthesis of DNA) with Taq DNA polymerase.

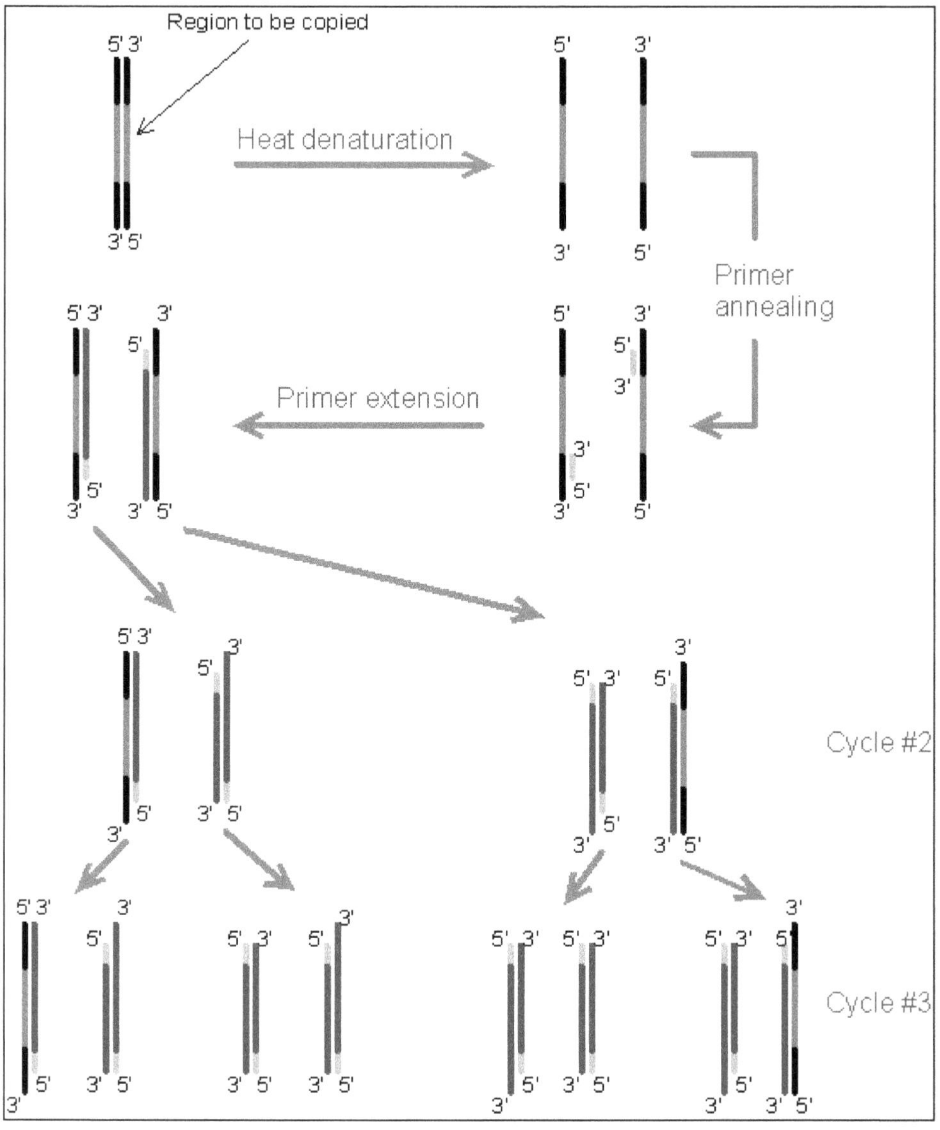

3. **Primer Extension (or) synthesis:** It involves heating the mixture to 72°C at which a special polymerase synthesizes the new DNA strand by using the original strand as the template starting by the primer utilizing their 3′OH free ends and continuing in the 3′ direction.
   - ✰ A typical thermal cycle might be as follows:
   - ✰ Heat denaturation at 94°C for 20 seconds
   - ✰ Primer annealing at 55°C for 20 seconds
   - ✰ Primer extension at 72°C for 30 seconds

- Average time for each cycle is approximately 4-5 minutes, considering the fact that heating and cooling between each stage also have to be considered.
- Initially these steps at three different temperatures were carried out in separate water baths but nowadays a thermal cycler is used (a machine that automatically changes the temperature at the correct time for each of the stages and can be programmed to carry out a set number of cycles).
- After the introduction of thermocycler, each cycle of replication can be completed in less than 5 minutes.
- At the end of each cycle the number of copies of desired segment becomes twice the number present at the end of previous cycle.
- Thus at the end of 'n' cycles 2^n copies of the segments are expected.

Application

- Amplification of small amounts of DNA for further analysis by DNA fingerprinting.
- DNA cloning for sequencing.
- The analysis of ancient DNA from fossils.
- The isolation of a particular gene of interest from a tissue sample.
- Generation of probes: large amount of probes can be synthesized by this technique.
- DNA based phylogenetic studies.
- Functional analysis of gene
- Diagnosis of hereditary diseases
- The detection and Diagnosis of infectious diseases
- It can be used to determine the sex of embryos
- Analysis of mutations: Deletions and insertions in a gene can be detected by differences in size of amplified product.
- Diagnosis of monogenic diseases (single gene disorders): For pre-natal diagnosis, PCR is used to amplify DNA from foetal cells obtained from amniotic fluid. PCR has also proved very important in carrier testing.
- Detection of microorganisms: Especially of organisms and viruses that are difficult to culture or take long time to culture or dangerous to culture.
- The PCR has even made it possible to analyze DNA from microscope slides of tissue preserved years before.
- Detection of microbial genes responsible for some aspect of pathogenesis or antibiotic resistance.
- Crucial forensic evidence may often be present in very small quantities, *e.g.* one human hair, body fluid stain (blood, saliva, semen). PCR can generate sufficient DNA from a single cell.

Limitations of PCR

- ☆ PCR is an extremely sensitive technique but is prone to contamination from extraneous DNA, leading to false positive results.
- ☆ Another potential problem is due to cross-contamination between samples. It is for this reason that sample preparation, running PCR and post-amplification detection must be carried out in separate rooms.
- ☆ Concentration of Mg is very crucial as low Mg^{2+} leads to low yields (or no yield) and high Mg^{2+} leads to accumulation of nonspecific products.
- ☆ Non-specific binding of primers and primer-primer dimmer formation are other possible reasons for unexpected results.
- ☆ Reagents and equipments are costly, hence can't be afforded by small laboratories.

Explain in detail about the types or variants of PCR?

Types of PCR

Multiplex PCR

It is an adaptation of PCR which allows simultaneous amplification of many sequences.

- ☆ It uses a number of pairs of primers to allow analysis of a number of fragments in a single sample *i.e.* several pairs of primers annealing to different target sequences.
- ☆ It is used for diagnosis of different diseases in the same sample.
- ☆ It can detect different pathogens in a single sample.
- ☆ It can be used to identify exonic and intronic sequences in specific genes and determination of gene dosage.
- ☆ Multiplex PCR has also been used for analysis of microsatellites and SNPs.
- ☆ Multiplex Ligation-dependent Probe Amplification (MLPA) permits multiple targets to be amplified using only a single pair of primers, avoiding the resolution limitations of multiplex PCR.

Nested PCR

This PCR increases the sensitivity due to small amounts of the target are detected by using two sets of primers, involving a double process of amplification.

- ☆ In the first PCR, one pair of primers is used to generate DNA products, which may contain products amplified from non target areas.
- ☆ The products from the first PCR are then used as template in second PCR, using one (hemi nesting) or two different primers nested within the first set.
- ☆ Therefore, specificity of first PCR product is verified with second one, thus increasing the specificity.

- ☆ It reduces the risk of unwanted products from primers binding to incorrect regions.
- ☆ It is often more successful in specifically amplifying long DNA products than conventional PCR, but it requires more detailed knowledge of the sequence of the target.
- ☆ The disadvantage of this PCR is the probability of contamination during transfer from the first amplified product into the tube in which the second amplification will be performed.
- ☆ But this contamination can be controlled using primers designed to anneal at different temperatures or by adding ultra pure oil to make a physical separation of two mixtures of amplification.

Real Time PCR

Real time PCR or quantitative PCR (qPCR) or quantitative real time PCR (QRT PCR) is an adaptation of the PCR method to quantify the number of copies of nucleic acids during PCR by tagging DNA or RNA using fluorescent probes.

- ☆ It is mainly used to quantify DNA or cDNA, determining gene or transcript numbers present within different samples.
- ☆ It offers advantages such as speed in the result, the reduced risk of contamination and the ease in the handling technology.
- ☆ It uses fluorescence detection systems which are generally of two types. Intercalating agents (*e.g.* SYBR green fluorescing agents) and labeled probes with fluoroprobes (*e.g.* Taqman probes).

Semiquantitative PCR

This PCR allows an approximation to the relative amount of nucleic acids present in a sample.

- ☆ In this optical density of amplified product can be calculated by densitometer after electrophoresis and ethidium bromide staining of the product.
- ☆ The disadvantage of the technique is possibility of non specific hybridizations, generating unsatisfactory results.

Reverse Transcriptase PCR (Rt PCR)

This amplification technique designed to amplify RNA (especially mRNA) sequence through synthesis of cDNA by reverse transcriptase enzyme.

- ☆ In this, PCR is preceeded by a reaction using reverse transcriptase that converts RNA to cDNA, and then amplification of cDNA retaining the original RNA sequence.
- ☆ This is mainly achieved by using Tth polymerase (obtained from *Thermus thermophilus*) enzyme, which has reverse transcriptase activity.
- ☆ It is mainly useful for studying of gene expression *in vitro*.

- ☆ It can be used to get the sequence of an RNA transcript, so that prediction of transcription start and termination sites is possible.
- ☆ It can also used for diagnosis of RNA viruses, as well as for evaluation of antimicrobial therapy.

Inverse PCR

It amplifies DNA next to a known sequence by using primers placed in the reverse direction to the normal position.

- ☆ In this, first target DNA is subject to series of restriction enzyme digestions. And then circulating the resulting fragments by self ligation.
- ☆ Then primers are digested to be extended out word from the known segment, it helps in amplification of the rest of the circle.
- ☆ It is mainly useful in studying of sequences on either side of various genomic inserts.

Asymmetric PCR

It is a method of PCR, which amplifies one strand of the target DNA.

- ☆ In this thermocycling is carried out with a limiting amount or leaving out one of the primers.
- ☆ When the limiting primer becomes depleted, replication increases arithmetically through extension of the excess primer.
- ☆ The main application of this technique is sequencing and hybridization probing to generate one strand as product.

Linear after the Exponential PCR (LATE PCR)

It is an modification method of asymmetric PCR, which creates single strands of DNA.

- ☆ This technique uses a limiting primer with a higher melting temperature (T_m) than the excess primer to maintain reaction efficiency as the limiting primer concentration decreases mid reaction.

Assembly PCR

It is also known as polymerase cycling assembly (PCA). This technique mainly used to build long pieces of DNA by performing PCR on a pool of long oligonucleotides with overlapping primers *i.e.* to assemble two or more pieces of DNA into one piece.

- ☆ This technique involves an initial PCR with primers that have an overlap and second PCR using the products as the template that generates the final full length products.
- ☆ This technique is mainly substitute for ligation based assembly.

Hot Start PCR

This technique is performed by heating the reaction components to the DNA melting temperature before adding the polymerase *i.e.* to inactivate the Taq polymerase until the reaction starts.

- ☆ It mainly helps to prevent non specific amplification at lower temperatures.
- ☆ It can also achieved by using specialized reagents that inhibit the polymerase activity at ambient temperature, either by the binding of antibody or by the presence of covalently bound inhibitor and this only dissociate after a high temperature activation step.

Touchdown PCR (TD PCR)

In this amplification reaction begins with an annealing temperature higher than optimum and reduces it each cycle to the optimum or touchdown temperature.

- ☆ In this, in early cycles annealing temperature is usually 3-5°C above standard T_m of the primers used, while in later cycles it is similar amount below the T_m.
- ☆ Because of initial higher annealing temperature, it leads to greater specificity for primer binding.
- ☆ While the lower temperatures permit more efficient amplification at the end of the reaction so it improves the outcomes.

Colony PCR

This technique mainly used to screen bacterial colonies directly by PCR after transformation with a plasmid vector.

- ☆ It is mainly to screen for correct DNA vector constructs.

Digital PCR

This amplification technique mainly useful for simultaneous amplification of thousands of samples, each in a separate droplet within an emulsion.

Long Range PCR

In this PCR reaction, mainly involves a mixture of polymerases to amplify longer stretches of DNA.

Suicide PCR

This technique mainly used in paleogenetics or in the studies where avoiding false positives to improve the specificity of the amplified fragment.

- ☆ In this, use any primer combination only once in a PCR, which should never have been used in any positive control PCR reaction.
- ☆ And also primers should always target a genomic region never amplified before in the lab using this or any other set of primers.

Variable Number of Tandem Repeat PCR (VNTR PCR)

This amplification reaction targets mainly the areas of the genome that exhibit length variation.

- ☆ In this analysis of genotypes mainly involves sizing of the amplification products by gel electrophoresis.
- ☆ It is the main basis for DNA fingerprinting *i.e.* by analysis of short tandem repeats.

AFLP PCR

It involves the digestion of DNA into fragments using restriction enzymes, amplification of the fragments using PCR and then analysis of the fragments using gel electrophoresis.

Allele Specific PCR

This amplification technique mainly employs allele specific primers that are designed to match a mutation.

- ☆ In this, primers are chosen mainly from polymorphic area with mutations located at its 3′end.
- ☆ During the reaction, mismatch primer will not initiate replication, whereas matched primers will initiate the replication.
- ☆ So the appearance of an amplification product indicates the genotype.

Inter sequence-specific PCR (ISSR- PCR)

It amplifies DNA between sequences that are repeated throughout the genome.

- ☆ In this, primers are selected from segments repeated throughout the genome to produce a unique finger print of amplified product lengths.
- ☆ *e.g.* Alu PCR, in this primer from a commonly repeated segment called Alu- sequence, so that it helps to amplify the sequences adjacent to these repeats.

Thermal Asymmetric Interlaced PCR (TAIL PCR)

This amplification reaction mainly uses a nested pair of primers with differing annealing temperatures to isolate the unknown sequences flanking a known area of the genome.

Ligation Mediated PCR

This amplification technique uses small DNA oligonucleotide linkers or adaptors that are first ligated to fragments of the target DNA, than amplification by target fragments by annealing the PCR primers to the linker sequences.

- ☆ It is mainly useful in DNA sequencing, DNA foot printing and genome walking.

Methylation Specific PCR

This amplification reaction is mainly used to identify the pattern of DNA methylation at cytosine-guanine (CpG) islands in genomic DNA.

- ☆ In this, first DNA is treated with sodium bisulfate, it helps to convert unmethylated cytosine base to uracil, which is complementary to adenosine in PCR primers.
- ☆ Two amplification reactions are carried out on bisulfate treated DNA: one primer set anneals to DNA with cytosine (corresponds to methylate cytosine), and the other set anneals to DNA with uracil (corresponds to unmethylated cytosine).
- ☆ It is mainly used in Real time PCR, which provides quantitative information about the methylation state of a given CpG island.

High Fidelity PCR

This amplification technique employs enzymes with proof reading activity (with 3′-5′ exonuclease activity). *e.g.* pfu DNA polymerase obtained from *Pyrococcus furiosus.*

- ☆ This mainly helps in decrease in error rates (mismatch bases) compared to Taq DNA polymerase.

Explain in brief about the following?

1. MALDI-TOF Mass spectrometry
2. PCR- ESI Mass spectrometry

MALDI-TOF Mass Spectrometry (Matrix Assisted Laser Desorption/Ionization – Time of flight mass spectrometry)

It is mainly used for determining the precise mass of peptides.

- ☆ In this, the sample for MALDI is uniformly mixed in a large quantity of matrix. The matrix absorbs the UV light and converts it to heat energy. A small part of matrix heats rapidly and is vaporized, together with the sample.
- ☆ In this method, peptides are first mixed with an organic acid and then dried onto a ceramic or metal slide.
- ☆ The sample is then blasted with a laser which causes the peptides to be ejected from the slide in the form of an ionized gas in which each molecule carries one or more positive changes.
- ☆ The time it takes for them to reach the detector is determined by their mass and charge.
- ☆ Large peptides move more slowly, and more highly charged peptides move more quickly.
- ☆ The precise mass is determined by analysis of those peptides with a single charge.

PCR- Electrospray Ionization Mass spectrometry (PCR – ESI/MS):

- ☆ During past two decades, genetic detection methods such as real time PCR and peptide nucleic acid fluorescent hybridization co-existed with traditional microbiological methods are using for detection of pathogens.
- ☆ But now, PCR-ESI/MS offers a new approach to identify pathogens directly from clinical specimen and the potential to identify genetic evidence of undiscovered pathogens and genetic changes that may occur to alter sequences used as targets in existing molecular assays.
- ☆ The PCR-ESI/MS assays contain a variety of purposefully designed primers sets that integrate common conserved and variable sequences found among various classes of organism.
- ☆ In this, proper amplification relies on the genetic similarities in microbial genomes.

  e.g. Bacteria have highly conserved sequences in several chromosomal locations, including the universally conserved regions of ribosomal RNA, other noncoding RNAs, and essential protein – encoding genes.
- ☆ Because PCR-ESI/MS is semiquantitative, a relative intensity of multiple organisms can be determined, allowing for selection of a treatment that manages multiple organisms. However, there are limitations for identification of mixed infections.
- ☆ One more limitation is subsequent contamination of future samples during processing because this instrument is an open platform, there is potential for instruments and workspace to be exposed to amplicons.
- ☆ Additional concern for PCR-ESI/MS is the expected cost and the feasibility of non preference facilities acquiring the technology.

Answer the following questions.

1. **What is microarray technique? Give the principle on which it is based?**
2. **Write the features and applications of microarray technology?**
3. How microarray technology can be used to differentiate a cancerous tissue from normal tissue?

The microarray technology is a tool for exploring the genome in a systematic and comprehensive fashion, to survey DNA and RNA variation.

- ☆ DNA microarray technologies have several names like DNA chips or genomic chips or gene array or biochip technology.
- ☆ It involves the microscopic array of single stranded DNA molecules immobilized on solid surface by biochemical analysis.
- ☆ DNA chips are derived from either cDNA (cDNA microarrays) or prepared by photolithography (short oligonucleotides array).

Principle

Hybridization of an unknown sample to an ordered array of immobilized DNA molecules of known sequences produces a specific hybridization pattern that can be analyzed with a given sample.

- ☆ Each cDNA or sample RNA is tagged with a fluorescent dye because each microchip contains a standard set of DNA sequences *i.e.* probes.
- ☆ By using autoradiography, laser scanning fluorescence detection devices enzyme detection system the hybridization pattern can be quantitatively or qualitatively analyzed.

Characteristic Features of DNA Chips/Microarrays

1. Parallelism: Microarray analysis allows parallel acquisition and analysis of massive data. It allows meaningful comparison between genes or gene products represented in microarray.
2. Miniaturization: Microarray analysis involves miniaturization of DNA probes and reaction volumes, thus reducing time and reagent consumption.
3. Speed: Microarray analysis is highly sensitive and allows rapid data acquisition.
4. Multiplexing: Multiple samples can be analyzed in a single assay.
   - ☆ The labeling and detection methods that involve multicolour fluorescence allows comparisons of multiple samples in a single DNA chip.
   - ☆ Multiplexing also increases the accuracy of comparative analysis by eliminating complicative factors such as chip to chip variation and discrepancies in reaction conditions.
5. Automation: Manufacturing technology permits the mass production of DNA chips and automation leads to proliferation of microarray assays by ensuring their quality, availability and affordability.
6. Combinational synthesis: Using this strategy, a set of all '4' oligonucleotides of the length 'k' nucleotides can be generated in '4k' synthesis cycles.

Applications

1. Functional genomics: Microarray for gene expression analysis provides an integrated platform for functional genomics.
2. SNP and point mutations: Genotyping individuals using SNPs needs only positive or negative assay, permitting easier automation.
   - ☆ The approach used for this purpose relies on the capacity to distinguish a perfect match from a single base mismatch.
3. DNA sequencing/Resequencing: This involves manufacturing the sequencing DNA chips that contain a complete set of immobilized

oligonucleotides of a particular size and hybridization of the target DNA of unknown sequence on to these DNA chips.

- ☆ Identification and analysis of the overlapping oligomers that form perfect duplexes with the DNA of interest permits reconstruction of the target DNA sequence.

4. Proteomics: Protein linkage maps can also be created using genomic sequence information.
5. Agricultural biotechnology:
   - ☆ Transgenic plants can also be rapidly analyzed using microarray on DNA chips.
   - ☆ Expression patterns under different environmental conditions can be predict at gene level.
   - ☆ To study polymorphism and to develop molecular markers tagged to specific economic traits for marker assisted selection in crop improvement.
   - ☆ Microarrays used to collect data on expression under different conditions.

DNA microarrays are used for gene expression patterns. Their use is explained in comparing gene expression patterns of a cancerous tissue and corresponding normal tissue.

Steps

1. mRNA is separately isolated from the cancerous as well as normal tissues.
2. RNA is not very stable, it is quickly degraded, so the mRNA can be copied in to cDNA using reverse transcriptase enzyme.
3. The cDNA is labeled by using suitable fluorescent dyes.
4. For convenience, each of the two cDNA preparations is labeled with a different fluorescent dyes, so they will produce fluorescence of different colours. The labeled cDNA is called probe.
5. Since we wish to compare gene expression profiles of cancerous and normal cell types, use both the cDNA probes simultaneously for hybridization with same DNA array.
6. After hybridization, the DNA array will have following four types of spots.
   - ☆ cDNA is present only in probe 1: These spots will fluoresce to generate one colour. (in fact, these colours are generated by the scanner and are, therefore, called false colours, they are not actual colours generated by the fluorescence).
   - ☆ cDNA is present only in probe 2: These spots will fluoresce and yield a different colour.

- ☆ cDNA is present in both probe 1 and 2. These spots will produce both the types of fluorescence and as a result, will produce a different colour.
- ☆ cDNA is absent from both the probes: These spots will produce no fluorescence and will appear as blank.

7. In this study, two different fluorescent dyes are used. The data on fluorescence image of spots are obtained on a separate channel of scanner for each of the two fluorescent dyes.

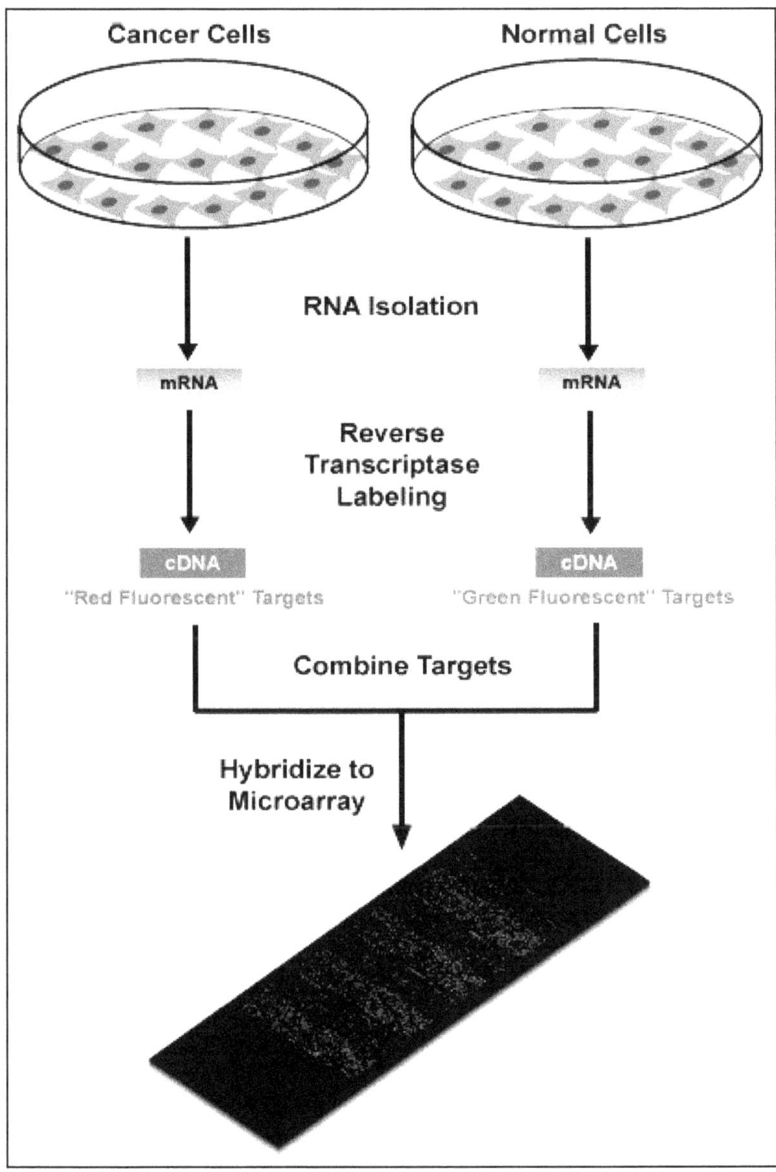

8. The spots are false coloured for each fluorescent dye, combined and displayed by the scanner.
9. The intensities of the images can be quantified using suitable software.
10. An analysis of the above microarray will provide information about the following:
    - ☆ The genes that are expressed in both normal and cancerous cells.
    - ☆ mRNA that were present in the cancerous cell types.
    - ☆ The frequency of cDNA sequence in the probe will be comparable to that of the mRNA in the cell.

Define gene cloning? What are its approaches? And write the steps involved in gene cloning.

Gene Cloning

Gene cloning (or) DNA cloning is defined as insertion of fragment of DNA representing a gene into a cloning vector and subsequent propagation of DNA molecule in a host organism.

- ☆ It is defined as isolation and amplification of an individual gene sequence by immersion of that sequence into a bacterium where it can be replicated.
- ☆ The basic aim of cloning genes to produce multiple copies of DNA to isolate and characterize specific genes for their use in basic and applied research.

There are two approaches for gene cloning

a. Cell based cloning
b. PCR based gene cloning

Among these two approaches, PCR is able to multiply many copies of gene within few hours, whereas it takes about weak to obtain the same by cell based cloning. In PCR, there as a limit to the length of DNA sequence that copied, 5 kb can be copied easily, but this is shorter than the length of many genes. Whereas cell based cloning useful for multiplication of long genes.

Basic events involved in gene cloning are

- ☆ Isolation and synthesis of gene of interest
- ☆ Insertion of isolated gene DNA fragment into a suitable vector to produce rDNA.
- ☆ Introduction (of) rDNA (vector) into a suitable organism/cell called host. This process is known as transformation. Generally, E. coli is used for initial cloning
- ☆ Selection of transformed host cell by rDNA

- ☆ Multiplication of rDNA within the host cell to produce a number of identical copies of cloned gene.
- ☆ A rDNA molecule is produced by joining together two (or) more DNA segments usually originating from different organisms.
- ☆ The rDNA molecule produced which contains the coding region from one organism joined to regulatory sequence from another organism and such a gene is called chimeric gene.
- ☆ The capability to produce rDNA molecules has given human beings the power and opportunity to create novel gene functions to suit specific needs.

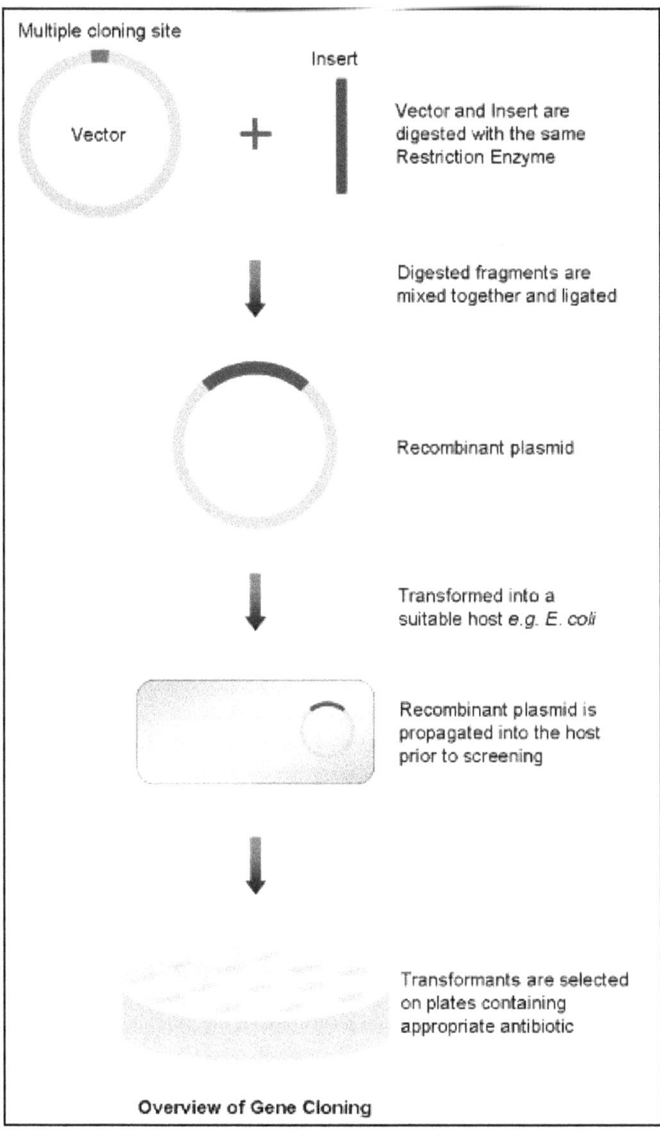

Overview of Gene Cloning

Write the difference Between PCR and Gene Cloning

Sl.No.	Parameter	PCR	Gene Cloning
1.	Manipulation	*In vitro*	Step *in vitro* and *in vivo* last
2.	Selectivity of specific segments from complex DNA	1st step	Last step
3.	Biological reagents require	DNA segment 2 primers DNTP technique polymers	Restriction enzymes ligases, Vector DNA, host cells
4.	Automation	Yes	No
5.	Labour intensive	No	Yes
6.	Error probability	Less	More
7.	Applications	More	Less
8.	Cost	Less	More
9.	Users skill	Not required	Required
10.	Time required for conducting a experiment	4 hrs	2-4 days

What is meant by Recombinant DNA technology? And explain the various enzymes used for DNA manipulation?

Recombinant DNA Technology

Recombinant DNA technology is the revolutionary technology which enables to transfer genes from any organism to another. This is feasible technology to transfer within or between species or taxa and even unrelated taxa

- The scope of this technology is capable to use gene pool-4 and overcome the barrier of cross compatibility.
- Recombinant DNA molecule used as a vector (plasmid, phage and virus) into which the desired DNA fragment has been inserted to enable its cloning in an appropriate host.

Recombinant DNA is product for following purpose.

- To obtain a large number of copies of specific DNA fragments
- To obtain large quantities of protein related to gene of interest.
- To integrate gene into chromosome

A cultivar developed by the rDNA procedure is called transgenic cultivar (or) genetically modified (GM) cultivar. The basic steps of all rDNA technologists are as follows.

- DNA of interest extracted from source organism through restriction enzyme.
- Desired DNA fragment is inserted into cloning vector.
- rDNA transferred and maintained in host through transformation.

- ☆ Host identified with cloned transgene.
- ☆ Cloned transgene can be manipulated to give protein product (expression of transgene).

Enzymes for DNA Manipulation

The enzymes available for DNA manipulation fall into four broad categories.

1. *DNA polymerases*

Which are enzymes that synthesize new polynucleotides complementary to an existing DNA or RNA template *i.e.* template dependent DNA polymerase.

Important Features of Template Dependent DNA Polymerase

i. To initiate DNA synthesis, these must be a short, double stranded region to provide a 3′end on to which the enzyme will add new nucleotide. *i.e.* short, synthetic olegonucleotide, usually about 20 nucleotides in length, which acts as a primer for DNA synthesis.

ii. Template – dependent DNA polymerase is multifunctional, being able to degrade DNA molecules as well as synthesize them.

  a. A 3′→5′ exonucleae activity enables the enzyme to remove nucleotides from the 3′ end of the strand that has just synthesized *i.e.* proof reading activity.

  b. A 5′→3′ exonuclease activity is less common

Types of DNA Polymerases

Polymerase	*Description*	*Main Use*
DNA polymerase I	Unmodified *E. coli* enzyme	DNA labeling
Klenow polymerase	Modified version of *E. coli* DNA polymerase I	DNA labeling
Sequenase	Modified version of phage T7 DNA polymerase I	DNA sequencing
Taq polymerase	*Thermus aquaticus* DNA polymerase I	PCR
Reverse transcriptase	RNA-dependent DNA polymerase, obtained from various retroviruses	cDNA synthesis

2. *Nucleases*

Which degrade DNA molecules by breaking the phophodiester bonds that link one nucleotide to the next nucleotide.

- ☆ Nucleases have a broad range of activities but most are either exonucleases, removing nucleotides from the ends of DNA/RNA molecules or endonucleases, making cuts at internal phosphodiester bonds.
- ☆ Nucleases are specific for DNA and some for RNA, some work only on double-stranded DNA and others only on single-stranded DNA *etc.*

Nucleases	*Description*	*Main Use*
Restriction endonucleases	Sequence-specific DNA endonucleases, from many sources	Restriction – many applications
Nuclease SI	Endonuclease specific for single-stranded DNA and RNA, from the fungus *Aspergillus oryzae*	Transcript mapping
Deoxyribonuclease I	Endonuclease specific for double-stranded DNA and RNA from *E. coli*	Nuclease footprinting

- ✰ Restriction endonucleases, which play a central role in all aspects of recombinant DNA technology.

Restriction Endonucleases

A, Restriction endonuclease is an enzyme that binds to a DNA molecule at a specific sequence and makes a double-stranded cut at or near that sequence.

There are three types of restriction restriction endonucleases (Type I, II and III).

- ✰ With Type I and III, there is no strict control over the position of the cut relative to the specific sequence in the DNA molecule that is recognized by the enzyme.
- ✰ Type II enzymes do not suffer from this disadvantage because the cut is always at the same place, either within the recognition sequence or very close to it. *e.g.* EcoRI cuts DNA only at the hexanucleotide sequence 5′-GAATTA-3′.
- ✰ Digestion of DNA with a Type II enzyme therefore gives a reproducible set of fragments whose sequences are predictable if the sequence of the target DNA molecule is known.
- ✰ Restriction enzymes cut DNA in two different ways. Many make a simple double-stranded cut, giving a blunt or flush ends. *e.g.* Alu1 enzyme.
- ✰ Many enzymes cut the two DNA strands at different position, usually two or four nucleotides apart, so that the resulting DNA fragments have short-stranded overhangs at each end. These are called sticky or cohesive ends because base pairing between them can stick the DNA molecule back together again. *e.g.* Sau 3A1, EcoR1 and PstI enzymes.

3. *DNA Ligases*

Which join DNA molecules together by synthesizing phosphodiester bonds between nucleotides at the ends of two different molecules, or at the two ends of a single molecule.

- ✰ The most widely used DNA ligase is obtained from *E. coli* cells infected with T4 bacteriophage. Its natural role is to synthesize phosphodiester bonds between unlinked nucleotides present in one polynucleotide of a double-stranded molecule.

☆ If the two molecules have complementary sticky ends, and the ends come together by random diffusion events in the ligation mixture, then transient base pairs might form between the two over-hangs.

☆ If molecules blunt ended, then they cannot base pair to one another, not even temporarily, and ligation is a much less efficient process, even when the DNA concentration is high and pairs of ends are in relatively close proximity.

4. End-modification Enzymes

1. Terminal deoxynucleotidyl transferase is a template-independent DNA polymerase, because it is able to synthesize a new DNA polynucleotide without base pairing of the incoming nucleotides to an existing strand of DNA or RNA. Its main role in recombinant DNA technology is in homopolymer tailing.
2. Two other end-modification enzymes are alakaline phosphatase and T4 polynucleotide kinase

   ☆ Alkaline phosphatase, which removes phosphate groups from the 5′ ends of DNA molecules, which T4 polynucleotide kinase performs the reverse reaction to alkaline phosphatase *i.e.* adding phosphates to 5′ ends. These enzymes are used during complicated experiments, but its main application is in the end-labeling of DNA molecules.

Conversion of Blunt Ends to Sticky Ends

1. Short, double stranded molecules called linkers or adapters are attached to the blunt ends. They contain a recognition sequence for a restriction endonuclease and so produce a sticky end after treatment with the appropriate enzyme.
2. Creation of sticky ends is by homopolymers tailing, in which nucleotides are added one after the other to the 3′ terminus at a blunt end. The enzyme involved is called "terminal deoxy nucleotidyl transferase". *e.g.* Poly (G) tail, which would enable the molecule to pair to other molecules that carry poly (C) tails.

What is meant by Restriction modification system? Explain in brief about its types and applications of Restriction modification system in molecular biology?

The Restriction modification system (RM system) is used by bacteria. Bacteria have restriction enzymes also called restriction endonucleases, which cleave double stranded DNA at specific points in to fragments, which are then degraded further by other endonucleases.

☆ By this prokaryotic organisms protect themselves from foreign DNA introduced by infectious agent such as bacteriophage.

Types of Restriction Modification System

These are five kinds of RM system: type I, type II, type III, type IV and 'Bcg-like' like restriction modification system.

Type I RM System

- ☆ It is the most complex of the known systems.
- ☆ It consists of three polypeptides: R (restriction), M (modification) and S (specificity).
- ☆ The M_2S complex is active as a protective methyltransferase, but the intact M_2SR_2 heteropentamer is required for restriction.
- ☆ The S subunit determines the specificity of both restriction and methylation.
- ☆ In this, cleavage occurs at variable distances from the recognition sequences, so discrete bands are not easily visualized by gel electrophoresis.

Type II RM System

- ☆ The type II RM systems are at the opposite extreme from the type I systems, being the most structurally simple of the restriction modification system.
- ☆ In this, instead of working as a complex, the methyltransferase and endonuclease are encoded as two separate proteins and act independently.
- ☆ The methyltansferase acts as a monomer, methylating the duplex one strand at a time while the endonuclease acts as a homodimer, which facilitates the cleavage of both strands.
- ☆ Thus both proteins recognize the same recognition site, and therefore compete for activity.
- ☆ In this, cleavage occurs at a defined position close to or within the recognition sequence, thus producing discrete fragments during gel electrophoresis.

Type III RM System

- ☆ DNA sequence specificity of type III system is provided by the methyltransferase subunit, which is different from the case in type I system, and the type III recognition sequences are asymmetric but not bipartite.
- ☆ In this, M protein is independently active *i.e.* methylates on its own, while R protein is active as a complex with M protein.
- ☆ The hetrodimer formed by R and M proteins competes with itself by modifying and restricting the same reaction. This results in incomplete digestion.

Type IV RM System

- ☆ This system has asymmetrical recognition sequences and shifted cleavage positions like the type IIS enzymes. (When the sites of cleavage were determined, it was found that both DNA strands were cleaved off to one side of the recognized sequence – hence the name S *i.e.* shifted cleavage).

- ☆ In this methyltransferase activity intrinsic to the endonuclease protein methylates one strand of the recognition sequence.
- ☆ This system also includes a second methyltransferase, which protectively methylates both strands of the recognition sequence.

'Bcg-like' RM System

- ☆ The 'Bcg-like' restriction modification system is named for their archetype, *Bcg* I from *Bacillus coagulans.*
- ☆ So far no body has suggested calling them 'type V' thogh in structural and functional terms might be justified.
- ☆ Like type IV RM system, this system has methyltransferase and endonuclease fused into a single polypeptide.
- ☆ Unlike type IV system, the Bcg-like system has a second B subunit that provides specificity.
- ☆ The cleavage pattern of Bcg-like system is unique they make a pair of double strand breaks, one pair to each side of the recognition sequence, then removing the recognition sequence as a short double stranded oligonucleotide.

Applications

1. Cloning: RM systems can be cloned in to plasmids and selected because of the resistance provided by the methylation enzyme.
   - ☆ Once the plasmid begins to replicate, the methylation enzyme will be produced and methylate the plasmid DNA, protecting it from a specific restriction enzyme.
2. Restriction Fragment Length Polymorphism (RFLP): When wide and mutants are analyzed by digestion with different restriction enzymes, the gel electrophoretic products vary in length, because mutant genes will not be cleaved in a similar pattern as wild type for presence due to the presence of mutations that render the restriction enzymes non-specific to the mutant sequence.
   - ☆ It is useful to analyze the composition of DNA is regarded to presence or absence of mutations that affect the cleavage specificity.

What is meant by vector? And mention the properties of good vector?

Vector

A vector is a DNA molecule that has the ability to replicate in an appropriate host cell, and into which the DNA insert is integrated for cloning. Therefore a vector must have a origin of DNA replication (ori) that functions efficiently in the concerned host cell.

The vector is a vehicle (or) carrier which is used for cloning foreign DNA in bacteria. The cloning vehicles are called vectors. Any extra-chromosomal small genome, *e.g.* Plasmid, phage and virus may be used as vector.

Properties of a Good Vector

- It should be able to replicate autonomously that is independent of the replication of host chromosome
- It should be easy to isolate and purify
- It should be easily introduced into the host cells.
- The vector should have suitable marker genes that allow easy detection or/and selection of the transformed the host cell. *e.g.* Genes for ampicillin and tetracycline resistance.
- The cells transform with recombination DNA should be indentifiable (or) selectable from those transformed by the unaltered vector.
- A vector should contain unique target sites for as many restriction enzymes as possible into which the DNA insert can be integrated. When expression of the DNA insert is desired, the vector should contain suitable regulatory elements like promoter, operator, and ribosome binding sites.

What are cloning and expression vectors? And explain in detail about the different types of vectors used in cloning?

Cloning Vectors

Vectors used for propagation of DNA inserts in a suitable host are called cloning vectors.

Expression Vectors

Vectors designed for the expression *i.e.* production of the protein specified by the DNA insert, are known as expression vectors. The gene carried by the expression vector is efficiently transcribed and translocated by the host cell.

The most commonly used vectors for cloning are plasmid vectors, cosmid vectors, phagemid vectors, shuttle vectors, yeast vectors, bacterial artificial chromosome vectors (BAC).

Plasmid Vectors

A plasmid is a DNA molecules other than bacterial chromosome that is capable of independent replication and transmission.

- Plasmids multiply at the same rate as that of bacterial cell multiply at a rate independent of that of host cell.
- Such that even upto 1000 copies per cell are obtained using the plasmid as vectors.
- The length of the DNA segments to be cloned is 10-15 base pairs. Any segments of DNA to be cloned is called as DNA insert

1. pBR - 322 - Ideal Plasmids Vector

The name pBR denotes 'p' signifies plasmid BR is from Boliver Rodriguez (1977) the two initials of the scientist who developed pBR-322 which is the most widely used plasmids.

Features

- It contains 4362 base pair (4.3kb: 10 base pairs) of double stranded DNA, its entire base sequence is known.
- It contains col-E1 replication origin with relaxed replication control.
- It has two selectable markers *i.e.* two antibiotic resistance genes, tetr for tetracycline and ampr for ampicillin.
- It has several unique recognition sites for 12 different restriction endonuclease enzymes located within the tetr and ampr genes.
- These plasmids can take up a DNA insert of 10 kb (kilo base pairs) length.
- The presence of restriction sites within the markers tetr and ampr permits an easy selection for cells transformed with the recombinant pBR-322.
- Insertion of the DNA fragment into the plasmid using restriction enzyme (*PstI or PvuI*) places the DNA insert within the gene ampr, this makes ampr non functional.
- Bacterial cells containing such a recombinant PBR 322 will be unable to grow in the presence of ampicillin, but will grow on tetracycline.

When restriction enzyme *Bam HI* and *sal I* is used, the DNA insert is placed within the gene tetr making it non – functional. Bacterial cells possessing such a recombinant PBR – 322 will therefore, grow on ampicillin but not on tetracycline. This allows an early selection of a single bacterial cell having recombinant p PBR 322 from among other types of cells.

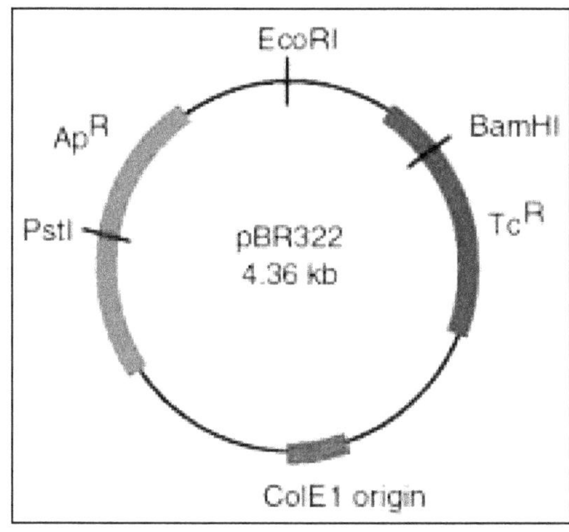

2. *pUC Vector/Plasmid*

The name pUC is derived because it was developed in the University of California (UC) by Messings and his colleagues. This vector is a derivative of pBR322 and is much smaller 2.7kb. It has all the essential parts of pBR322. *e.g.* The ampicillin resistant genes and Col E ori of replication.

The second scorable marker is due to *E. coli* gene lac Z encoding the fragment of ß-galactosidase, which aplits lactose into glucose and galactose. When *E. coli* cells are transformed by these pUC vectors there will be

- Cells which are transformed but non recombinant these cells are ampr and able to synthesize ß-galactosidase when an inducer of lac operon IPTG and X-gal (5 Bromo 4 chloro 3 Indolyl ß-D galactopyranoside) a substrate for ß-galactosidase was added in the media, it will be broken down by the enzymes to give deep blue colonies
- Cells with recombinant pUC vectors – these cells are ampr but unable to synthesize ß-galactosidase. When IPTG and X-gal are put in the media it will give white colonies because the substrate cannot be broken down. Recombinants can therefore be selected by the colour of the colonies and there will be no need of replica plating as is done incase of pBR322

  Cells plated + agar + ampicillin + IPTG + X-gal = Non recombinant -Blue colonies

  Cells plated + agar + ampicillin + IPTG (Isopropyl thio-galactoside) + X-gal = Recombinant-white colonies

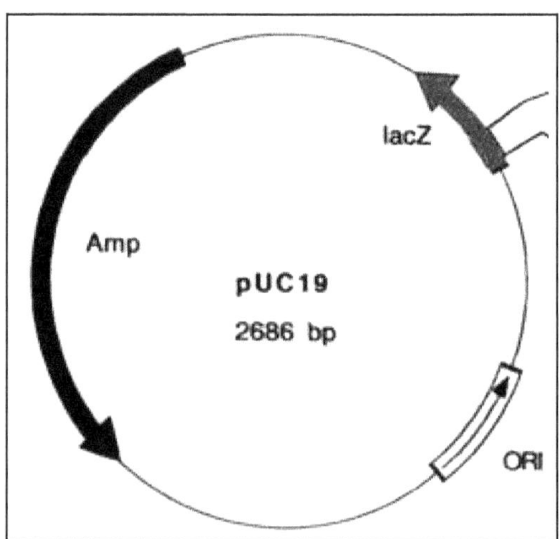

3. *Ti plasmid and Ri plasmid*

The Ti plasmid and Ri plasmid is a large conjugative plasmid or megaplasmid of about 200 kb. These are commonly used plasmids for transformation of plant cells.

- *Agrobacterium tumefaciens* has the Ti plasmid (ca 200 kb) while *Agrobacterium rhizogenes* has the Ri plasmid have similar general features and can be interchanged between the two species.

☆ These are plant pathogenic gram negative soil bacteria to cause crown gall (*A. tumefaciens*) and hairy root (*A. rhizogenes*) diseases of dicot plants. They infect plant cells near wounds, usually at the crown of roots at the soil surface.

☆ These plasmids naturally transfer a part of their DNA, the T – DNA into the host plant genome, which makes *Agrobactrium* a natural genetic engineer.

☆ The T – region (ca 23 kb) which contain genes for opine metabolism and phytohormone independence, this region is transferred into host cells and is integrated into their genome. They have another region, called 'vir' region, which produces, an endonuclease essential for the excision transfer of T-region into plant cells.

☆ The genes to be transferred are placed within T DNA and contain eukaryotic regulatory sequences. As a result these genes are expresses only in plant cells. They are not expressed in the Agrobacterium.

Bacteriophage Vectors

Bacteriophages are viruses that infect bacteria. These are usually called phages. The phages are constructed from two basic components.

i. Capsid (or) protein coat within which the nucleic acid genome is enclosed.
ii. Nucleic acid genome which is packaged with in the capsid phages can be both DNA and RNA phages (MS2)

The DNA phages can be double standard (T2 T4 T6), (or) single standard (X 174 and M-13).

RNA virus – MS2

Several bacteria phages are used as cloning vectors, the most commonly used bacteria phage vectors being (lambda) phage, M-13 phages.

Phage vectors present two advantages over plasmid vectors

1. They are more efficient than plasmids for cloning of large DNA fragments. The length of the DNA segments is to be cloned is upto 25 kb.
2. It is easier to screen a large number of phage plaques than bacterial colonies for the identification of recombinant pla ques/clones.

1. Bacteriophage vectors (Lambda)

The genome (48.502 base pairs) contains an origin of replication and genes for head and tail proteins and enzymes of DNA replication, lysis and lysogeny, and single stranded protruding cohesive ends of 12 bases at its 51 ends.

☆ These two cohesive ends are referred to as cos sites (sticky (or) cohesive ends) these cohesive ends enable DNA to form a circular molecule when it is injected into the *E. coli* cells.

☆ The central part (Nonessential segment) of the chromosome may be excised with restriction enzyme and replace with foreign DNA.

Advantages

- ✰ Large size DNA fragment upto 25 kb can be cloned as compared to plasmid vector which can take up 10kb
- ✰ Large number of phage plaques can be easily screened
- ✰ They are more efficie nt than plasmids for cloning of large DNA fragments

2. *Phage M-13 Vectors*

They are derived from the 6.4 kb genome of the E. *coli* filamentous bacteriophage M13.

- ✰ This phage has a single stranded linear DNA genome in phase particles which converts into double stranded circular DNA molecule in the host cells.
- ✰ It contains origin of replication and a scorable marker gene lac-Z that compliments the gal host giving blue colonies.
- ✰ On transformation only white (or) clear plaques are obtained, thus permitting easy selection of recombinant plaques.

Cosmid Vectors

Cosmids are essentially plasmids that contain a minimum of 250bp of DNA including *cos* site and sequences needed for binding of and cleavage by terminase

A typical cosmid has

- ✰ Origin of replication
- ✰ Selectable markers from a plasmid ampr and tetr
- ✰ Cloning sites (unique restriction sites) and Cos site (The sequence yielding cohensive ends) of which are essential for efficient packing of -DNA into virus particle which infect host cells. Packaged cosmids infect host cells like -particle but inside the host they replicate and propagate like plasmids.

The typical features of cosmids are as follows:

i. Cosmids can be used to clone DNA inserts upto 45kb

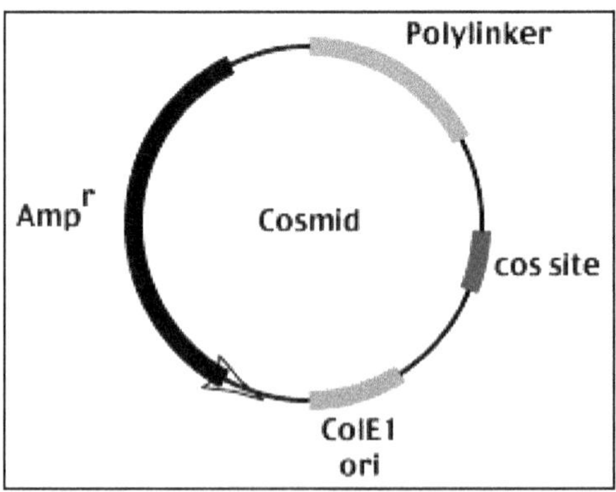

ii. They can be packaged into -particles which infect host cells, which is many fold more efficient than plasmid transformation.

iii. Selection of recombinant DNA is based on the procedure applicable to the plasmid making up the cosmid.

iv. Finally, these vectors are amplified and maintained in the same manner as the contributing plasmid

e.g. Cosmid PHC-79 and cosmid PJB-B

Phagemid Vectors

Plasmids + Bacteriophage, a plasmid vector that contains the original from the page, in addition to that of plasmid.

These are artificially constructed vectors containing the characters of plasmids and bacteriophages. These consist of socrable and selectable markers for antibiotic resistance.

e.g. PUC-118, PUC-119

Yeast Vector

Yeast is an Eucaryotic which can grow as a single cell and produce colonies on agar plate. The vectors used in yeast are

1. Plasmid vectors
2. ARS (Autonomously Replicating sequences) vectors.
3. Minichromosome vectors
4. Yeast artificial chromosomes (YAC). Among these the last type of vectors are most commonly used

Yeast Artificial Chromosome Vectors

These are developed by David Burke *et al.*, in 1987. These are linear vectors that behave like a Yeast chromosome hence they are called yeast artificial chromosome vectors.

A typical YAC contains the following functional elements from yeast. *e.g.* pYAC 3

i. An ARS sequence for replication

ii. A CEN4 sequence – centromeric sequence function

iii. A TEL sequence (Telomeric sequence) at two ends for protection from exonuclease action.

iv. One (or) two selectable marker genes *e.g.* TRP 1 and URA 3

v. SUP4, a selectable marker into which the DNA insert is integrated.

vi. And necessary sequences from *E. coli* plasmid for selection and propagation in *E. coli* YAC's are used for cloning very large (1000-2000kb) DNA segments used for mapping of complex Eukaryotic chromosomes.

The problem faced with YAC's is the recombination between the copies of inserts and more particularly deletions in DNA vectors.

Bacterial Artificial Chromosome (BAC)

- ☆ Artificial chromosomes are circular or linear vectors that are stably maintained in, 1 to 2 copies per cell.
- ☆ Bacterial artificial chromosome is another cloning vector system in *E.coli* developed by Melsimon and his colleagues have the origin of replication 'Ori'S' of *E.coli* 'F' factor which allows a strict copy number control at 1/2 copies per cell.
- ☆ The low copy number helps to maintain the DNA inserts without any change arising from recombination between the copies of DNA inserts and avoids any counter selection that may arise due to over expression of cloned genes.
- ☆ These vectors are used to clone the DNA inserts up to 300kb.
- ☆ These vectors are able to maintain in stable state and extensively used in analysis of large genomes but the main disadvantage of BAC vectors is somewhat laborious construction of BAC libraries.

Phasmid Vectors

- ☆ These vectors are shortened linear genomes containing DNA replication and lytic functions plus the cohesive ends of the phage.
- ☆ Their middle non-essential segement is replaced by a linerized plasmid with intact replication nodule.
- ☆ A phasemid vector contains several tandem copies of the plasmid to make it longer than 38 kb, the minimum size needed for packaging in particles

What is meant by binary vector? Enlist the key steps involved in the construction of binary vectors?

Binary Vector/Shuttle Vector

It is a standard tool designed for transformation of higher plants mediated by *Agrobacterium tumefaciens*.

- ☆ These vectors replicate in two different species *i.e.* both *E. coli* and *A. tumefaciens*. Therefore it contains both origins of replication, one specific for each host species, as well as genes necessary for their replication, which will not provided by the host cells.
- ☆ It also consists of borders of T- DNA, multiple cloning sites, selectable marker genes, reporter genes, and other accessory elements that can improve the efficiency of vector system.

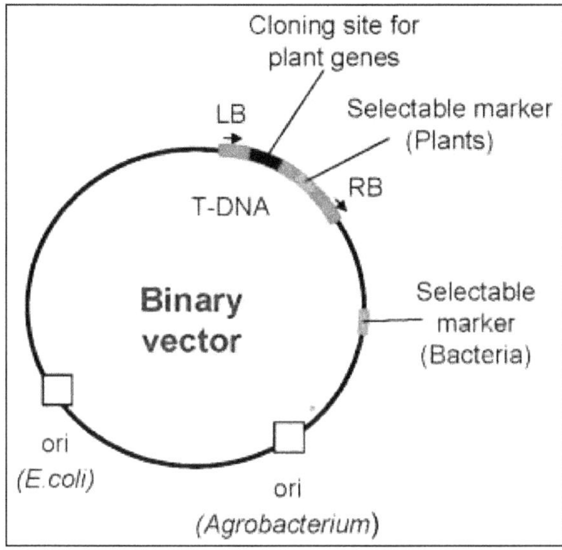

Steps Involved in Construction of Binary Vectors

1. Obtain plasmids and other DNA fragments necessary for construction of vectors.
2. Combine the bacteria selectable marker and the plasmid replication functions for *E. coli*
3. Insert the plasmid replication functions for *A. tumefaciens*
4. Insert the plasmid mobilization functions.
5. Insert the right border, left border and multiple cloning sites to give the empty vector.
6. Construct the expression unit of the selectable marker gene separately.
7. Insert the unit in to the empty vector to give the selection vector.
8. Construct the expression unit of the reporter gene separately.
9. Insert the unit into the selection vector (vector with plant selectable marker) to give the reporter vector (vector with both a reporter gene and a selectable marker gene).

What are the different methods of gene isolation? Describe the essential steps involved in identification of target gene from any plant species using map based cloning method?

Gene/DNA fragment of interest can be isolated by any of the following methods:

1. From genomic libraries
2. From cDNA libraries
3. Chemical synthesis

4. Transposon tagging
5. Map based cloning
6. Polymerase chain reaction

Map Based Cloning Approach

Positional cloning or map based cloning can be used to identify any gene, given an adequate map of the region of the chromosome in which it is located.

- ✰ This strategy is used when no information is available about the gene or its product.
- ✰ This involves the identification of gene using molecular maker linkage analysis and isolation through chromosome walking using linked molecular markers like RFLP, RAPD, AFLP and SSRs *etc.*

Steps

1. Fine mapping of target trait.
2. Identification of tightly linked flanking DNA markers.
3. Physical mapping to know the approximate distance between marker and gene in base pairs.
4. Screening of large fragment genomic DNA library such as YAC, using flanking DNA markers as a probe.
5. Chromosome landing, walking, if needed.
6. Identification of large fragment DNA clones spanning the target gene region.
7. Screening of cDNA library using either the identified DNA clones or the exon sequences contained in the cloned DNA fragment as probe.
8. Identification and characterization of cDNA clone showing positive signal.
9. Transformation of crop genotype using cDNA clone.
10. cDNA that confers the expected trait phenotype in the transformed genotype is the target gene.

Explain in brief about the following?

1. **Chromosomal walking**
2. **Chromosomal jumping**

Chromosome Walking

Positional cloning is accomplished by mapping the gene of interest, identifying an RFLP or other molecular marker near the gene and then walking or jumping along the chromosome until the gene is reached.

Major steps involved here are:

1. Prepare the restriction map of a clone

2. Sub clone the restricted fragment
3. Screen genomic library with sub clone as probe
4. Prepare restriction map for new clone
5. Sub clone restricted fragment
6. Rescreen library with new sub clone as probe
7. Repeat 1-3 steps for as many cycles as needed to reach the gene of interest

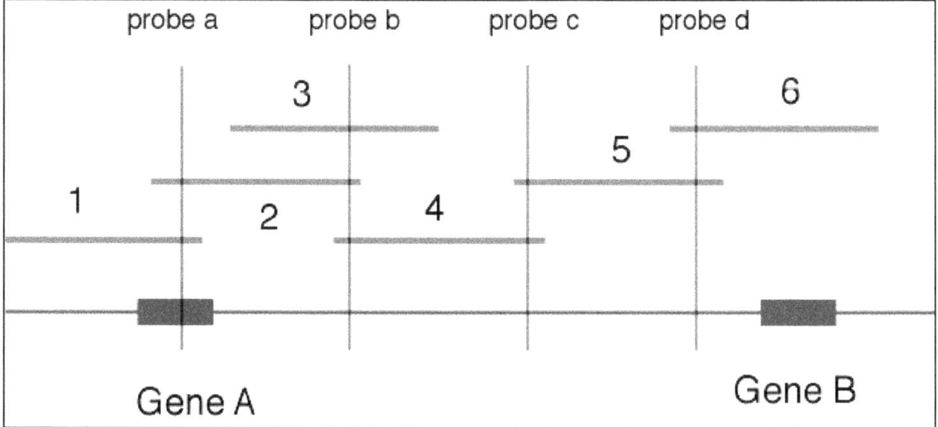

- ✰ Chromosomal walks are initiated by the selection of molecular markers (RFLP) or known gene clone close to the gene of interest and use of this clone as a hybridization probe to screen a genomic library for overlapping sequences.
- ✰ Chromosome walking is very difficult in species with large genome and abundance of dispersed repetitive DNA.
- ✰ Chromosome walking is easier in organisms such as *Arabidopsis thaliana* and *Ceanorabditis elegans* which have small genomes and little dispersed repetitive DNA.

Chromosomal Jumping

When a distance from the closest molecular marker to the gene of interest is large, a technique called Chromosome jumping can be used to speed up an otherwise long walk.

In this, main steps involved are:

1. Digestion of long DNA fragment using restriction enzymes.
2. After restriction, circularize the small DNA fragments
3. Cleave the DNA on either side of the junction
4. Use the cleaved fragments as probes for jumping along the chromosome.
   - ✰ Each jump over a distance of 100 kb or more.

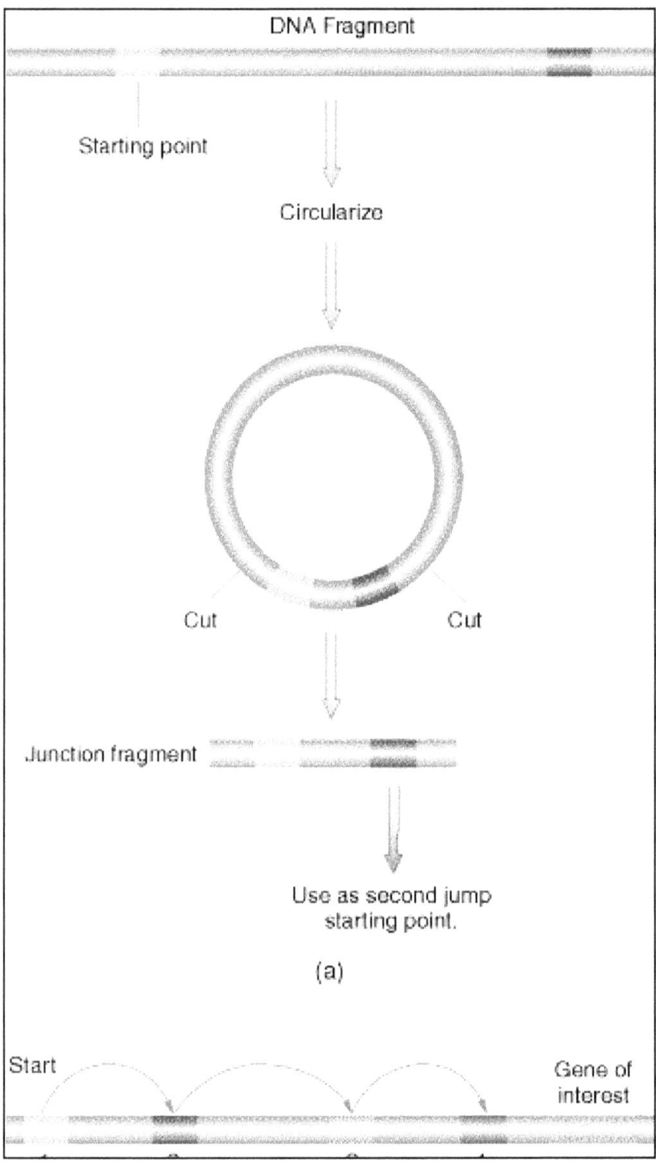

- ☆ Chromosome jumping as a short cut method for replacing long chromosomal walks.
- ☆ This procedure can also be used to jump over repetitive DNA sequences that block chromosomal walks.

5

Mutations and *In vitro* Mutagenesis

Write short notes on the following?

1. **Define mutation? And explain in brief about the molecular basis of mutations?**
2. **Frame shift mutations**
3. **Suppressor mutations**
4. **Base analogs**

Mutation

Mutation can be defined as a sudden heritable change in the character of an organism which is not due to either segregation or recombination.

Molecular Basis of Mutations

The term mutation is presently used to cover only those changes which alter the chemical structure of the gene at molecular level. Such changes are commonly referred to as "point mutations".

- ☆ Point mutations involve a change in the base sequence of a gene which results in the production of a mutant phenotype.
- ☆ Point mutations can be subdivided into the following three classes on the basis of molecular change associated with them.

  1. Base substitution 2. Base deletion and 3. Base addition

1. **Base Substitution:** When a single base in a DNA molecule is replaced by another base it is known as base substitution. This can be of two types.
   a. **Transition:** Replacement of a purine by another purine or a pyrimidine by another pyrimidine. (or) The substitution of a purine by another purine or of a pyrimidine by another pyrimidine base in DNA or RNA is known as transition. (A G or C T).
   b. **Transversion:** Replacement of a purine by a pyrimidine and vice versa. (or) The substitution of a purine by a pyrimidine or of a pyrimidine by a purine in DNA or RNA is known as transversion. (A or G C or T or U).
2. **Base deletion:** In base deletion, one or more bases are altogether deleted.
3. **Base addition:** There is insertion of one or more bases.

If the number of bases added or deleted is not a multiple of three, a frameshift mutation is obtained, as the reading frame in such case is shifted from the point of addition or deletion onwards.

Frame Shift Mutations

A mutation produced by either deletion or addition of bases that are not in multiples of three as a result, reading frame changes down stream of the site of addition or deletion.

- ☆ If addition and deletion will occurs at same site, it leads to restoration of normal reading frame *i.e.* double frame shift mutations.
- ☆ Generally, frame shift mutations arise mainly by error during replication or due to acridine dyes.
- ☆ Frame shift mutations result in nonfunctional proteins, so these are more deleterious than mutations produced by base substitutions.
- ☆ Frameshifting is associated with specific tRNAs in two circumstances:
  i. Some mutant tRNA supressors recognize a "codon" for four bases instead of the usual three bases.
  ii. Certain "slippery" sequences allow a tRNA to move a base up or down mRNA in the A site *i.e.* programmed frameshifting.

Suppressor Mutations

A mutation, which suppresses the phenotypic expression of another mutation: usually in another gene or in the same gene at a different site.

- ☆ It is a second mutation that alleviates or reverts the phenotypic effects of an already existing mutation.
- ☆ Intragenic suppression results from suppressor mutations that occur in the same gene as the original mutation.
- ☆ Intergenic (also known as extragenic) suppression relieves the effects of a mutation in one gene by a mutation somewhere else within the genome.

- ✰ Intergenic suppression is useful for identifying and studying interactions between molecules, such as proteins.

Base Analogs

Base analogs have structures similar to the normal bases and are incorporated in to DNA during replication to exert their mutagenic effects.

- ✰ The two most commonly used base analogs are 5-bromouracil and 2-aminopurine.
- ✰ The pyramidine 5-bromouracil is a thymine analog, the bromine at the 5th position is similar to the methyl group at the 5th position in thymine.
- ✰ This bromine at this position changes the charge distribution and increases the frequency of tautaomeric shifts.
- ✰ In its more stable keto form, 5-bromouracil pairs with adenine. After a tautomeric shift to its enol form, it can pairs with guanine. So it will cause an A:T!G:C transition.
- ✰ 2-aminopurine acts in a similar manner but is incorporated in place of a adenine or guanine.

Define *in vitro* mutagenesis? And explain in brief about the various techniques available for performing *in vitro* mutagenesis?

***In vitro* mutagenesis** method in which a mutation is produced in a segment of cloned DNA. The DNA is then inserted into a cell or organism, and the effects of the mutation are studied.

- ✰ It is mainly used to make specific and intentional changes to the DNA sequence of a gene and any gene products.
- ✰ It is more specifically referred to as site-directed mutagenesis or oligonucleotide-directed mutagenesis.
- ✰ Modifications introduced in to DNA molecules can be classified in to three groups:
  - i. Deletion, the elimination of a given number of nucleotides from a segment of DNA.
  - ii. Insertion, the addition of a given number of nucleotides to a segment of DNA.
  - iii. Substitution, the exchange of certain nucleotides with others without changing the length of the affected segment of DNA.

Many different techniques are available for performing *in vitro* mutagenesis.

Classical in vivo Mutagenesis

- ✰ Earlier mutagenesis was done using radiation like X-rays or chemical mutagens like aminopurine, nitrosoguanidine, and bisulfate *etc.*

☆ In 1973, Charles Weismann used nucleotide analogue N-hydroxycytidine for mutagenesis, which induces transition of GC to AT. But these methods are not specific as site site directed *in vitro* mutagenesis.

Classical Site Directed *in vitro* Mutagenesis

☆ In 1971, Clyde Hutchinsosn and Marshall Edgell showed that it is possible to produce mutants with small fragments of phage Öx174 and restriction nucleases.

☆ In 1978, Hutchinson and Michael Smith, given more flexible approach to site-directed mutagenesis using oligonucleotides in a primer extension method with DNA polymerase *i.e.* oligonucleotide mediated mutagenesis.

☆ This oligonucleotide mediated mutagenesis is mainly based on the simple concepts like

i. A synthetic nucleotide encoding the desired mutation is annealed to the target region of the wild-type template DNA where it serves as primer for initiation of DNA synthesis *in vitro*.

ii. Extension of the oligonucleotide by a DNA polymerase generates a double stranded DNA that carries the desired mutation.

iii. The mutated DNA is than inserted at the appropriate location of the target gene, and the mutant protein is expressed.

☆ Throughout the 1980's protocols for oligonucleotide-mediated mutagenesis typically involves the:

a. Design and synthesis of mutagenic nucleotides.

b. Hybridization of mutagenic nucleotides to single-stranded target DNA cloned in to bacteriophage or phagemid vector. Use of such single stranded templates eliminates competition between the mutagenic oligonucleotide and a complementary strand of DNA.

c. Extension of hybridized oligonucleotide by DNA polymerase in the presence of all four deoxynucleoside triphosphates (dNTPs).

d. Formation of closed circular DNA by ligation with DNA ligase.

e. Transfection of susceptible bacteria.

f. Screening of clones by hybridization to the mutagenic oligonucleotide primer for those carrying the desired mutation.

g. Preparation of single stranded DNA from the mutagenized clones.

h. Confirmation by DNA sequencing that the target DNA in the mutagenized clones carries the desired mutation.

i. Recovery of the mutated fragment of the DNA.

j. Substitution of the mutagenized fragment for the corresponding segment of wild-type DNA.

☆ In classical site directed *in vitro* mutagenesis, the proportion of clones containing the desired mutation varies from 0.1 per cent to >50 per cent depending upon the efficiency of the mutagenic oligonucleotide.

- It is extremely efficient and reliable method of mutagenesis over the years, but now a days site directed mutations are more easily generated by using PCR based methods.

PCR Mediated Site Directed Mutagenesis

In this, amplification with mutated oligonucleotide primers produces a fragment containing the desired mutation in sufficient quantity to be separated from the original and unmutated fragments by gel electrophoresis.

- In this, primers designed with mutations that can introduce small sequence changes, and primer extension or inverse PCR can be used to achieve longer mutagenic regions.
- Of the many variants of PCR-based mutagenesis, two methods have long durability and robustness. Those are overlap extension mutagenesis and megaprimer mutagenesis.

Overlap Extension Mutagenesis

- In this, two overlapping DNA fragments are amplified in separate PCRs.
- The mutation of interest is constructed in the regions of overlap and is present in both amplified fragments.
- The overlapping fragments are mixed and, in third PCR, are amplified into a full-length DNA using two primers that bind to extremes of the two initial fragments.
- The method is most effective, but it requires two mutagenic primers, two flanking oligonucleotides, and three PCRs to construct a mutation.

Megaprimer Based Mutagenesis

It is the simplest and cost effective method of PCR-based mutagenesis currently available.

- This method involves two rounds of PCR that employ two flanking primers and one internal mutagenic primer containing the desired base substitutions.
- The flanking primers can be complementary to sequences in the cloned gene or to adjacent vector sequences.
- The mutagenic primer is always oriented towards the nearer of the two flanking primers so that the length of the megaprimer is kept to a minimum.
- In this, the first round of PCR is performed with the mutagenic primer, the nearer of the flanking primers, and a wild-type DNA template.
- The product of this reaction, a double stranded megaprimer, is purified and used in a second PCR together with the reverse flanking primer.

- ✰ The product is a double stranded DNA that contains the mutation and whose size is equal to the distance between the two outside flanking primers.
- ✰ If the flanking primers contain restriction sites for ease of cloning, the product of the second-round PCR can be used to reconstruct a mutagenized version of the target gene.
- ✰ The chief problem with this megaprimer method is the time-consuming and laborious purification step. Purification is required to remove residual primers from the amplified megaprimer synthesized in the first PCR.
- ✰ This simple solution to this difficulty is to use flanking primers with significantly different melting temperatures (T_m) to prime the two PCRs.

Advantages of PCR-Based Site Directed Mutagenesis

- ✰ Speed and ease.
- ✰ High rates of recovery of mutants.
- ✰ Availability of commercial kits.
- ✰ No need of cloning bacteriophage vectors.
- ✰ Ability to use double stranded DNA templates and to introduce mutations at almost any site.
- ✰ Use of high temperatures to reduce the ability of the template DNA to form secondary structures that lower the efficiency of the extension reaction on single stranded DNA templates.

Disadvantages of PCR-Based Site Directed Mutagenesis

- ✰ Relatively high rate of errors in PCR products that often contain unwanted mutations in addition to desired alterations. But this problem can be overcome by:
  - i. Limiting the number of amplification cycles and
  - ii. Instead of *Taq* polymerase, using of thermostable DNA polymerases, such as *Pfu* and *vent*, which carry a 3′-5′ exonuclease activity. So that introduction of unwanted nucleotides at the 3′ termini of amplified product can also be solved.
- ✰ Large number of primers and amplification reactions required for each mutagenesis experiment. *e.g.* Megaprimer mutagenesis.
- ✰ Requirement to optimize the conditions for PCR for each new set of primers and/or template.
- ✰ High frequency of unmutagenized clones resulting from contamination of the PCR amplified DNA with the parental, wild type DNA used as template in the PCR. But this can be solved by using a restriction enzyme such as *Dpn*I to digest the plasmid selectively.

- Inefficiency of amplifying DNA fragments longer than 2-3 kb by standard PCR.

How recombination occurs in bacterial, fungal and viral genomes? And also explain how this recombination permits the estimation of genetic distance in these organisms?

Recombination is nothing but the formation of new combinations of genes, either naturally, by crossing over or independent assortment, or in the laboratory by direct manipulation of genetic material.

Recombination in Bacteria

In bacteria, recombination occurs mainly by three mechanisms like, transformation, conjugation and transduction.

Transformation

Transformation is the process by which bacteria pick up DNA from their environment.

- Transformation was first discovered by Griffith in 1928 in *Diplococcus pneumonia* and in 1944 Avery *et al.*, showed that DNA was responsible for transformation.
- For successful transformation, bacteria must be competent. *i.e.* bacteria should have appropriate enzymes (the transformation machinery) required to transport the exogenous DNA in to the cell.
- By transformation, it is possible to estimate map distance by co-transformation (simultaneous transfer of two genes, it is quite frequent if they are closely linked).
- If two genes are closely linked so that they are frequently located in the same DNA fragment, similar double transformants are expected.
- If two genes are located far apart, they will frequently occur on different DNA fragments, therefore, double- transformation can occur only when the two specific fragments are taken up by the same recipient cell, and two independent recombinant events occur.
- So by analyzing the frequency of single and double transformation events, one can determine whether or not two genes arc liked.

$$\text{Co-transformation index} = \frac{\text{Number of double cross over recombinants}}{\text{Number of single cross over recombinants} + \text{Number of double cross over recombinants}}$$

Conjugation

It is a mating process involving bacteria.

- It involves transfer of genetic information from one bacterial cell to another, and requires physical contact between the two bacteria involved.

- It was first discovered by J. Lederberg and E. Tatum in 1946 in *E. coli.*
- The contact between the cells is via a protein tube called as F or sex pilus, which is also the conduit for the transfer of the genetic material.
- Basic conjugation involves two strains of bacteria F^+ and F^-. The difference between these two strains is the presence of fertility factors (or F factors) in the F^+ cells.
- Genetic transfer in conjugation is from an F^+ cells to an F^- cell, and the genetic material transferred is the F factor itself.
- Occasionally, the F factor integrates in to a random position in the bacterial chromosome, when this happens; the bacterial cell is called Hfr instead of F^+.
- Hfr bacteria are still able to initiate conjugation with F^- cells, but the outcome is completely different from conjugation involving F^+ bacteria.
- By conjugation, the location of gene in the chromosome is usually determined by interrupted mating technique.
- In this technique, the mating between Hfr and F^- cells in interrupted at regular intervals after mating initiation and recombinants for the genes are scored in the progeny.
- Matings are easily interrupted by vigorously agitating the cell suspension in a blender as this breaks the conjugation bridge between Hfr and F^- cells.
- The minimum time (*i.e.* duration of mating before it was interrupted) required for recombinants of the concerned genes to appear, gives the approximate location of the gene in the linkage map.
- The exact location of gene is still determined through transduction mapping or by three point test cross.

Transduction

It involves the exchange of DNA between bacteria using bacterial viruses (bacteriophage) as an intermediate.

- It was first time discovered by N.Zinder and J. Lederberg in 1952 in bacterium "*Salmonella typhemurium*".
- There are two types of transduction, generalized and specialized transduction, which differ in their mechanism and in the DNA that gets transferred.
- In generalized transduction, a random or nearly random segment of the bacterial chromosome is transferred by the virus.
- In specialized transduction, restricted set of bacterial genes is transferred to another bacterium.
- By transduction, the extent of linkage is estimated as the frequency of cotransduction *i.e.*, two genes being transduced together through a single transducing particle.

- ☆ The closer are the two genes, the higher will be the frequency of their transduction.

Recombination in Fungi

In general, fungi are haploid (n), and hence the product of meiosis of zygotes can be directly analyzed.

- ☆ In fungi, the spores produced are arranged as tetrads resulting from one complete meiotic division.
- ☆ In some fungi, spores produced are arranged in the precise sequence in which they are produced called ordered tetrads. *e.g. Neurospora*
- ☆ And in others, the spores are not present in the same sequence in which they were produced called unordered tetrads. *e.g.* Yeast.
- ☆ Of these, only ordered tetrads permit the estimation of recombination frequency between gene and its centromere.

In case of ordered tetrads, the recombination frequency between gene and its centromere will be calculated by using the formula:

$$\text{Recombination frequency (per cent)} = \frac{\frac{1}{2}\text{(Number of asci showing second division segregation)}}{\text{Total number of asci}} \times 100$$

- ☆ Here, arrangement of spores is mainly depends on presence or absence of crossing over between gene and its cetromere.
    a. If there is no crossing over, it is called first division segregation and it results in A: a (4:4) spore arrangement.
    b. If there is crossing over, it is called second division segregation and it results in any of four types of spore arrangements like 2A:2a:2A:2a, 2a:2A:2a:2A, 2A:4a:2A and 2a:4A:2a.

In case of unordered tetrads, the detection of linkage and the estimation of recombination frequency between genes is based on the classification of a large number of tetrads produced by the appropriate heterozygote into different classes. This is called as tetrad analysis.

a. Parental ditypes (PD), *e.g.* Tetrads having AB and ab spores, here the chromosome carrying genes A and B will move to one pole, while those carrying genes a and b will move to the opposite pole during anaphase I.
b. Non parental ditypes (NPD), *e.g.* Tetrads having Ab and aB spores, here chromosomes having genes A and b will move to one pole, and those having a and B to the opposite pole during anaphase I.
c. Tetratypes (T), *e.g.* Tetrads contains four different types of spores representing all the four of the genotypes like AB, ab, aB and Ab.

- ☆ If the two genes are linked, parental ditype tetrads will be more frequent, non parental ditype tetrads will be less frequent, while tetratype tetrads will have an intermediate frequency.
- ☆ In contrast, if two genes are segregating independently, parental ditype and nonparental ditype tetrads will be more frequent and their frequencies will be comparable, while tetratype tetrads will be relatively infrequent.

In case of these unordered tetrads, the recombination frequency between gene and its centromere will be calculated by using the formula:

$$\text{Recombination frequency (per cent)} = \frac{\frac{1}{2}\text{(No. of tetratype asci)} + \text{No. of nonparental ditype asci}}{\text{Total number of asci}} \times 100$$

Recombination in Viruses

The phenomenon of recombination in viruses was first described by A.D. Hershey and R. Rotman in 1945.

- ☆ Viral chromosomes undergo recombination during their reproduction in host cells.
- ☆ For this, host cell must be simultaneously infected by two different strains/ genotypes of a virus in order to permit their chromosomes to undergo recombination.
- ☆ This is achieved by infecting host cells with a mixture of the two selected virus strains. In such cells, chromosomes of the two strains may undergo recombination to yield recombinant progeny phages.

  e.g. A recombination between a^+b and ab^+ chromosomes would yield the wild type a^+b^+ and ab progeny.
- ☆ By this, linkage maps of several viruses have been prepared and found to be circular. *e.g.* T_2 and T_4 phages. Recombination among the r^{II} mutants of T_4 phage was extensively studied by Benzer for developing the fine structure of r^{II} locus.

6

Plant Molecular Biology

Answer the following questions.

1. **Explain in brief about the mitochondrial and chloroplast genome?**
2. **Describe briefly the process of chloroplast transformation. And give its advantages over nuclear transformation?**

Mitochondrial and Chloroplast Genome

Both mitochondrion and chloroplast are semi autonomous organelles with DNA.

Mitochondrial DNA

- ☆ Mitochondrial DNA is called a mt DNA which is double stranded, circular and not covered by membrane.
- ☆ Size of it varies between species: Human mt DNA has 16 Kb and Plant cell has 200 Kb – 2500 Kb.
- ☆ It exists in dozens of copies in each mitochondrion. In Trypanosomes, each mitochondrion contains 1000 of copies of mt DNA arranged into a complex interlinked structure called Kinetoplast.
- ☆ Animal mitochondrial genome contains 37 genes that encode 13 proteins, 22 tRNAs, and 2 rRNAs. The 13 mitochondrial genes code for protein subunits of the enzyme complexes of the oxidative phosphorylation system, which helps mitochondria to act as the powerhouses of the cell.
- ☆ In this, genes transcribed in a polycistronic manner *i.e.* a large mRNA molecule with instruction to make many proteins simultaneously.

- ☆ This DNA mainly AT rich with lack of introns except some organisms like yeast mt DNA.

Chloroplast DNA

- ☆ Chloroplast DNA is called as ctDNA, cp DNA or plastome.
- ☆ Its genome size varies from 120 to 170 kb of DNA.
- ☆ It exists in multiple copies. Each chloroplast with several nucleoid regions containing 8-10 rings of DNA molecules. The number of DNA copies in mature chloroplast is 15-20.
- ☆ Chloroplast genome includes ~100 genes, 46-90 protein coding genes, 4 rRNA and 30 tRNA genes.
- ☆ In this also, majority of genes transcribed in a polycistronic manner.
- ☆ This DNA mainly AT rich.
- ☆ Many chloroplast genes contain introns, which form two general classe.
    a. Introns of tRNA genes, which are usually located in the anticodon loop and are similar to those found in yeast nuclear tRNA genes.
    b. The introns present in protein encoding genes are similar to those of mitochondrial genes.

Chloroplast Transformation

It is nothing but introduction of foreign genes in to the chloroplast genome. Choroplast transformation was first achieved in *Chlamydomonas reinharbdtii*, single cell green algae in 1988.

- ☆ Chloroplast transformation can be used in the production of transgenic plants with herbicide resistance, insect resistance, viral resistance, fungal resistance, abiotic and biotic stress tolerance, production of biopharmaceuticals *etc.*
- ☆ Biolistic/Particle bombardment method involves the introduction of *E. coli* plasmids containing a gene of interest and marker gene into chloroplasts or plastids.
- ☆ The insertions of foreign genes into plasmid DNA occur by homologous recombination via the sequences flanking at the insertion site.
- ☆ First successful chloroplast transformation was performed in *Chlamydomonas reinhardtii* by particle bombardment method. It involves simple operation and high transformation efficiency makes it a favourable way for plastid or chloroplast transformation.
- ☆ PEG- mediated and Agrobacterium- mediated transformation method was also employed in the early days.
- ☆ After the first chloroplast transformation in *Chlamydomonas reinhardtii*, the stable plastid transformation has been established in higher plants

like tobacco, Arabidopsis, rape, lesquerella, rice, potato, lettuce, soybean, cotton, carrot and tomato.

- ✰ However, plastid transformation is routinely performed only in tobacco because of higher efficiency of transformation in tobacco than other plants.

Advantages of Chloroplast Transformation

Chloroplast transformation offers several advantages over nuclear transformation.

1. Risk of transgene escape: Chloroplast genome is maternally inherited and there is rare occurrence of pollen transmission. It provides a strong level of containment and thus reduces the escape of transgene from one cell to other.
2. Expression level: It exhibits higher level of transgene expression and thus higher level of protein production due to the presence of multiple copies of chloroplast transgenes per cell.
3. Homologous recombination: Chloroplast transformation involves homologous recombination and is therefore precise and predictable.
   - ✰ This minimizes the insertion of unnecessary DNA that accompanies in nuclear genome transformation.
   - ✰ This also avoids the deletions and rearrangements of transgene DNA, and host genome DNA at the site of insertion.
4. Gene silencing or RNA interference does not occur in genetically engineered chloroplasts.
5. In chloroplast transformation, absence of position effect due to lack of a compact chromatin structure and efficient transgene integration by homologous recombination.
6. In chloroplast transformation, because of ability to form disulfide bond formation and folding human proteins results in high level production of biopharamaceuticals in plants.
7. In chloroplast transformation, multiple transgene expression is possible due to polycistronic mRNA transcription.
8. High level of expression and engineering foreign genes without the use of antibiotic resistant genes makes this compartment ideal for the development of edible vaccines.
9. Choloplast is originated from Cyanobacteria through endosymbiosis. It shows significant similarities with the bacterial genome. Thus, any bacterial genome can be inserted in to chloroplast genome.
10. Foreign proteins observed to be toxic in the cytosol are non-toxic when accumulated within transgenic chloroplasts as they are compartmentalized inside chloroplast.

Define photosynthesis? How are C_4 plants superior to C_3 plants with respect to photosynthesis? What strategies have been used towards engineering C_4 pathway in C_3 crops?

Photosynthesis

Photosynthesis is a vital physiological process where in the chloroplast of green plants synthesizes sugars by using water and carbon dioxide in the presence of light.

- Photosynthesis literally means synthesis with the help of light *i.e.* plant synthesize organic matter (carbohydrates) in the presence of light.
- Photosynthesis is sometimes called as carbon assimilation (assimilation: absorption into the system).
- This is represented by the following traditional equation.

$$6CO_2 + 12H_2O \xrightarrow[\text{Green pigments}]{\text{Light}} \underset{\text{Carbohydrates}}{6O_2 + C_6H_{12}O_6}$$

C_3 Plants

The majority of terrestrial plants, including many important crops such as rice, wheat, barley, soybean, and potato, assimilate atmospheric CO_2 directly through the C_3 photosynthetic pathway, also known as the Calvin cycle, and these are classified as C_3 plants.

- The enzyme of primary CO_2 fixation in this pathway, ribulose 1,5-bisphosphate carboxylase/oxygenase (Rubisco), reacts not only with CO_2 but also with O_2, leading to photorespiration, which essentially wastes assimilated carbon.
- Under current atmospheric conditions, potential photosynthesis in C_3 plants is suppressed by oxygen by as much as 40 per cent.
- The extent of suppression further increases under stress conditions such as drought, high light, and high temperature, through a decline of the CO_2 concentration inside leaves due to closure of stomata.

C_4 Plants

C_4 plants such as maize, sorghum, and sugarcane have evolved a novel biochemical mechanism to overcome photorespiration. In addition to the C_3 pathway, they use the C_4 photosynthetic cycle to elevate the CO_2 concentration at the site of Rubisco and thus suppress its oxygenase activity.

- Leaves of C4 plants have two types of photosynthetic cells, the mesophyll cell (MC) and bundle sheath cell (BSC).
- While all the photosynthetic enzymes are confined in MCs in C3 plants, they are localized in MCs and/or BSCs in C4 plants.
- In addition, C4 plants show extensive venation, with a ring of BSCs surrounding each vein and an outer ring of MCs surrounding the bundle sheath. This unique leaf structure, known as Kranz anatomy.

C_4 plants - agronomically desirable traits like

a. Higher photosynthetic capacity/high carbon assimilation
b. Higher growth rate and bio mass production
c. High nutrient and water use efficiency
d. Biofuel production.

- The main feedstocks for bioethanol are sugarcane (*Saccharum officinarum*) and maize (*Zea mays*), both of which are C_4 grasses, highly efficient at converting solar energy into chemical energy, and both are food crops.
- As many C_4 plants have high light, water and nitrogen use efficiency, as compared with C_3 species, they are ideal as feedstock crops.
- In this way, photosynthesis may be exploited to generate liquid transportation fuels. For example, more than 50 per cent of Brazil's transport fuel is comprisedof bioethanol (Pohit *et al.*, 2009) made from fermenting sugar extracted from sugar cane (*Saccharum* hybrid)

Strategies for Engineering of C_4 Pathway in C_3 Plants

Approaches for improvement of photosynthetic efficiency in C_3 plant involves

1. Genetic manipulation of Rubisco
2. Over expression of C_4 enzymes in C_3 plants

Genetic Manipulation of Rubisco

One major objective for Rubisco manipulation has been to alter the discrimination between CO_2 and O_2 (*i.e.* to alter the specificity factor, t).

- Mutagenesis in *in vitro* has been used to make changes to DNA encoding both large and small subunits of Rubisco
- In tobacco changing leucine 335 for valine decreased the both specificity and carboxylation rates to 25 per cent of the wild type (Whitney, 1999).
- Specific factor for mutant with a two amino acid extension (aspertate-lysine) extension at the C-terminus was at least 10 per cent higher than the wild type chlamydomonas (Zhu *et al.*, 1998).
- Recently Ishikawa *et al., 2011* introduced the small subunit (RbcS) of high catalytic turnover rate (kcat) Rubisco from the C_4 plant sorghum (*Sorghum bicolor*) significantly enhances kcat of Rubisco in transgenic rice (*Oryza sativa*).
- Decresing of photorespiratory losses can also possible by plants in which chloroplastic glycolate can be converted directly to glycerate in the chloroplast by introducing genes of the *E.coli* glycolate catabolic pathway in to *Arabdiopsis thaliana* (Kebeish *et al.*, 2007).

Over Expression of C_4 Enzymes in C_3 Plants

Previously, attempts have been made to transfer C_4 traits to C_3 plants by conventional hybridization between C_3 and C_4 plants. However, this approach was available only in several plant genera such as *Panicum, Moricandia, Brassica, Atriplex,* and *Flaveria.* Moreover, most C_3-C_4 hybrids showed infertility due to abnormal chromosome pairing and/or genetic barriers. Recent developments in plant genetic engineering have enabled us to introduce the desired genes encoding C_4 enzymes into C_3 plants. In the past several years a variety of "C_4 transgenic" C_3 plants have been produced.

But once we over express C4 enzymes in C3 plants it leads to that cell specific over expression ie mesophyll cell over expression and bundle sheath cell expression for example PEPC enzyme overexpress in mesophyll cell only.

Over Expression of Phosphoenolpyruvate Carboxylase (PEPC)

- ✰ The efficient fixation of Atmospheric CO_2 by PEPC-prerequisite for establishing C_4 cycle in C_3 plants, we need to express this enzyme in C_3 plants.
- ✰ PEPC gene isolated from maize (Izui *etal.*,1986) and overexpressed into tobacco (Hudspeth *et al.*, 1992).
- ✰ Resulting transgenic plants showed PEPC activity-2 fold more, but no significant increase in CO_2 assimilation.
- ✰ Over expression of PEPC – Rice: Over expression of PEPC gene from maize into rice leads to 110 fold increased activity of PEPC.
- ✰ Overexpression of two bacterial PEPC genes from *Escherichia coli* and *Corynebacterium glutamicum* under results in 5-fold increase in PEPC activity in potato leaves.
- ✰ Combined over expression of PEPC with PPDK in rice plants results in 35 per cent increase in photosynthetic efficiency and 25 per cent increase in grain yield (Ku *et al.*, 2001) it also causes decreased rate of photorespiration, increased rate of CO_2 assimilation.

What is meant by respiration? Explain in brief about aerobic and anaerobic respiration?

The cellular oxidation or break down of carbohydrates into CO_2 and H_2O, and release of energy is called as respiration. It is a reverse process of photosynthesis. In respiration, the oxidation of various organic food substances like carbohydrates, fats, proteins *etc.* may take place. Among these, glucose is the commonest.

$$C_6H_{12}O_6 + 6O_2 \rightarrow 6CO_2 + 6H_2O + \text{Energy (686 kcal)}$$

Degradation of organic food for the purpose of releasing energy can occur with or without the participation of oxygen. Hence, respiration can be classified into two types; aerobic and anaerobic respiration.

Aerobic Respiration

Aerobic respiration takes place in the presence of oxygen and the respiratory substrate gets completely oxidized to carbon dioxide and water as end products.

$C_6H_{12}O_6 + 6O_2 \rightarrow 6CO_2 + 6H_2O$ + Energy (686 kcal)
(Glucose)

This type of respiration is of common occurrence and it is often used as a synonym of respiration.

Anaerobic Respiration

It takes place in the absence of oxygen and the respiratory substrate is incompletely oxidized. Some other compounds are also formed in addition to carbon dioxide. This type of respiration is of rare occurrence but, common among microorganisms like yeasts.

$C_6H_{12}O_6 \rightarrow 2C_2H_5OH + 2CO_2$ + 56 kcal
(Glucose) (Ethanol)

Explain in detail about glycolysis/Embden – Meyer Hof – Paranas (EMP) pathway of energy production?

Glycolysis can take place even in the absence of O2. One molecule of the 6 carbon compound, glucose is broken down through a series of enzyme reactions into two 3-carbon compounds, the pyruvic acid. Glycolysis takes place in the cytoplasm and it does not require oxygen. Hence it is an anaerobic process.

1. Glucose molecules react with ATP molecules in the presence of the enzyme hexokinase to form glucose -6- phosphate.

   Glucose + ATP → Glucose -6- phosphate + ADP

2. Glucose-6-phosphate is isomerised into fructose-6-phosphate in the presence of phospho hexose isomerase.

   Fructose + ATP → Fructose -6- phosphate + ADP

3. Fructose-6-phosphate reacts with one molecule of ATP in the presence of phosphor hexo kinase forming fructose 1, 6-disphosphate.

   Fructose – 6- phosphate + ATP → Fructose -1,6- biphosphate + ADP

4. Fructose 1, 6 diphosphate is converted into two trioses, 3-phospho glyceraldehydes and dihydroxy acetone phosphate in the presence of aldolase.

   Fructose -1,6- biphosphate → 3-phospho glyceraldehyde+ DHAP

5. 3-phosphoglyceraldehyde reacts with H_3PO_4 and forms 1,3-diphospho-glyceraldehyde where, the reaction is non-enzymatic.

6. 1, 3-Diphosphoglyceraldehyde is oxidized to form 1,3- diphosphoglycerate in the presence of triose-phosphate dehydrogenase and coenzyme NAD^+. The NAD^+ acts as hydrogen acceptor and reduced to $NADH^+ + H^+$ in the reaction.

Glyceraldehde -3- phosphate + NAD + Pi → 1,3- diphosphoglycerate + NADH

7. 1,3-Diphosphoglycerate reacts with ADP in the presence of phosphoglyceric transphorylase (kinase) to form 3 phosphoglyceric acid and ATP.

   1,3- diphosphoglycerate + ADP → 3, Phosphoglycerate + ATP

8. 3, Phosphoglycerate → 2, Phosphoglycerate acid is isomerized into 2,phosphoglyceric acid in the presence of the enzyme, phospho glycero mutase

   3, Phosphoglycerate → 2, Phosphoglycerate

9. 2 phosphoglyceric acid is converted into 2-phosphoenolpyruvic acid in the presence of enolase.

   2, Phosphoglycerate → Phosphoenol pyruvate + H_2O

10. 2 phospho enol pyruvic acid reacts with ADP to form one molecule each of pyruvic acid and ATP in the presence of pyruvate kinase.

    Phosphoenol pyruvate + ADP → Pyruvate + ATP

☆ Glycolysis or EMP pathway is common in both aerobic and anaerobic respiration.

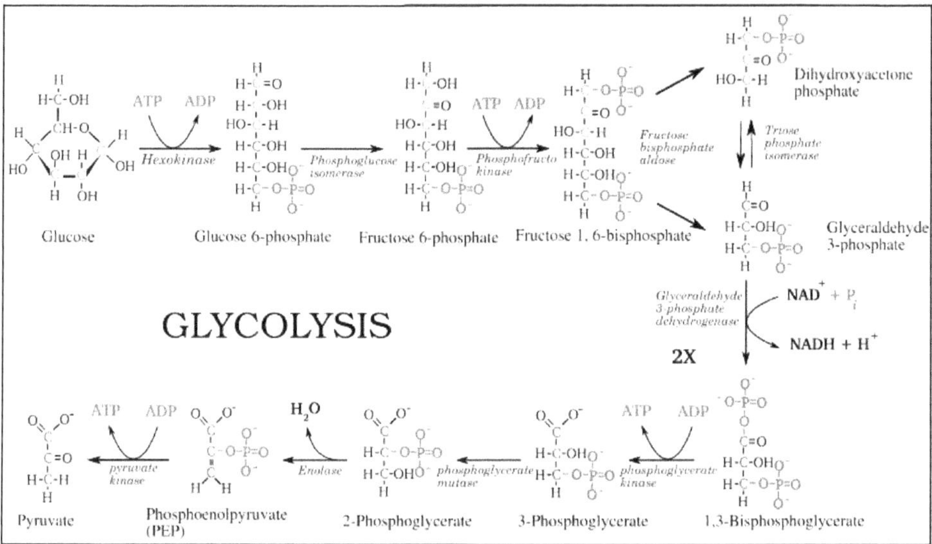

The overall glycolytic process can be summarized as follows

$C_6H_{12}O_6$ + 2ATP + 2NAD + 4ADP+$2H_3PO_4$ → 2 $CH_3COCOOH$ + 2ADP + $2NADH_2$ + 4 ATP + Pyruvic acid

☆ Thus there is a gain of 4-2 = 2 ATP molecules per hexose sugar molecule oxidized during this process.

☆ Besides this, 2 molecules of reduced coenzyme $NADH_2$ are also produced per molecule of hexose sugar in glycolysis.

☆ During aerobic respiration, these two $NADH_2$ are oxidized via the electron transport chain to yield 3 ATP molecules each. Thus 6 ATP molecules are formed.

What is meant by Citric acid cycle? And explain in detail about the various reactions involved in krebs cycle?

Krebs' Cycle/Citric Acid Cycle/TCA Cycle

The pyruvic acid produced in glycolysis enters into Krebs' cycle for further oxidation. Krebs' cycle is also known as citric acid cycle or Tri carboxylic acid (TCA) cycle. This aerobic process takes place in mitochondria where necessary enzymes are present in matrix.

1. Pyruvic acid reacts with CoA and NAD and is oxidatively decarboxylated. One molecule of CO_2 is released and NAD is reduced. *i.e.* Pyruvic acid is converted into acetyl CoA in the presence of enzyme pyruvate dehydrogenase.

   Pyruvic acid + CoA + NAD $\rightarrow$ Acetyl – COA + CO_2 + $NADH_2$

2. Acetyl-CoA condenses with oxaloacetic acid in the presence of condensing enzyme and water molecule to form citric acid. CoA becomes free.

   Acetyl CoA + Oxaloacetic acid $\rightarrow$ Citric acid + CoA+ H_2O

3. Citric acid is dehydrated in the presence of aconitase to form cis – aconitic acid

   Citric acid $\rightarrow$ Cis – Aconitic acid + H_2O

4. Cis-aconitic acid reacts with one molecule of water to form Isocitric acid

   Cis-aconitic acid + H_2O $\rightarrow$ Isocitric acid

5. Iso-citric acid is oxidized to oxalo succinic acid in the presence of Isocitric dehydrogenase. NADP is reduced to $NADPH_2$ in the reaction.

   Isocitric acid + NADP $\rightarrow$ Oxalo succinic acid + $NADPH_2$

6. Oxalo succinic acid is decarboxylated in the presence of oxalo succinic decarboxylase to form α - ketoglutaric acid and a second molecule of CO_2 is released.

   Oxalosuccinic acid $\rightarrow$ α-ketoglutaric acid + CO_2

7. α - ketoglutaric acid reacts with CoA and NAD in the presence of α - ketoglutaric acid dehydrogenase complex and is oxidatively decarboxylated to form succinyl CoA and a third mole of CO_2 is released. NAD is reduced in the reaction.

   α-keto glutaric acid + CoA + NAD $\rightarrow$ Succinyl-CoA + CO_2 + $NADH_2$

8. Succinyl CoA reacts with water molecule to form succinic acid. CoA becomes free and one molecule of GDP (Guanosine diphosphate) is phosphorylated in presence of inorganic phosphate to form one molecule of GTP.

Succinyl-CoA + H_2O + GDP + ip → Succinic acid + GTP

☆ GTP may react with ADP to form one molecule of ATP

GTP + ADP → ATP + GDP

9. Succinic acid is oxidized to fumaric acid in the presence of succinic dehydrogenase and co enzyme FAD is reduced in this reaction.

   Succinic acid + FAD → Fumaric acid + $FADH_2$

10. One mole of H_2O is added to Fumaric acid in the presence of fumarase to form malic acid.

    Fumaric acid + H_2O → Malic acid

11. In the last step, malic acid is oxidized to oxaloacetic acid in the presence of malic dehydrogenase and one molecule of coenzyme *i.e.* NAD is reduced.

    Malic acid + NAD → Oxaloacetic acid + $NADH_2$

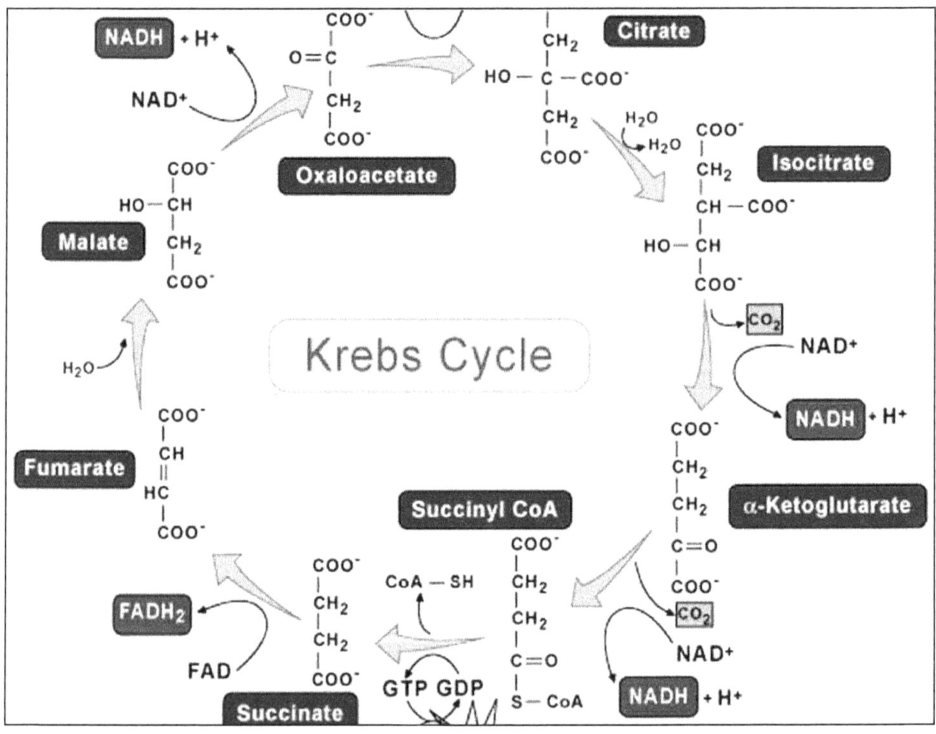

Explain in brief about the following?

1. **Pentose Phosphate Pathway**
2. **Oxidative phosphorylation**

Pentose Phosphate Pathway (PPP)/Hexose Mono Phosphate (HMP) Shunt/ Phosphogluconate Pathway/Warburg and Dicken's Pathway

Pentose phosphate pathway (also called the phosphogluconate pathway and the hexose monophosphate shunt) is a metabolic pathway parallel to glycolysis that generates NADPH and pentoses (5-carbon sugars). While it does involve oxidation of glucose, its primary role is anabolic rather than catabolic.

- The pentose phosphate pathway occurs in the cytoplasm outside the mitochondria and it is an alternative pathway to glycolysis and Kreb's cycle.
- The presence of some compounds like iodoacetate, fluorides, arsenates *etc.* inhibit some steps in glycolysis and that leads to the alternate pathway.
- This pathway was discovered by Warburg and Dicken (1938).
- This pathway does not produce ATP but it produces another form of energy called reducing power in the form of NADPH.
- It is not oxidized in the electron transport system but, it serves as hydrogen and electron donor in the biosynthesis of fatty acids and steroids.
- The pentose phosphate pathway consists of two distinct phases. In the first phase, hexose is converted into pentose and in the second phase, pentose is reconverted in to hexose.
- In the process, oxidation of glucose 6 phosphate leads to the formation of 6 phosphogluconic acid (pentose phosphate). Since glucose is directly oxidized without entering glycolysis, it is called as direct oxidation.

  6 Glucose 6 phosphate +12 NADP $\rightarrow$ 5 Glucose 6 Phosphate + 12 $NADPH_2$+ 6 CO_2
- It provides ribose sugars for the synthesis of nucleic acids and is also required for shikimic acid pathway.
- Although ATP is not produced, NADPH is produced and serves as hydrogen and electron donor in the biosynthesis of fatty acids and steroids.
- The pathway is also called as Phosphogluconate pathway as the first product in this pathway is phosphogluconate.

Oxidative Phosphorylation

Oxidative phosphorylation is the metabolic pathway in which the mitochondria in cells use their structure, enzymes, and energy released by the oxidation of nutrients to reform ATP.

- Oxidative phosphorylation is the process in which ATP is formed as a result of the transfer of electrons from NADH or $FADH_2$ to O_2 by a series of electron carriers.
- One molecule of ATP with 7.6 Kcal.energy is synthesized at each place when electrons are transferred from
  - Reduced NADH2 or NADPH2 to FAD
  - Reduced cytochrome b to cytochrome c
  - Reduced cytochrome a to cytochrome a3

- Thus, oxidation of one molecule of reduced NADH2 or NADPH2 will result in the formation of 3 ATP molecules while the oxidation of FADH2 lead to the synthesis of 2 ATP molecules.
- According to the most recent findings, although in eukaryotes terminal oxidation of mitochondrial NADH/NADPH results in the production of 3 ATP molecules but that of extra mitochondrial NADH/NADPH yields only 2 ATP molecules.
- Therefore, the two reduced coenzyme molecules (NADH) produced per hexose sugar molecule during Glycolysis will yield only 4 ATP molecules instead of 6 ATP molecules.
- Complete oxidation of a glucose molecule (hexose sugar) in aerobic respiration results in the net gain of 36 ATP molecules in most eukaryotes.
- One glucose molecule contains about 686 Kcal. Energy and 38 ATP molecules will have 273.6 Kcal energy.
- Therefore about 40 per cent (273.6/686) energy of the glucose molecule is utilized during aerobic breakdown and the rest is lost as heat. Since huge amount of energy is generated in mitochondria in the form of ATP molecules, they are called as *Power Houses of the cell.*

Write the differences between the following?

1. **Respiration and Photosynthesis**
2. **Oxidative phosporylation and Photophosporylation?**
3. **Photorespiration and Dark respiration**

Sl.No.	*Respiration*	*Photosynthesis*
1.	It is catabolic process resulting in the destruction of stored food	It is an anabolic process resulting in the manufacture of food.
2.	Light is not essential for the process	Light is very much essential
3.	Oxygen is absorbed in the process	Oxygen is liberated
4.	Carbon dioxide and water are produced	Carbon dioxide is fixed to form carbon containing compound
5.	Potential energy is converted into Kinetic energy	Light energy is converted into chemical energy
6.	Glucose and oxygen are the raw materials	Carbon dioxide and water are the raw materials
7.	Energy is released during respiration and hence it is an exothermic process.	Energy is stored during photosynthesis and hence it is an endothermic process
8.	Reduction in the dry weight	Gain in the dry weight
9.	Chlorophyllous tissues are not necessary	Chlorophyllous tissues are essential forthe process

Sl.No.	Oxidative Phosphorylation	Photophosphorylation
1.	It occurs during respiration	Occurs during photosynthesis
2.	Occurs inside the mitochondria (inner membrane of cristae)	Occurs inside the chloroplast (in the thylakoid membrane)
3.	Molecular O_2 is required for terminal oxidation	Molecular O_2 is not required
4.	Pigment systems are not involved	Pigment systems, PSI and PSII are involved
5.	It occurs in electron transport system	Occurs during cyclic and non cyclic electron transport
6.	ATP molecules are released to cytoplasm and used in various metabolic reactions of the cell	ATP molecules produced are utilized for CO_2 assimilation in the dark reaction of photocynthesis

Sl.No.	Photorespiration	Dark/Mitochondrial Respiration
1.	It occurs in the presence of light	It occurs in the presence of both light and dark.
2.	The substrate is glycolate	The respiratory substrate may be carbohydrate, fat or protein.
3.	It occurs in chloroplast, peroxisome and mitochondria	The process occurs in the cytoplasm and mitochondria
4.	It occurs in temperate plants like, wheat and cotton (mainly in C_3 plants)	It occurs in C_4 plants (maize and sugar cane)
5.	It occurs in the green tissues of plants	It occurs in all the living plants (both green and non green)
6.	The optimum temperature is 25- 35ºC	It is not temperature sensitive
7.	This process increases with increased CO_2 concentration	This process saturate at 2-3 percent O_2 in the atmosphere and beyond this concentration there is no increase.
8.	Hydrogen peroxide is formed during the reaction	Hydrogen peroxide is not formed.
9.	ATP molecules are not produced	Several ATP molecules are produced.
10.	Reduced coenzymes such as $NADPH_2$, $NADH_2$ and $FADH_2$ are not produced.	Reduced coenzymes such as $NADPH_2$, $NADH_2$ and $FADH_2$ are produced.
11.	One molecule of ammonia is released	No ammonia is produced per molecule of CO_2 released.
12.	Phosphorylation does not occur	Oxidative phosphorylation occurs.

Explain how the urea cycle linked to TCA cycle?

- ✰ The urea cycle consist of five enzymatically controlled steps that are catalyzed by carbamyl phosphate synthetase, ornithine transcarbamylase, argininosuccinate synthetase, argininosuccinase, and arginase, respectively.
- ✰ The complete cycle is present in physiological meaningful levels in the liver of terrestrial vertebrates, and in man represents the sole mechanism for ammonia disposal. The formation of carbamyl phosphate and the synthesis of argininosuccinate are potential limiting steps in urea biosynthesis but

substrate and not enzymes levels are rate-limiting under physiological conditions.

- ☆ In the adult, urea cycle enzymes change as a unit, and are largely influenced by dietary protein content. The urea cycle is closely linked to the citric acid cycle deriving one of its nitrogens through transamination of oxalacetate to form asparate and returns fumarate to that cycle.
- ☆ The biosynthesis of urea demands the expenditure of energy but less than 20 per cent of the energy derived from metabolism of gluconeogenic amino acids is required for ureogenesis. Embryological development of the urea cycle in the tadpole and in mammalian fetal liver therefore permits use of amino acids as new sources of energy to meet oxidative demands for continuing growth.

What is meant by protein targeting? And explain the mechanism of targeting proteins in both prokaryotic and eukaryotic cells?

Protein targeting or **protein sorting** is the biological mechanism by which proteins are transported to the appropriate destinations in the cell or outside of it. Proteins can be targeted to the inner space of an organelle, different intracellular membranes, plasma membrane, or to exterior of the cell via secretion.

Objectives

1. To understand how proteins find their destination in prokaryotic and eukaryotic cells
2. To know how proteins are bio-recycled

Targeting in Prokaryotic Cells

- ☆ In bacterial cells, the targeting decision is relatively straightforward *i.e.* whether protein destined to be an intracellular protein or an extracellular one.
- ☆ Secreted proteins contain a signal sequence. This is a short (6-30) stretch of hydrophobic amino acids, flanked on the N-terminal side by one or more positively charged amino acids such as lysine or arginine, and containing neutral amino acids with short side-chains (such as glycine or alanine) at the cleavage site.
- ☆ As proteins with signal sequences are synthesized, they are bound by the SecB protein. This prevents the protein from folding.
- ☆ SecB delivers the protein to the cell membrane where is secreted through a pore formed by the SecE and SecY proteins.
- ☆ Secretion is driven by the SecA ATPase.
- ☆ After the protein has been secreted, the signal sequence is removed by a membrane bound leader peptidase.

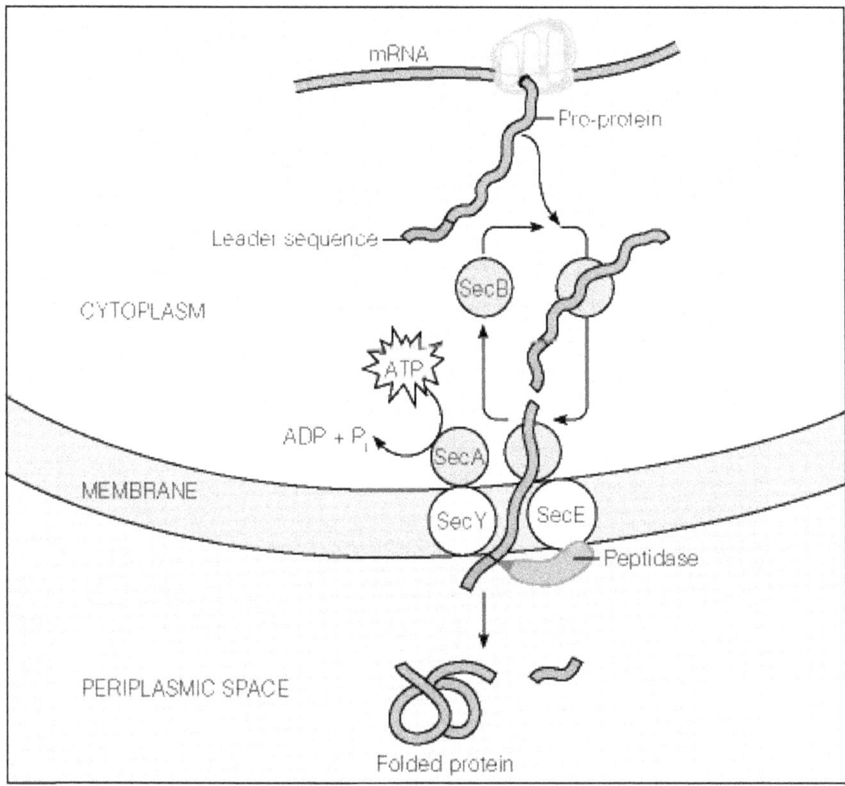

Targeting in Eukaryotic Cells

In eukaryotic cells, the situation is more complex. Extracellular proteins can be targeted for secretion or to the cell membrane, or to one of the many internal

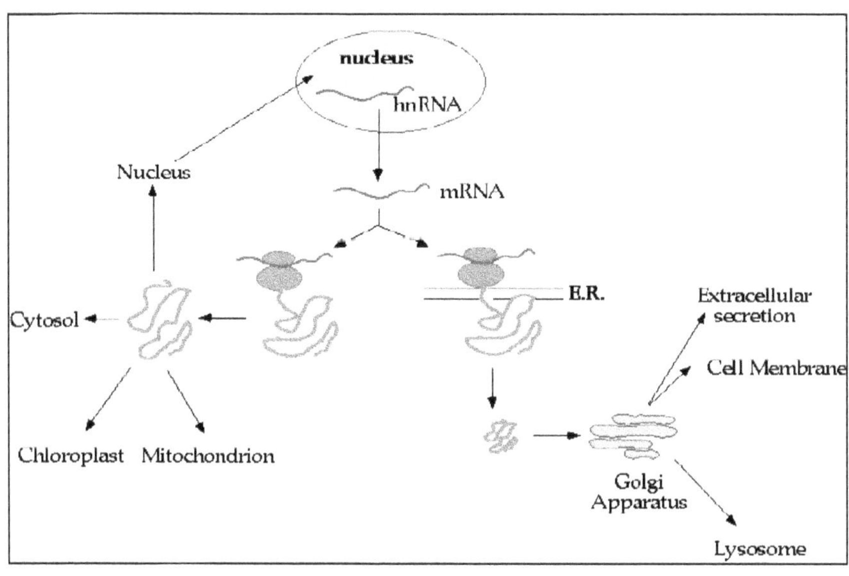

organelles such as the lysosome. Intracellular proteins can be targeted for the cytoplasm, to the nucleus or to special organelles such as the mitochondrion.

Protein secretion in eukaryotic cells also involves a signal sequence.

The SRP Cycle

- ☆ The signal recognition particle (SRP) associates with ribosomes that are in the process of translating the mRNA for a secretory protein.
- ☆ The protein has a signal at the N-terminus.
- ☆ Subsequently, the ribosome-bound SRP interacts with the SRP-receptor a component of the ER membrane.
- ☆ Finally, SRP recycles to associate with another ribosome, and translation continues with the secretory protein transversing the membrane through a channel called the translocon.
- ☆ Proteins that must be targeted to the nucleus have a nuclear localization signal (NLS). Once common type of signal is a series of five or so closely spaced positively charged amino acids.

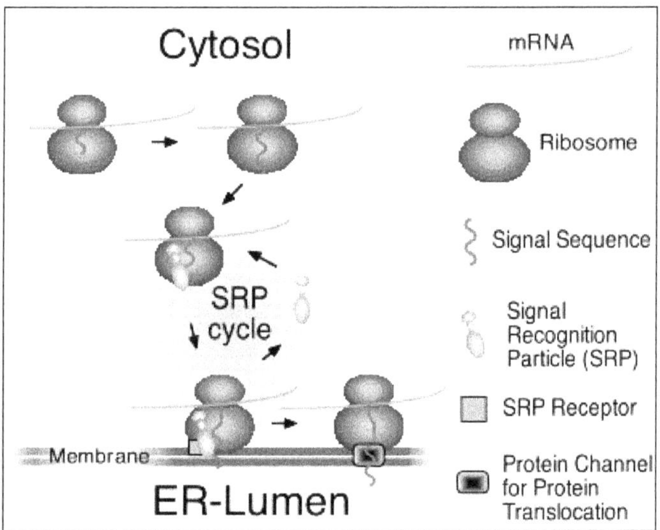

Write short notes on the following?

1. **Protein turn over**
2. **Targeted protein degradation**

Protein Turn Over

Protein lifetimes must also be regulated. Some proteins are needed for only very short times and could be harmful is present for too long. Others are needed all the time and it would be unnecessarily wasteful to keep re-synthesizing them.

The lifetime of proteins in eukaryotes appears to be determined by the nature of the N-terminal amino acid. Some amino acids (*e.g.* Ala, Cys, Gly, Met, Pro, Ser, Thr, Val) stabilize proteins (at least in yeast); others (*e.g.* Arg, His, Ile, Leu, Lys, Phe, Trp, Tyr) destabilize proteins.

Targeted Protein Degradation:

For constant working of cell, it needs to remove

- ☆ Incorrectly synthesized proteins (with errors in amino acid sequence)
- ☆ Damaged proteins (*i.e.* oxidative damage)
- ☆ Cell-cycle specific proteins
- ☆ Other signaling proteins which are no longer necessary

Mechanisms of Protein Degradation

1. **Protein degradation via lysosomes:** Lysosomes are acidic vesicles that contain about 50 different enzymes involved in degradation:
   i. Proteases (cathepsins): cleave peptide bonds
   ii. Phosphatases: remove covalently bound phosphates
   iii. Nucleases: cleave DNA/RNA
   iv. Lipases: cleave lipid molecules
   v. Carbohydrate-cleaving enzymes: remove covalently bound sugars from glycoproteins
      - ☆ Lysosomes often secrete their contents into the extracellular medium via exocytosis.
      - ☆ Lysosomes can also target damaged organelles in a process called autophagy.
      - ☆ Sometimes, lysosomes are triggered to rupture inside a cell, resulting in autolysis, also called apoptosis or programmed cell death.
2. **Protein degradation via ubiquitin labeling of surplus proteins:**
   - ☆ Ubiquitin (a small 76-residue protein) is attached to the protein
     - ❒ First, an *activating enzyme* attaches itself to the carboxy terminus of free ubiquitin in an ATP-dependent process.
     - ❒ Then, the activated ubiquitin is transferred onto a second enzyme which at the same time recognizes damaged proteins.
     - ❒ The activated ubiquitin is then *covalently* linked to lysine residues on the surface of the damaged protein.
   - ☆ These *ubiquitin-tagged* proteins are now recognized by specific *proteases* in the cytosol which in turn cleave and degrade the tagged protein.
   - ☆ These proteases are combined in a very large protein complex called the *proteasome*.

What is meant by protein engineering? Explain its approaches and applications?

Protein Engineering

Protein engineering is the design of new enzymes or proteins with new or desirable functions. It is based on the use of recombinant DNA technology to change amino acid sequences.

There are two general strategies for protein engineering, 'rational' protein design and directed evolution. These techniques are not mutually exclusive; researchers will often apply both.

Rational Design

In rational protein design, the scientist uses detailed knowledge of the structure and function of the protein to make desired changes.

- ✰ In general, this has the advantage of being inexpensive and technically easy, since site-directed mutagenesis techniques are well-developed.
- ✰ However, its major drawback is that detailed structural knowledge of a protein is often unavailable, and, even when it is available, it can be extremely difficult to predict the effects of various mutations.

Directed Evolution

In directed evolution, random mutagenesis is applied to a protein, and a selection regime is used to pick out variants that have the desired qualities. Further rounds of mutation and selection are then applied.

- ✰ This method mimics natural evolution and, in general, produces superior results to rational design.
- ✰ An additional technique known as DNA shuffling mixes and matches pieces of successful variants in order to produce better results.
- ✰ This process mimics the recombination that occurs naturally during sexual reproduction.
- ✰ The advantage of directed evolution is that it requires no prior structural knowledge of a protein, nor is it necessary to be able to predict what effect a given mutation will have. Indeed, the results of directed evolution experiments are often surprising in that desired changes are often caused by mutations that were not expected to have that effect.
- ✰ The drawback is that they require high-throughput, which is not feasible for all proteins. Large amounts of recombinant DNA must be mutated and the products screened for desired qualities. The sheer number of variants often requires expensive robotic equipment to automate the process. Furthermore, not all desired activities can be easily screened for.

Applications of Protein Engineering

So far a variety of protein engineering applications have been reported. These applications range from biocatalysis for food and industry to environmental, medical and nanobiotechnology applications.

1. *Food and Detergent Industry Applications*

- ☆ Particularly, the enzymes used in food industry were emphasized as an important group of enzymes, the industrially important properties of which could be further improved by protein engineering. Those properties include thermostability, specificity and catalytic efficiency.
- ☆ An important application area of protein engineering regarding food industry is the wheat gluten proteins.
- ☆ Their heterologous expression and protein engineering has been studied using a variety of expression systems, such as *E.coli*, yeasts or cultured insect cells.
- ☆ Food industry makes use of a variety of food-processing enzymes, such as amylases and lipases, the properties of which are improved using recombinant DNA technology and protein engineering.

2. *Environmental Applications*

Environmental applications of enzyme and protein engineering are also another important field.

- ☆ Early reports on enzyme and cell applications in industry and in environmental monitoring.

  e.g. Environmental biosensors,
- ☆ Genetic methods and strategies for designing microorganisms to eliminate environmental pollutants, which includes gene expression regulation to provide high catalytic activity under environmental stress conditions.
- ☆ The importance of microbial strains and their enzymes in bioremediation and biotransformation applications.

  e.g. Molybdenum hydroxylases, enzymes that catalyze the initial bacterial hydroxylation of a N-heteroaromatic compound, and 2,4- dioxygenases that play a role in the bacterial quinaldine degradation.
- ☆ Protein engineering of oxygenases, an important group of enzymes with high selectivity and specificity, which enable the microbial utilization and biodegradation of organic, toxic compounds.
- ☆ Petroleum biorefining is also an important environmental application area, where new biocatalysts are required.

  e.g. Protein engineering, isolation and study of new extremophilic microorganisms, genetic engineering developments are all promising advances to develop new biocatalysts for petroleum refining.
- ☆ Microbial bioplastics, or polyhydroxyalkanoates (PHAs), are also an important research area in environmental biotechnology.

☆ Another important environmental application of protein engineering involves fungal enzymes. Particularly peroxidases isolated from fungi can transform xenobiotics and many pollutants.

3. *Medical Applications*

Medical applications of protein engineering are also diverse.

☆ The use of protein engineering for cancer treatment studies is a major area of interest.

☆ Advances in protein engineering and recombinant DNA technology were expected to increase the use of pretargeted radioimmunotherapy.

☆ Recently, the term "modular protein engineering" has been introduced for emerging cancer therapies. Treatment strategies based on targeted nanoconjugates to be specifically directed against target cells are becoming increasingly important.

☆ Additionally, multifunctional and smart drug vehicles can be produced at the nanoscale, by protein engineering. These strategies could be combined to identify and select targets for protein-based drug delivery

☆ Protein engineering applications for therapeutic protein production is an important area, particularly for medicine. It was stated that protein engineering resulted in a second generation of therapeutic protein products with application-specific properties obtained by mutation, deletion of fusion. The third generation of such products were mentioned as "gene therapy".

☆ Protein engineering applications with antibodies are also diverse. Owing to advances in recombinant DNA technology, "antibody engineering" is possible. "Antibody modeling" studies to engineer antibody-like molecules and increase their stability and specificity are also common, particularly for humanization of antibodies of animal origin.

☆ Pharmacokinetic properties of antibodies have been improved by protein engineering and antibody variants of different size and antigen binding sites have been produced for the ultimate use as imaging probes specific to target tissues.

4. *Applications for Biopolymer Production*

☆ Protein engineering and macromolecular self assembly are utilized to produce peptide-based biomaterials, such as elastin-like polypeptides, silk-like polymers, *etc.*

☆ The ability of protein engineering to create and improve protein domains can be utilized for producing new biomaterials for medical and engineering applications.

e.g. The use of protein engineering to make new protein and peptide domains which enable advanced functional hydrogel formation. These domains include leucine zipper coiled-coil domains, the EF-band domains and elastin-like polypeptides.

5. *Nanobiotechnology Applications*

- The synthesis and assembly of nanotechnological systems into functional structures and devices has been difficult and limiting their potential applications for a long time.
- Combinatorial biology methods commonly applied in protein engineering studies, such as phage display and bacterial cell surface display technologies, are also used to select polypeptide sequences which selectively bind to inorganic compound surfaces, for ultimate applications of nanobiotechnology.
- Engineering protein and peptide building blocks to be used as molecular motors, transducers, biosensors, and structural elements of nanodevices, and the importance of proteins and peptides for the development of biocompatible nanomaterials, as well as the impact of computational techniques in this field have been well recognized.
- Another interesting nanotechnology application is the use of amyloid fibrils as structural templates for nanowire construction.

6. *Applications with Redox Proteins and Enzymes*

- Improvement of redox proteins and enzymes by protein engineering is also an important application field.
- Protein engineering of redox-active enzymes pointed out two emerging areas of protein engineering of redox-active enzymes: novel nucleic acid-based catalyst construction, and intra-molecular electron transfer network remodeling.

7. *Applications with Various Industrially Important Enzymes*

Protein engineering applications with a variety of industrially important enzymes were reported. These include nitrilases, aldolases, microbial beta-D-xylosidases *etc.*

- Other protein engineering examples with industrially and/or pharmacologically important enzymes include studies on cholesterol oxidase, cyclodextrin glucanotransferases, human butyrylcholinesterase, microbial glucoamylases, lipases of different origins, phospholipases.
- Studies on extremozymes, enzymes isolated from extremophilic species, revealed their different structural and functional characteristics which could be exploited for biotechnological applications and improved further by protein engineering.

8. *Other Applications*

- These binding proteins of non-Ig origin are called "affibody binding proteins". With their high affinity, these proteins have been used in many different applications such as diagnostics, bioseparation, functional inhibition, viral targeting, and *in vivo* tumor imaging or therapy.
- Inteins are protein splicing elements that are involved in a variety of applications such as protein purification, protein semisynthesis, *in vivo* and *in vitro* protein modifications.
- The use of intein tags for protein purification in plants with high protein production could potentially enable industrial production of pharmaceutically important proteins.
- "Insertional protein engineering" applications are also becoming important, particularly for biosensor studies.
- "Zinc finger protein engineering" is another approach that has been used in gene regulation applications.
- Applications of protein engineering in enzymatic biofuel cell design are also becoming increasingly important. Particularly, obtaining biofuels from lignocellulosic resources is a challenge, as the enzyme hydrolysis efficiency of lignocellulose is low which increases the costs of biofuels.
- Thus, protein engineering methods have been used to improve the performance of lignocellulose-degrading enzymes, and biofuels-synthesizing enzymes.
- Protein engineering is also applied to obtain an efficient electrical communication between biocatalyst(s) and the electrode by rational design and directed evolution, within the frame of biocatalyst engineering.
- "Virus engineering" is another emerging field, where the virus particles are modified by protein engineering. Viruses have many promising applications in medicine, biotechnology and nanotechnology. They could be used as new vaccines, gene therapy and targeted drug delivery vectors, molecular imaging agents and as building blocks for electronic nanodevices or nanomaterials construction.
- "Protein cysteine modifications" are also important protein engineering applications. As cysteine modifications in proteins cause diversities in protein functions, cysteine thiol chemistry has been applied for *in vitro* glycoprotein synthesis.
- Cyclotides are important proteins that have recently been popular for protein engineering applications. They are plant proteins made up from small disulfide-rich peptides and are exceptionally stable to thermal, chemical or enzymatic degradation. This property of cyclotides makes them valuable molecular templates for many protein engineering and drug design applications.

Protein engineering applications cover a broad range, including biocatalysis for food and industry, as well as medical, environmental and nanobiotechnological applications. With advances in recombinant DNA technology tools, "omics" technologies and high-throughput screening facilities, improved methods for protein engineering will be available, which would enable easy modification or improvement of more proteins/enzymes for further specific applications.

What is meant by metabolic engineering? Explain its approaches and applications in Agriculture?

Metabolic Engineering

It is the practice of optimizing genetic and regulatory processes within cells to increase the cells production of certain substance.

These processes are chemical networks that use a series of biochemical reactions and enzymes that allow cells to convert raw materials into molecules necessary for the cells survival.

- ☆ It mainly targets the regulatory networks in a cell to efficiently engineer the metabolism rather than direct deleting or overexpressing the genes that encode for metabolic enzymes.
- ☆ It is the rational redesign of cellular metabolism for industrial applications. It is mostly commonly targeted towards production of industrial biochemicals like fuels, pharmaceuticals, flavours and fragrances etc but can also be focused on catalysis of useful processes like bioconversion or development of biological products like blood products.

Approaches for Metabolic Engineering

Metabolic engineering involves the adjustment of metabolic and regulatory processes to improve desired cellular behaviours such as the production of proteins and chemicals.

- ☆ But cellular metabolic and regulatory networks are often large and complex, the construction and analysis of computational models of these networks can be useful for identifying current network states and evaluating the effects of network perturbations on desired phenotypes.

Descriptive Approaches for Identifying Intracellular Metabolic States

A variety of experimental measurements can be used to quantify the state of metabolic and regulatory networks which includes metabolic flux analysis, gene/protein expression, metabolite concentration, enzyme activity, uptake and secretion rates and transcription factor-DNA binding assays *etc.*

Metabolic Flux Analysis (MFA)

It is an experimental fluxomic technique used to examine production and consumption rates of metabolites in a biological system.

- It mainly works by employing stoichiometric models of metabolism and mass spectrometry methods with isotopic mass resolution, the transfer of moieties containing isotopic tracers from one metabolite into another can be elucidated, and finally information about the metabolic network thus derived.
- It is critical for identifying pathway bottlenecks and elucidating network regulation in biological systems especially in engineered cells with nonnative metabolic capacities.

Metabolite and Gene/Protein Expression Measurements

These include the measurements of metabolite concentrations and gene or protein expression by using mass spectroscopy for hundreds of metabolites in a single condition.

- These studies also help to elucidate metabolite fluxes and their regulation in addition to metabolite concentrations.

Predictive Approaches for Improving Cellular Phenotypes

These approaches mainly used to identify which environmental and/or genetic perturbations would improve cellular phenotypes, such as the production of desired chemicals. These include pathway based approaches, optimization based approaches and kinetic modeling approaches.

Path way Based Approaches

These pathways mainly help to derive minimal media requirements of an organism and assess the robustness and redundancy of key metabolic pathways.

- The methods to identify relevant metabolic pathways include extreme currents, elementary modes, extreme pathways and minimal generators.
- Of these, elementary modes are mainly used for determining the metabolic impact of gene knockouts and the required pathways for producing desired chemicals.
- These elementary flux modes also have been successfully used to engineer strains with a variety of desired phenotypes, including sugar coutilization, ethanol and carotenoid production *etc.*

Optimization Based Approaches

These approaches mainly used to identify mutations that would improve desired phenotypes.

- These approaches include flux balance analysis (FBA), minimization of metabolic adjustment (MOMA), regulatory on/off mechanism (ROOM), regulated flux balance analysis (rFBA) and probabilistic regulation of metabolism (PROM).

Kinetic Modeling Approaches

Earlier two models do not take enzyme kinetics into account, so they are unable to predict how changes in kinetic properties, enzyme and metabolic concentrations would affect fluxes through metabolic pathways.

- ✰ Kinetic models are mainly needed to make these predictions since they capture the dependence of fluxes on metabolite and enzyme concentrations.
- ✰ The classical model for elucidating parameters responsible for control of metabolic fluxes is metabolic control analysis (MCA).

Applications of Metabolic Engineering in Agriculture

1. **Nitrogen fixation in plants:** It can be achieved by either expression of all 18 components of the nitrogensae biosynthetic apparatus in plants or by engineer a rhizosphere symbiosis in between a nitrogen fixing microorganism and a plant host.
2. **Crop plants with altered nutrient content:** Golden rice proves the concept that the nutrient content of a crop plant can be improved by metabolic engineering.
3. **Engineering crops for biofuel production:** Here, plants could be engineered to break down its own cellulose on demand, releasing fermentation ready sugars for biofuel production.
4. **Enhancement of photosynthetic efficiency:** It is achieved by installation of C_4 photosynthesis in C_3 plants or by altering the rubisco enzyme that catalyze the fisrt key step in CO_2 fixation as a part of calvin cycle.

Define and explain in detail about the plant hormones?

Plant Hormones

Plant hormones are chemicals that regulate plant growth. These are also known as the growth factors, growth substances, plant growth regulators or phytohormones.

On the basis of the function there are five main types of plant hormones includes auxins, gibberellins, cytokinins, ethylene and abscisic acid.

Auxin

It is the plant growth regulators that mainly help in rooting besides its minor functions.

- ✰ This growth substance was found out by Charles Darwin in his book called *The Power of Movement of Plants*.
- ✰ Auxins are synthesized on the root and shoot tips and are produced relatively in lesser amount in the roots.
- ✰ Tryptophan, an amino acid, is the precursor of auxin.

- ☆ Auxins produced at the shoot are transported to root through parenchyma.
- ☆ Low concentration of auxin results in the growth whereas high concentration will inhibit the growth of the same organ.
- ☆ Auxin is active in free state and can be easily extracted.
- ☆ Bound auxin is inactive and is meant for storage purpose like IAA- Inositol, IBA- Alanine.
- ☆ Auxins are weak organic acids.
- ☆ Some auxin occur naturally in plants and are called the natural auxin while many are synthetic:
  a. Natural auxin: It was first isolated from human urine by Kogl and Haagensmit. It was named as auxin-a. But later Kogl obtained another auxin called auxin-b (auxentriolic acid) from corn germ oil and heteroauxin from human urine. Heteroauxin, also known as indole-3-acetic acid (IAA) is the best known natural auxin. Indole butyric acid is another natural auxin.
  b. Synthetic auxin: *e.g.* naphthalene acetic acid (NAA), phenoxy acetic acid (PAA), 2, 4- dichlorophenoxy acetic acid (2, 4-D), 2,4,5-trichlorophenoxy acetic acid (2,4,5-T), naphthalene acetamide (NAAM).
- ☆ Anti-auxin: Inhibits auxin activity *e.g.* TIBA (triidobenzoic acid) and PCIB (p-chlorophenoxy isobutyric acid).
- ☆ Auxin is used for Avena test culture, split pea test and root growth inhibition test.

Functions of Auxin

- ☆ Cell enlargement: Auxin causes cell enlargement by solubilisation of carbohydrates, loosening of cell wall micro-fibrils, synthesis of new wall material and increase in respiration.
- ☆ Prevention of lodging: Lower inter-nodes of the stem of cereals are long and weak. As a result the plant bends down or droops. Application of NAAM prevents lodging.
- ☆ Apical dominance: In many plants, apical bud suppresses the growth of lateral buds. This condition is known as apical dominance. A plant with strong apical dominance has little or no branching like in sunflower.
- ☆ It is responsible for the photo-tropism and geotropism.
- ☆ It promotes root initiation in callus and stem cutting.
- ☆ Delay of abscission of leaves by preventing formation of abscission layer. But promotes this in older leaves and fruits.
- ☆ It induces parthenocarpy (production of fruit without fertilization).
- ☆ Increases the number of female flowers.
- ☆ Activity of cambium is promoted by auxin.

- ☆ Healing of injury in plants.
- ☆ Promotes xylem differentiation.

Gibberllins

It is the second major plant hormone and it was first known by a Japneses farmer Konishi (1898) and working was found out by Kurosawa in 1926.

- ☆ It was first extracted from the ascomycetous fungus *GibberellaFujikuroi*, the casual organisms of "foolish seedling of rice" or commonly called *bakanae of rice*.
- ☆ Anti-gibberellins: Phosphon D, amo-1618, CCC(chloro choline chloride) and maleic hydrazide (MH).
- ☆ Gibberellins are used in induction of alpha-amylase in barley endosperm test, dwarf maize test and dwarf pea test.
- ☆ Its precursor is acetyl CoA.
- ☆ These are found in most of the groups of plants like algae, fungi, mosses, ferns *etc.*
- ☆ These are produced at the apices of young leaves, embryo, buds etc and are transported through xylem.
- ☆ Chemically these are terpenes (lipids) and are weak acids.
- ☆ These have gibbane ring skeleton.
- ☆ More than 100 types of gibberellins have been identified but GA3 form is most common.

Functions of Gibberellins

- ☆ In bolting: In rosette forming plants, inter-node growth is poor but large leaves appear to rise arise in tufts. The inter-nodes suddenly elongate and the stem becomes normal just before flowering. This is called bolting.
- ☆ Parthenocarpy: Gibberellins have been found to be effective in inducing pathenocarpy in tomatoes, apple *etc.*
- ☆ Breaking of dormancy: These can effectively break the dormacy of potato tuber, winter buds and seeds of many trees.
- ☆ Increase in fruit size: These increase the number and fruit size of grapes.
- ☆ Production of male flowers: These induce production of male flowers on genetically female plants.
- ☆ Inter-nodal elongation: It helps in the elongation of the stem but not in roots.
- ☆ It helps in the germination of the seed.
- ☆ It is used to delay the ripening in plants like citrus.

Cytokinins

Skoog (1964) noted that in general "the term Cytokinin is universally used for substance which promotes cell division and exerts other growth regulatory functions as kinetin".

- ☆ It is found at the tissues where rapid growth is taking place like growing fruits, root tips and shoot buds.
- ☆ Chemically they are purines derived from tRNA.
- ☆ The first naturally occurring cytokinin was identified from the young maize (*Zea mays*) and hence it was named as zeatin.
- ☆ Cytokinin is also present in the coconut water (liquid endosperm).
- ☆ It is used in tissue culture, chlorophyll preservation test and cell division test.

Functions of Cytokinins

- ☆ Cell division: These are found in a higher amount where rapid division is going on.

  Morphogenesis: Cytokinins promotes cell division and in the presence of auxin, it promotes cell division even in the meristematic tissues. In tissue culture, mitotic divisions are accelerated when both auxin and cytokinin are present. The ratio of high cytokinin and low auxin promote shoot buds in tissue culture.
- ☆ Apical dominance: Cytokinin and auxin acts antagonistically in the control of apical dominance.
- ☆ Delay in senescence: It delay the senescence of plant organs by controlling protein synthesis and mobilization of resources. This phenomenon is called the Richmond Lang effect. They helps to produce chloroplast in leaves. These are also called anti-ageing hormones.
- ☆ Flowering: It helps in inducing flowering in certain species of plants like Lemna and Wolffia.
- ☆ It promotes phloem transport.
- ☆ It also promotes accumulation of salts in the cells.
- ☆ It promotes production of female flowers.
- ☆ It increases the resistance to low and high temperature and diseases.

Abscisic Acid

It was discovered by Addicott and Wareing separately. They named it abscisin II and dormin respectively. In 1967 it was decided to call it as abscisic acid (ABA).

- ☆ It is major inhibitor of growth in plants and is antagonistic to all the three growth promoters, especially GA.
- ☆ Its precursor is violaxanthin (a xanthophyll in chloroplast).

- It is used in cotton or bean explant test.
- It is chemically a dextro-rotatory cis-sesquiterpene.
- In liverworts and algae, a compound called lunularic acid has similar activities as ABA.

Function of Abscisic Acid

- It hastens the formation of abscission layer and senescence.
- Transpiration: It helps in closing of stomata by causing potassium ions to leave the guard cells during periods of water shortage or drought and hence is also known as stress hormone.
- It promotes bud dormancy in seeds during winters.
- Seed dormancy: It induces seeds dormancy hence is named dormin. Thus it helps the seed to withstand desiccation and other unfavorable factors.
- It inhibits cambial activity.
- Flowering: it induces flowering in some short day plants like strawberry.
- It plays an important role in seed development, maturation and dormancy.
- It induces synthesis of carotenoids in green oranges making them yellow.

Ethylene

It is the only gaseous growth hormone. It is produced by almost all the organs but maximum production occurs in ripening fruits and senescent leaves.

- High concentration of auxin leads to the formation of ethylene.
- Ethylene is used for sprouting of storage organs like rhizome and tubers.
- It is used in the triple response test.
- It can be placed in the group of both stimulator and inhibitor but largely is growth inhibitor.
- It is the only hormone which is gas at normal temperature and pressure.
- Since it is a gas, it can move in air too. Hence, any fruit kept with ripened fruit gets ripened too.
- Its precursor is methionine.

Functions of Ethylene

- It helps in ripening of fruits.
- Inhibition of stem elongation and stimulation of transverse growth by causing increase in the girth of the plant and promotes horizontal growth.
- It promotes apogeotropism in roots.
- It increases the speed of senescence.
- It also induces abscission of leaves.

- It helps in breaking the dormancy of storage organs and initiates germination in peanut seeds.
- Root initiation: Low concentration of ethylene induces rooting and growth of lateral roots and root hairs.
- Flowering: It is used to initiate flowering and synchronizing fruit set in pineapples.
- It helps in production of female flowers in a plant which is genetically male.

Other identified plant growth regulators include:

Brassinosteroids

These are a class of polyhydroxysteroids, a group of plant growth regulators.

- Brassinosteroids have been recognized as a sixth class of plant hormones, which stimulate cell elongation and division, gravitropism, resistance to stress, and xylem differentiation.
- They inhibit root growth and leaf abscission.
- Brassinolide was the first identified brassinosteroid and was isolated from extracts of rapeseed (*Brassica napus*).

Salicylic Acid

It activates genes in some plants that produce chemicals that aid in the defense against pathogenic invaders.

Jasmonates

These are produced from fatty acids and seem to promote the production of defense proteins that are used to fend off invading organisms.

- They are believed to also have a role in seed germination, and affect the storage of protein in seeds, and seem to affect root growth.

Plant Peptide Hormones

This encompasses all small secreted peptides that are involved in cell-to-cell signaling.

- These small peptide hormones play crucial roles in plant growth and development, including defense mechanisms, the control of cell division and expansion, and pollen self-incompatibility.

Polyamines

These are strongly basic molecules with low molecular weight that have been found in all organisms studied thus far.

- They are essential for plant growth and development and affect the process of mitosis and meiosis.

Nitric oxide (NO): It serves as signal in hormonal and defense responses (*e.g.* stomatal closure, root development, germination, nitrogen fixation, cell death, stress response).

- ☆ NO can be produced by a yet undefined NO synthase, a special type of nitrite reductase, nitrate reductase, mitochondrial cytochrome oxidase or non enzymatic processes and regulate plant cell organelle functions (*e.g.* ATP synthesis in chloroplasts and mitochondria).

Strigolactones

Its main implication in the inhibition of shoots branching.

Karrikins

These are not plant hormones because they are not made by plants, but are a group of plant growth regulators found in the smoke of burning plant material that have the ability to stimulate the germination of seeds.

Some Common Applications in Agriculture/Crop improvement include

- ☆ The high percentage of germination of sown seeds in the field has a bearing on the output. Pretreatment of seeds with IAA, NAA, GA *etc.* has been found to be very effective not only in increased the percentage of germination but also in the total yield of the crop plants. *e.g.* Use of GA or IAA greatly enhances the growth of plants and total area of leaf surfaces.
- ☆ Plants can be multiplied by vegetative propagation. Hormones like NAA and IBA are very effective in inducing roots in stem cuttings. These hormones can also be used for grafting propagation.
- ☆ Use of hormones like IAA, NAA, IBA and Gibberellins ensures fruit setting and many of the fruits which develop from such hormone treatments are seedless, larger in size and sweeter in content.
- ☆ Hormones prevent premature falling of fruits, otherwise nearly 50-70 per cent of the set fruits fail to mature and most of them fall off because of the formation of abscission layer in their stalks.
- ☆ Preservation of agricultural products before marketing is another area in which hormones can be used effectively. *e.g.* Tubers, rhizomes, bulbs and such products sprout while they store. For this some synthetic hormones like NAA, maleic hydrazide can be used to preserve the said products for quite a period of time, thus one can improve the keeping quality of agricultural products.
- ☆ Hormones like GA can also be commercially exploited for inducing flowering in unseasonal periods. But one should know which hormone is effective on which plant: otherwise phytohormones fail to produce the desired results.
- ☆ Hormones can also be exploited for removal of weeds. *e.g.* Synthetic hormones like 2,4-D, 2,4,5-T can be used in paddy fields to destroy weeds.

☆ Hormonal concentrations play a significant role in culturing explants into undifferentiated callus and callus to differentiate into roots, shoots or the entire plant from eh callus.

☆ The callus cells can be further induced to develop into shoots, roots or both by providing auxins and cytokinins in a defined ratio. At high ratio of auxin to cytokinin callus produces only shoots, at lower ratio the callus induces only roots, but at an intermediate ratio both shoots and roots develop.

Answer the following questions.

1. **Define and explain in brief about the biofertilizers?**
2. **Define and explain in brief about the biological nitrogen fixation?**
3. **Define symbiosis? And explain in brief about the molecular mechanism behind the effective establishment between plant and bacteria?**
4. **What is the enzyme involved in biological nitrogen fixation? And explain in brief about structural features of it?**
5. **Explain in brief about the genetics of nitrogen fixation?**

Biofertilizers

Biofertilizers are those substances that contain living microorganisms and they colonize the rhizosphere of the plant and increase the supply or availability of primary nutrient and/or growth stimulus to the target crop.

☆ Biofertlizers are an environmentally friendly substitute for harmful chemical fertilizers.

☆ They transform organic matter in to nutrients that can be used to make plants healthy and productive.

☆ Types of biofertilizer available are

a. Nitrogen fixing biofertilizer: Rhizobium, Azotobacter, Azospirillum, Bradyrhizobium.
b. Phosphorus solubilising biofertilizer (PSB): Bacillus, Pseudomonas, Aspergillus.
c. Phosphorus mobilizing biofertilizer: Mycorrhiza
d. Plant growth promoting biofertilizer: Pseudomonas

☆ Steps involved in production of biofertilizers are

a. Isolation of microbes from the soil
b. Laboratory screening of microbes for plant growth
c. Greenhouse screening of microbes to promote growth in potted soil
d. Field screening of most effective microbe in cropped soil
e. Refinement of inoculums
f. Production of biofertilizer

☆ Mode of action of biofertilizer

a. They fix nitrogen in the soil and the root nodules of the legumes crop and make it available to the plant.
b. They solubilise the insoluble form of the phosphate like tricalcium, iron and aluminium phosphate into the available form.
c. They produce hormones and anti metabolites which promote root growth.
d. They also decompose the organic matter.
e. When biofertilizers are applied to the seed and the soil they increases the availability of the nutrient to the plant and increases the yield up to 10-20 per cent without producing any adverse effect to the environment.

Nitrogen Fixation

The conversion of dinitrogen gas (N_2) to ammonia (NH_3) is called nitrogen fixation.

There are three ways that nitrogen fixation occurs in the biosphere includes

a. Lightening and other natural combustion process
b. The industrial Haber-Bosch process and
c. Biological nitrogen fixation
d. The catalytic efficiency of BNF, which proceeds under ambient conditions, stands in pronounced contrast to the Haber-Bosch process, and indeed, any other synthetic system, that either require high temperatures to get reasonable rates, or are only able to sustain a limited number of turnovers.

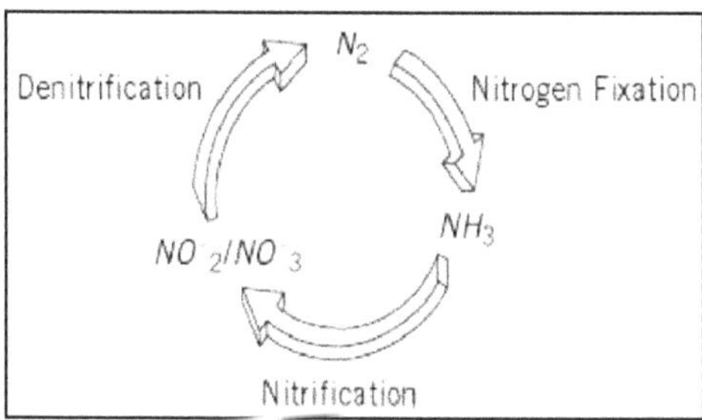

Biological Nitrogen Fixation (BNF)

The process of conversion of nitrogen to a combined form (either ammonia or nitrate) by prokaryotes is referred to as Biological nitrogen fixation.

☆ It was first discovered by Beijerinck in 1901.

☆ Ammonia is necessary for the formation of biologically essential, nitrogen-containing compounds such as amino acids and nucleic acids, a fixed source is necessary to sustain life on earth.

- ☆ Ammonia is also sequestered into sediments or reconverted to N_2 through the combined biological processes of nitrification and denitrification.
- ☆ BNF is carried out by a special class of prokaryotes. These prokaryotes include aquatic organisms like cyanobacteria.
- ☆ Most of the nitrogen fixing prokaryotes found in soil is free living in nature. They fix the nitrogen without the direct interaction with other organisms. *e.g. Azotobacter, Bacillus, Clostridium* and *Klebsiella*.
- ☆ Free living prokaryotes often behave as anaerobes during nitrogen fixation. They contribute to a very less amount of global nitrogen fixation due to scarcity of appropriate carbon sources.
- ☆ Associative types of microorganisms remain in close association with the rhizosphere region of members of family Poaceae (Rice, wheat, corn, oats, barley *etc.*).
- ☆ Associative nitrogen fixation can supply 20–25 per cent of total nitrogen requirements in rice and maize. *e.g. Azospirillum*
- ☆ Symbiosis is a naturally occurring phenomenon where the microbes and higher plants mutually and beneficially contribute to the process of nitrogen fixation.
- ☆ The symbiotic associations occur between the plants of family Leguminosae and bacteria of genera *Azorhizobium, Bradyrhizobium, Photorhizobium, Rhizobium* and *Sinorhizobium*.

Symbiosis

Symbiosis is nothing but mutual beneficial relationship between two different organisms living in close physical association, typically to the advantage of both.

- ☆ In BNF, the symbiotic association is results due to plant root nodule contain differentiated bacteria that specialize in nitrogen fixation.
- ☆ In this symbiosis, the plant provides the microbe carbon energy source through the process of photosynthesis whereas the microbe supplies the plant host a fixed nitrogen source through the process of nitrogen fixation.

Establishment of Symbiotic Relationship

Symbiotic relationship between a leguminous plant host and an associated rhizobium is a complicated process that is specific between a particular host and a particular symbiont.

- ☆ During initiation of the infection process, bacteria induce the formation of curls at the tip of plant root hairs which then envelop the invading bacteria.
- ☆ Then infection threads are formed within the root hair. Infection threads are tubular structures of plant origin that penetrate the root hairs and the root cortex through which the invading bacterium traverse.

- At or about the time infection thread formation occurs, cell division is also induced within the root cortex, resulting in the formation of a nodule primordium.
- When the infection thread contacts a newly divided primordial cell within the root cortex, the bacteria are released from the infection thread tip.
- Subsequent penetration of the released bacterial cell into the cytoplasm of the primordial cell occurs through a process of endocytosis resulting in the formation of a plant cell membrane (bacteroidal membrane) that surrounds the bacterium.
- Bacterial cell division and bacteroid membrane proliferation then occur, resulting in the host plant cell cytoplasm becoming filled with bacteria.
- These bacteria then differentiate into cells specialized for nitrogen fixation that are called bacteroids.

To establish an effective symbiosis, developmental and metabolic cooperation between the plant and microbe are necessary.

- This cooperation is accomplished through the reciprocal communication and control of gene expression between the two partners.
- During nodule development, signaling occurs through the action of nodulation (nod) gene expression by the bacterium and expression of nodulins (ENOD sequences) by the plant.
- The first of these signals is provided by flavanoid molecules, three-ring aromatic compounds derived from phenylpropanoid metabolism, which are produced by the plant and found in the root exudate.
- Flavanoids act as specific inducers of bacterial nodD gene expression.
- The nodD gene encodes a regulatory protein that controls the expression of other nod genes.
- The concerted action of other nod gene products, whose synthesis is induced by nodD gene expression, results in forming and releasing bacterial signal molecules called Nod factors.
- Nod factors are responsible for signaling the initial root hair curling event. Their core structures are composed of b1-4 linked acetylglucosamine residues of various length (usually one to four molecules).
- The specificity of a particular Nod factor is determined by its level of oligomerization and by certain chemical modifications of the oligosaccharide core, such as O-acylation or N-acylation at the nonreducing end of the oligosaccharide and sulfation at the reducing end.

Nitrogenase

Nitrogenase is the enzyme that catalyzes biological nitrogen fixation. The reaction usually depicted as

$$N_2 + 8H^+ + 8e^- + 16\ MgATP \rightarrow 2\ NH_3 + H_2 + 16\ MgADP + 16\ P_i$$

Component Proteins of Nitrogenase

- ☆ Nitrogenase consists of two component metalloproteins designated the iron (Fe-) protein and the molybdenum iron (MoFe-) protein, these have unique biochemical ability to catalyze the reduction of dinitrogen to ammonia.
- ☆ The MoFe- protein contains the active site for substrate reduction, and is organized as an $\alpha_2\beta_2$ tetramer (where α and β subunits are homologous) of molecular weight 240KD.
- ☆ Associated with this protein are 2 Mo, 30 Fe and 32 S organized into two copies of each of two extraordinary metalloclusters designated the FeMo-cofactor and the P- cluster.
- ☆ The FeMo- cofactor represents the site of substrate reduction, while the P- cluster is likely the initial acceptor of electrons from the Fe- protein.
- ☆ The Fe- protein mediates the coupling of ATP hydrolysis to electron transfer, and is the only known electron donor that can support substrate reduction by the MoFe- protein.
- ☆ The Fe- protein is a dimer of identical subunits (of molecular weight 60KD) that contains [4Fe:4S] metallocluster per dimer.
- ☆ In addition to this molybdenum containing nitrogenase, alternate nitrogenases also exist that are homologous to this system, but with the molybdenum almost certainly substituted by vanadium or iron.

Genetics of Nitrogen Fixation

The genetics of nitrogen fixation was initially studied in *Klebsiella pneumoniae.*

- ☆ The genes that encode the enzymes involved in the nitrogen fixation are referred as nif genes.
- ☆ The primary enzyme encoded by nif genes is nitrogenase complex.
- ☆ The genes of *K. pneumoniae* are a part of a complex regulon called nif gene regulon.
- ☆ The nif gene of *K. pneumoniae* spans 24kb of DNA and contains about 20 genes arranged in several transcriptional units and diverse array of structural genes and regulatory proteins are found in the nif gene regulon.
- ☆ Dinitrogenase is a complex protein consisting of α and β subunits. α subunit is the product of *nifD* gene while β subunit that of *nifK* gene.
- ☆ The *nifH* gene encodes for the protein dinitrogenase reductase which is a protein dimer made up of two identical subunits.
- ☆ Besides these three structural genes the complete assembly of nitrogenase requires products of nif genes that are involved in the synthesis of FeMo-co, in formation of FeS clusters and maturation of nitrogenase components.

- ✰ Fe-Mo cofactor is synthesized through participation of several genes including *nifN, nifV, nifB, nifQ, nifE, nifX, nifU,* nifS and *nifY*.
- ✰ Gene *nifS* and *nifU* play a part in the assemblage of Fe-S clusters.
- ✰ Products of *nifH, nifM, nifU* and *nifS* are required for maturation of Fe protein.
- ✰ *nifE* and *nifN* products have been proposed to function as scaffold for FeMo-co biosynthesis.
- ✰ The *nifB* gene product acts as iron and sulphur containing precursor of FeMo-co.
- ✰ The gene *nifQ* is a molybdenum-sulfur containing precursor of FeMo-co.
- ✰ The gene *nifV* encodes homocitrate synthase and is required for the synthesis of FeMo-co.
- ✰ The gene *nifW* is involved in stability of dinitrogenase and protect the protein from oxygen inactivation (Cheng, 2008).
- ✰ *Klebsiella* also contains the genes that mediate electron transport to nitrogenase.
- ✰ *nifF* encodes flavodoxin that transfers electrons to nitrogenase and *nifJ* encodes pyruvate oxidoreductase that transfers electrons to flavodoxin from the pyruvate.
- ✰ *nif A* encodes positive regulatory protein that serves to activate transcription of other genes while *nifL* acts as repressor of nitrogenase.
- ✰ For most free-living, nitrogen-fixing organisms, such as *K. pneumoniae*, nif-gene expression should be responsive to three environmental conditions.

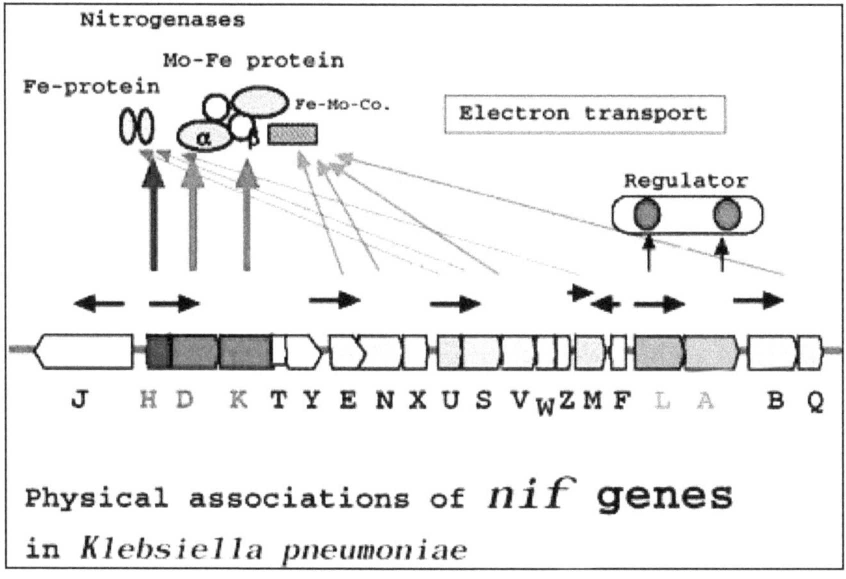

Physical associations of *nif* genes in *Klebsiella pneumoniae*

- ☆ These parameters include (1) the availability of a fixed nitrogen source, (2) whether or not oxygen is present, and (3) the energy charge status of the cell.
- ☆ The reasons that these conditions are important is that nitrogenase activity is both oxygen-sensitive and requires the intense consumption of metabolic energy.
- ☆ Thus, if oxygen is present, if a fixed nitrogen source is already available, and if cellular growth is limited by the availability of a carbon source rather than a nitrogen source, the expression of nif-genes would represent a waste of metabolic energy.
- ☆ In free-living nitrogen fixers that have alternative nitrogenases, the expression of nif-genes is additionally controlled by the availability of Mo or V.

Define nanotechnology? And explain how bionanotechnology differs from nanobiotechnology?

Nanotechnology

Nanotechnology is the study and application of extremely small things and can be used across all the other science fields, such as chemistry, biology, physics, materials science, and engineering *i.e.* engineering of functional system at molecular scale.

- ☆ In nanotechnology, a particle is defined as a small object that behaves as a whole unit with respect to its transport and properties.
- ☆ Particles are further classified according to diameter, ultrafine particles, or nanoparticles are between 1 and 100 nanometers in size.
- ☆ Bionanotechnology, nanobiotechnology and nanobiology are terms that refer to the insertion of nanotechnology and biology.

Bionanotechnology

Bionanotechnology refers to the study of how the goals of nanotechnology can be guided by studying how biological machines work and adapting these biological motifs into improving existing nanotechnologies or creating new ones.

- ☆ It mainly involves the development of techniques of nanotechnology might be guided by studying the structure and function of the natural nano molecules found in living cells.
- ☆ DNA nanotechnology or cellular engineering would be classified as bionanotechnology because they involve working with biomolecules on the nanboscale.
- ☆ Lipid nanotechnology is another major area of research in bionanotechnology, where physic-chemical properties of lipids such as their antifouling and self assembly are exploited to build nanodevices with applications in medicine and engineering.

- Bionanotechnology mainly involves the fields like nanotechnology in nutrition and medicine, nanotechnology in health and food technology applications and nanomaterial manufacturing *etc.*

Nanobiotechnology

Nanobiotechnology refers to the ways that nanotechnology is used to create devices to study biological systems.

- Many new technologies involving nanoparticles as delivery systems or as sensors would be examples of nanobiotechnology since they involve using nanotechnology to advance the goals of biology.
- Another example of nanobiotechnology is nanospheres coated with fluorescent polymers could become part of biological assays, and the technology might lead to particles which could be introduced into the human body to track down metabolites associated with tumors and other health problems.
- The possibility to exploit the structures and processes of biomolecules for novel functional materials, biosensors, biomolecules and medical applications has created the rapidly growing field of nanobiotechnology.

Define and explain in detail about the biofortification of food crops?

Biofortification

Breeding crops to enhance their nutritional composition, known as biofortification, is one potential strategy for addressing certain forms of undernutrition.

- Biofortification uses the very staple foods that the poor are already eating to deliver necessary micronutrients to them.
- It is an agricultural intervention targeted to rural areas where more than seventy-five percent of the poor in developing countries live, and where access to supplements, fortified foods and other urban-based interventions are limited.

Examples of biofortification projects include:

- Iron- biofortification of rice, beans, sweet potato, cassava and legumes
- Zinc- biofortification of wheat, rice, beans, sweet potato and maize
- Provitamin A carotenoid- biofortification of sweet potato, maize and cassava and
- Amino acid and protein- biofortification of sorghum and cassava.

There are more than twenty 'spillover' countries, which would also benefit from these biofortified varieties following their release in selected target countries.

Crop	*Nutrient*	*Countries of First Release*
Sweet potato	Provitamin A	Uganda, Mozambique
Bean	Iron	Rwanda, D.R. Congo
Pearl millet	Iron	India
Cassava	Provitamin A	Nigeria, D.R. Congo
Maize	Provitamin A	Zambia
Rice	Zinc	Bangladesh, India
Wheat	Zinc	India, Pakistan

Biofortified crop varieties are developed by plant breeding using selective breeding and/or genetic modification.

Selective Breeding

These programmes search for variation in the characteristic of interest, for example higher iron content, which existing varieties of the crop.

- ☆ This characteristic is then bred into cultivated varieties by crossing (deliberately inter-breeding) and selecting those individual plants with the desired characteristic.
- ☆ In selective breeding scientists use:
    a. Seed banks- collections of seeds usually collected in the past, which may have greater genetic variation than current varieties.
    b. Mutagenesis- a chemical or physical induction of genetic mutations used to generate new variation.
    c. Wide crosses- inter-breeding between a cultivated species and another, normally closely related species.

Genetic Modification

Genetic modification technology is used to transfer a single gene or several genes, from any species, into another. Researchers can control both when (at what point in development) and where a gene is switched on (expressed) in the plant to alter desired characteristics of a crop.

Frame Work for Biofortified Crops

Identifying target population and staple food consumption

↓

Setting nutrient target levels

↓

Screening and applied breeding technology

↓

Crop improvement

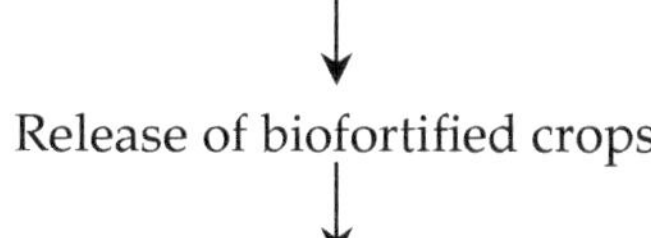

↓

Facilitates dissimilation, marketing and consumer acceptance

↓

Improved nutrient status of target population

The potential benefits of biofortification include:

- Scope – reaching rural communities without access to pharmaceutical supplements or fortified food and improving life-time nutritional status.
- Endurability – less susceptible to social and economic changes than short term interventions.
- Cost effectiveness – the potential to impact a large number of people at a low cost per person.

The possible limitations include:

- Narrow focus – increasing any single micronutrient in the diet is unlikely to address the whole problem.
- Allergenicity and toxicity – increasing the amount or incidence of certain plant products in the diet could have a negative impact on health for some people.
- Top-down approach – a technological solution alone will not address root causes of the problem, such as social inequality, lack of education and poverty.
- Lack of capacity – plant breeding is an ongoing exercise requiring continued effort and financial support, at a regional level, with local farmer engagement.

Define and explain in brief about the bioremediation?

Bioremediation

Bioremediation is the technology that uses microorganism metabolism to remove pollutants it uses relatively lowcost, low-technology techniques, which generally have a high public acceptance and can often be carried out on site.

- It is the process whereby organic wastes are biologically degraded under controlled conditions to an innocuous state, or to levels below concentration limits established by regulatory authorities.
- For bioremediation to be effective, microorganisms must enzymatically attack the pollutants and convert them to harmless products.
- As bioremediation can be effective only where environmental conditions permit microbial growth and activity, its application often involves the

manipulation of environmental parameters to allow microbial growth and degradation to proceed at a faster rate.

- Different techniques are employed depending on the degree of saturation and aeration of an area.
  a. *In situ* techniques are defined as those that are applied to soil and groundwater at the site with minimal disturbance. *e.g.* Phyto-remediation, Biosparging, Bioventing and Bioaugmentation *etc.*
  b. *Ex situ* techniques are those that are applied to soil and groundwater at the site which has been removed from the site via excavation (soil) or pumping (water). *e.g.* Landfarming, composting and Biopiles *etc.*

Advantages of Bioremediation

- It is a natural process; it takes a little time, as an acceptable waste treatment process for contaminated material such as soil.
- Bioremediation also requires a very less effort and can often be carried out on site, often without causing a major disruption of normal activities.
- This also eliminates the need to transport quantities of waste off site and the potential threats to human health and the environment that can arise during transportation.
- Bioremediation is also a cost effective process as it lost less than the other conventional methods that are used for clean-up of hazardous waste.
- It also helps in complete destruction of the pollutants, many of the hazardous compounds can be transformed to harmless products, and this feature also eliminates the chance of future liability associated with treatment and disposal of contaminated material.
- It does not use any dangerous chemicals. The nutrients added to make microbes grow are fertilizers commonly used on lawns and gardens.
- Because bioremediation changes the harmful chemicals into water and harmless gases, the harmful chemicals are completely destroyed.

Limitations of Bioremediation

- Bioremediation is limited to those compounds that are biodegradable. Not all compounds are susceptible to rapid and complete degradation.
- There are some concerns that the products of biodegradation may be more persistent or toxic than the parent compound.
- Biological processes are often highly specific. Important site factors required for success include the presence of metabolically capable microbial populations, suitable environmental growth conditions, and appropriate levels of nutrients and contaminants.
- It is difficult to extrapolate from bench and pilot-scale studies to full-scale field operations.

- ☆ Research is needed to develop and engineer bioremediation technologies that are appropriate for sites with complex mixtures of contaminants that are not evenly dispersed in the environment. Contaminants may be present as solids, liquids, and gases.
- ☆ Bioremediation often takes longer than other treatment options, such as excavation and removal of soil or incineration.
- ☆ Regulatory uncertainty remains regarding acceptable performance criteria for bioremediation.

Define Phytoremediation? And explain in brief about the mode of action of phytoremediation along with advantages and limitations?

Phytoremediation

Phytoremediation can be defined as "the efficient use of plants to remove, detoxify or immobilise environmental contaminants in a growth matrix (soil, water or sediments) through the natural biological, chemical or physical activities and processes of the plants".

- ☆ Plants are unique organisms equipped with remarkable metabolic and absorption capabilities, as well as transport systems that can take up nutrients or contaminants selectively from the growth matrix, soil or water.
- ☆ Phytoremediation involves growing plants in a contaminated matrix, for a required growth period, to remove contaminants from the matrix, or facilitate immobilisation (binding/containment) or degradation (detoxification) of the pollutants.
- ☆ Phytoremediation may take several years to clean up a site. The cleanup time will depend on several factors. For example, phytoremediation will take longer where:
  - a. Contaminant concentrations are high.
  - b. The contaminated area is large or deep.
  - c. Plants that have a long growing time are used.
  - d. The growing season is short.

There are several different types of phytoremediationmechanisms. These are:

1. **Rhizosphere biodegradation:** In this process, the plant releases natural substances through its roots, supplying nutrients to microorganisms in the soil. The microorganisms enhance biological degradation.
2. **Phyto-stabilization:** In this process, chemical compounds produced by the plant immobilize contaminants, rather than degrade them.
3. **Phyto-accumulation** (also called phyto-extraction): In this process, plant roots absorb the contaminants along with other nutrients and water. The contaminant mass is not destroyed but ends up in the plant shoots and leaves. This method is used primarily for wastes containing metals.

4. **Rhizofiltration** (Hydroponic Systems for Treating Water Streams): Rhizofiltration is similar to phyto-accumulation, but the plants used for cleanup are raised in greenhouses with their roots in water. This system can be used for *ex-situ* groundwater treatment
5. **Phyto-volatilization**: In this process, plants take up water containing organic contaminants and release the contaminants into the air through their leaves.
6. **Phyto-degradation:** In this process, plants actually metabolize and destroy contaminants within plant tissues.
7. **Hydraulic Control:** In this process, trees indirectly remediate by controlling groundwater movement. Trees act as natural pumps when their roots reach down towards the water table and establish a dense root mass that take up large quantities of water.

Advantages

- ☆ The cost of the phytoremediation is lower than that of traditional processes both *in situ* and *ex situ.*
- ☆ In this, the plants can be easily monitored.
- ☆ The possibility of the recovery and re-use of valuable metals in phytoremediation (by companies specializing in "phyto mining").
- ☆ It is potentially the least harmful method because it uses naturally occurring organisms and preserves the environment in a more natural state.

Limitations

- ☆ Phytoremediation is limited to the surface area and depth occupied by the roots.
- ☆ Slow growth and low biomass require a long-term commitment.
- ☆ With plant-based systems of remediation, it is not possible to completely prevent the leaching of contaminants into the groundwater (without the complete removal of the contaminated ground, which in itself does not resolve the problem of contamination)
- ☆ The survival of the plants is affected by the toxicity of the contaminated land and the general condition of the soil.
- ☆ Bio-accumulation of contaminants, especially metals, into the plants which then pass into the food chain, from primary level consumers upwards or requires the safe disposal of the affected plant material.

What is molecular farming? And explain the advantages of plants as expression systems for molecular farming?

Molecular Farming

It refers to the use of genetic engineering to insert genes that code for useful pharmaceuticals into host animals or plants that would otherwise not express those genes, thus creating a genetically modified organism. This is also known as Pharming or Molecular pharming or Biopharming.

- The first recombinant plant derived protein (PDP) was human serum albumin, initially produced in 1990 in transgenic tobacco and potato plants.
- Global demand for pharmaceuticals is at unprecedented levels, and current production capacity will soon be overwhelmed.
- Expanding the existing microbial systems, although feasible for some therapeutic products, is not a satisfactory option on several grounds.
- But the use of plants offers a number of advantages over other expression systems.

Expression System	*Yeast*	*Bacteria*	*Plant Virus*	*Transgenic Plants*	*Animal Cell Cultures*	*Transgenic Animals*
Cost of maintaining	Inexpensive	Inexpensive	Inexpensive	Inexpensive	Expensive	Expensive
Type of storage	-2.0ºC	-2.0ºC	-2.0ºC	Room temperature	Maintained under nitrogen gas	Not applicable
Gene size restriction	Unknown	Unknown	Limited	Not limited	Limited	Limited
Production cost	Medium	Medium	Low	Low	High	High
Protein yield	High	Medium	Very high	High	Medium to high	High
Therapeutic risk	Unknown	Yes	Unknown	Unknown	Yes	Yes

- Plants can be used in two ways: one way is to insert the desired gene into a virus that is normally found in plants, such as the tobacco mosaic virus in the tobacco plant. The other way is to insert the desired gene directly into the plant DNA to produce a transgenic plant.
- Like animals, plants are complex, multicellular organisms and therefore their process of protein synthesis is more similar to that of animals than those of bacteria or yeast, which are not capable of producing complex proteins.
- Another advantage of using plants is that proteins that are produced in a plant accumulate to high levels in its tissues.
- The use of plants avoids the risk of contamination with animal pathogens, such as viruses, that could be harmful to humans. No plant virus has been found to be pathogenic to humans.
- Purification of the desired product from plants is often easier than bacteria, which can be labour and cost intensive.

- Whereas transgenic plants or virus infected plants, can be grown on an agricultural scale requiring only water, minerals and sun light.
- Mammalian cell cultivation is an extremely delicate process that is also very expensive, requiring bioreactors that cost several hundred million dollars when production is scaled up to commercial levels.

Define biodiesel? And explain how biofuel production potential can be increased by using biotechnological tools?

Biodiesel

Biodiesel refers to vegetative oil or animal fat based diesel fuel consisting of long chain alkyl esters. Biodiesel is typically made by chemically reacting lipids like vegetative oil, soybean oil and animal fat with an alcohol producing fatty acid esters.

- The commonly used crops for biofuel production are likely maize, cashew nut, oats, lupin, kenaf, calendula, cotton, hemp, soybean, coffee, flax, hazelnuts, euphorbia, pumpkin seed, coriander, mustard seed, camelina, sesame, safflower, rice, tung tree, sun flower, cocoa, pea nut, opium, rape seed, olives, castor beans, jojoba, jatropa, avocado, coconut and oil palm *etc.*

Recent technology advancements in cellulosic biomass conversion technologies combined with high fossil fuel prices, have rekindled interest in the potential of cellulosic biofuels. Future biotechnology based developments in processing technology will likely include:

- Improved cellulose and hemicellulase production economics via microbe or plant based production system.
- Improved fermentation strains that efficiently utilize both hemicelluloses and cellulosic sugars.
- Consolidated bioprocessing microbes which combine the ability to breakdown cellulosic materials with the ability to efficiently ferment various sugars to ethanol.

The next horizon for biotechnology will be its impact on the development of improved biomass feedstocks for biofuels production. It includes:

- Marker assisted breeding, the use of genomics to generate DNA markers associated with specific plant phenotypes.
- Precision breeding, the use of recombinant DNA technology to reintroduce plant genes into a plant under different regulation.
- Transgenics, the ability to transfer genes between species.

The application of biotechnology to perennial grass species such as switch grass, sugar cane and *miscanthus* over the next several years will result in a more economically competitive and environmentally friendly biofuel feedstock, and will enable the biofuel industry to scale to a point where it can meet a significant percentage of global transportation fuel demand.

7

Molecular Markers and Genomics

What are molecular markers? How do they differ from biochemical markers? Explain in brief about the role of molecular markers in crop improvement?

Molecular marker is a DNA segment that is readily detected and whose inheritance can easily monitored. The use of molecular markers is based on the naturally occurring DNA polymorphism.

- A DNA marker is a small region of DNA showing sequence polymorphism in different individuals with in a species or among different species.
- There are three types of genetic markers (represents the genetic differences between individual organisms or species).
  1. Morphological markers, which themselves are phenotypic traits or characters.
  2. Biochemical markers, which include allelic variants of enzymes called isozymes.
  3. DNA or Molecular markers, which reveal sites of variation in DNA.
- Morphological markers are usually visually characterized phenotypic characters such as flower colour, seed shape, growth habits or pigmentation *etc.*
- Isozyme markers are differences in enzymes that are detected by electrophoresis and specific staining.

- ✰ The major disadvantage of morphological and biochemical markers are that they may be limited in number and are influenced by environmental factors or the developmental stage of the plant. But these are extremely useful to the plant breeders.
- ✰ DNA markers are the most widely used type of markers predominantly due to their abundance.
- ✰ DNA markers arise from different classes of DNA mutations such as substitution mutations (point mutations), rearrangements (insertion or deletion) or errors in replication of tandemly repeated DNA.
- ✰ DNA markers are selectively neutral because they are usually located in non-coding regions of DNA.
- ✰ Unlike morphological and biochemical markers, DNA markers are practically unlimited in number and are not affected by environmental factors and/or developmental stage of the plant.

Role of Molecular Markers in Crop Improvement

1. Construction of high density (saturated) genetic maps of crops to provide assessment of each segment of genome for economic use: on the basis of map position of the gene the DNA molecule rather than through its phenotypic effect.
2. Estimation of genetic diversity, DNA finger printing of crop varieties and genetic collections help to characterize genetic diversity for effective management and utilization germplasm resources.
3. Comparative gene mapping in different plant species to determine parallelism in the gene order (synteny) and evolutionary relationship.
4. Marker aided early generation selection of transgressive segregants to increase the speed and efficiency of developing new varieties.
5. Introgression of genes from wild germplasm.
6. Mapping of poly genes: QTLs.
7. Gene pyramiding: Assembly of many genes and alleles with similar effect from different loci is practically difficult through screening for molecular markers associated with each of such genes. The genes for different characters can also be assembled in to genotype with help of markers.
8. Map based cloning of genes

What are the properties of ideal marker system? And discuss the different types of molecular markers, their principle, advantages and disadvantages?

Ideal marker system should have these properties.

1. Easily available and identifiable
2. Associated with specific locus
3. Easy and rapid assay

4. Highly variable, polymorphic and reproducible
5. Selectively neutral to environmental conditions or management practices
6. Co-dominant inheritance and recurrent occurrence in genome
7. Cost and time effective

DNA markers may be broadly devided in to three classes based on the method of detection

i. Hybridization based *e.g.* RFLP
ii. PCR based *e.g.* RAPD, AFLP and SSR
iii. DNA sequence based *e.g.* SNP

Restriction Fragment Length Polymorphism (RFLP)

- ☆ First molecular marker developed in 1980 by Botestein and is based on southern hybridization technique.
- ☆ RFLP means variation in homologous fragment length of DNA on being digested with Restriction endonucleases.

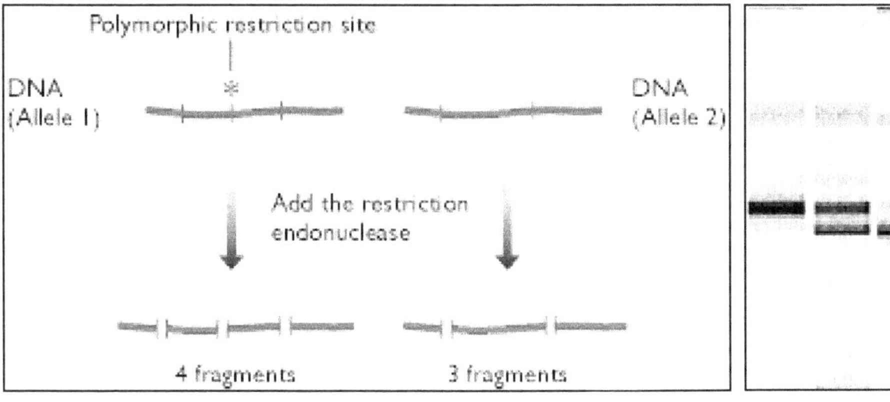

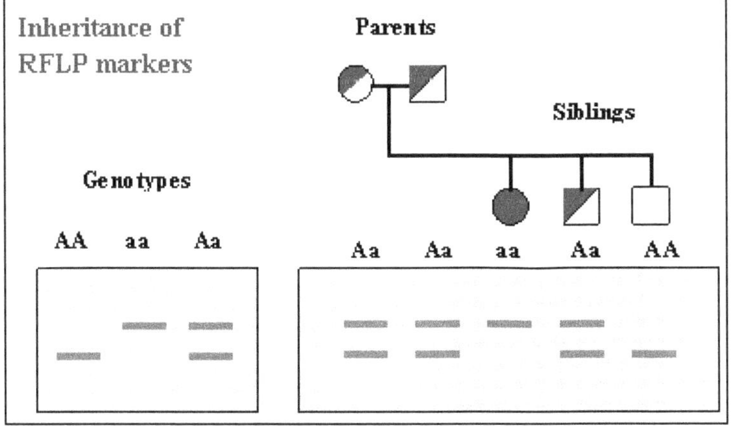

- In this, low copy/single copy number DNA sequence is used as probe. So the size of the DNA fragments homologous to the probe DNA depends upon the location of target sites.
- Base changes can change the sequence of target site and deletion and insertions can also change the location of target sites.

Advantages

- Transferable across the population
- Have good repeatability, highly reproducible and very robust
- Useful in comparative genome mapping
- Co dominant marker

Disadvantages

- The assay is tedious and time consuming
- Requires large quantity of DNA (20µl of DNA)
- Extremely limited utility in marker assisted selection due to very low assay efficiency.

Random Amplified Polymorphic DNA (RAPD)

- It is dominant marker based on PCR technique
- It employs a decamer primer of arbitrary sequence which is annealed to template DNA at 37°C
- Variation in RAPD profile in the form of presence or absence of bands results from variations in primer binding sites.
- The number of primer binding sites or number of fragments expected to be amplified = $2fN/16^{n}$, where N = Genome size (bp), n = primer size (bp) and 2f = probability of number primer binding sites including both the strands

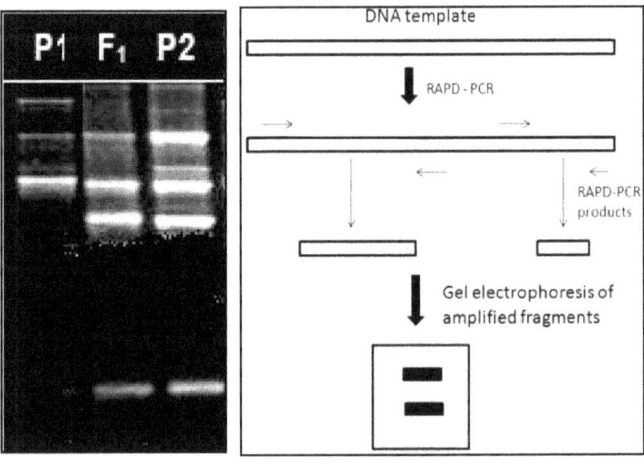

Advantages

- ☆ Requires small quantity of DNA (10ng/µl)
- ☆ Quality of DNA is not important requirement
- ☆ Needs limited investment in time and training

Disadvantages

- ☆ Non reproducible due to low annealing temperature. For this RAPD marker will convert in to SCAR marker by sequencing RAPD marker termini and design a longer (*e.g.* 24 nt) primers for specific amplification through PCR.
- ☆ SCAR marker is Co dominant marker

RAPD	SCAR
Primer size is 10 nucleotides (decamer)	Up to 24 nucleotides
Results are not consistent	Results are consistent
Single primer	Both forward and reverse primers are different
Size of fragment is small	Size of fragment is large
Number of fragments more	Only single fragment
Non reproducible and less robust	Reproducible and robust

Sequence Tagged Sites (STS)

- ☆ STS markers generated by unmodified PCR markers are designed based on cDNA sequence, random genomic DNA and ends of large genomic DNA fragment cloned in cosmid, λ and artificial vectors.

Advantages

- ☆ Has all advantages of PCR
- ☆ Very robust in nature

Disadvantages

- ☆ Considerable effort and time required to generate DNA sequence for designing primers.
- ☆ Fragments amplified by primers may not show length variation *i.e.* monomophic. So convert in to CAPS markers using restriction endonuclease enzymes.
  - ❑ CAPS will not work if there is no restriction site in sequence
  - ❑ CAPS marker is Co dominant marker

Amplified Fragment Length Polymorphism (AFLP)

- AFLP are based on both RFLP and PCR
- AFLP markers are generated by selective amplification of DNA fragments obtained by restriction enzyme digestion.
- High molecular weight DNA is digested with two restriction enzymes one hexa cutter (EcoRI) and one tetra cutter (Mse I)
- It provides the highest number of amplified products among all the DNA profiling systems.
- Dominant marker because many fragments, we just cannot find out its homologous band, so it is almost like dominant marker

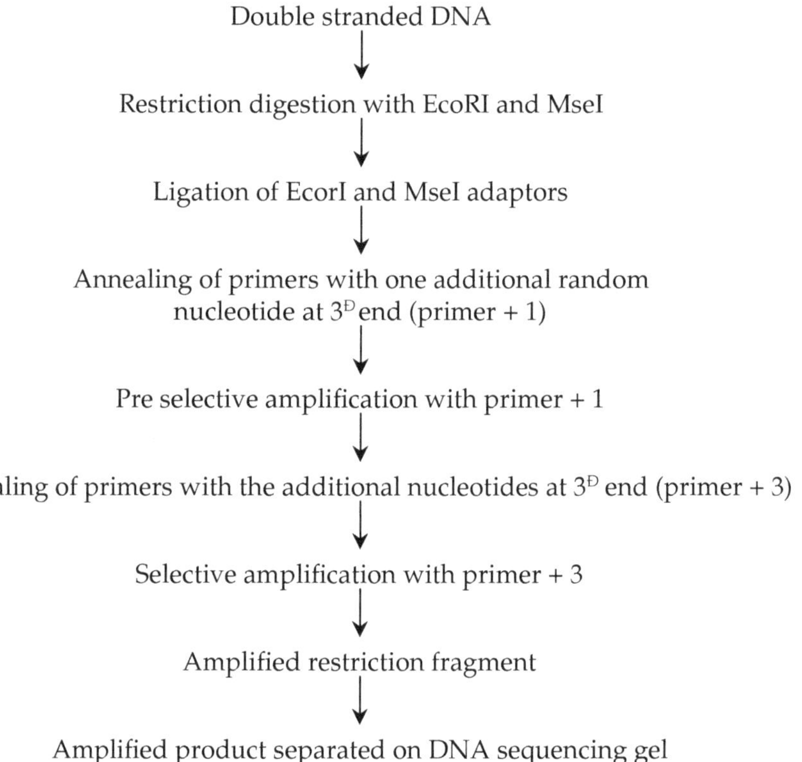

Advantages

- No prior sequence information needed
- Plenty in number
- Polymorphic information content is high
- Can be used in unsequenced crops
- Provide raw material for STS derivation

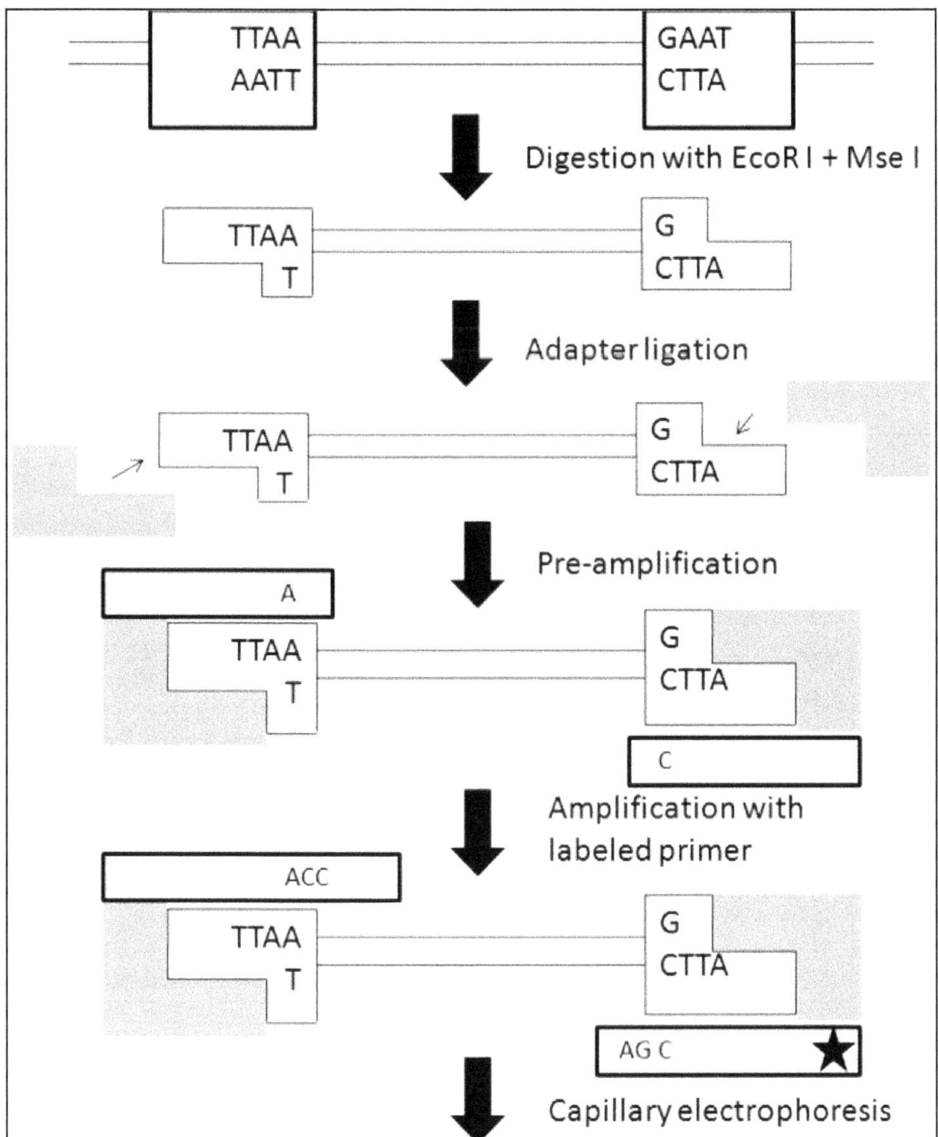

Disadvantages

- ✰ Lengthier and costlier than other PCR based markers
- ✰ Requires good quality DNA for complete digestion by enzyme because partial digestion of DNA results in non reproducible variation in DNA profile

Variable Number of Tandem Repeats (VNTR)/Simple sequence Repeats (SSR)/ Simple Sequence Length Polymorphism (SSLP)

- Variable number of tandem repeats of mono or di or tri or tetra or penta nucleotides dispersed throughout the genome.
- Variation due to number of repeats in different individuals at a given locus and these variation constitutes the alleles of VNTR loci.
- Classification:

Feature	*Microsatellites*	*Minisatellites*
DNA fragment	100-200 bp	0.2-2 kb
Tandem repeat sequence	2-9 bp	15-16 bp
Number of repeat units per locus	100 times	1000 times

  - In microsatellite marker, the fragment length of amplification is usually bigger than the expected size.
- Co dominant markers

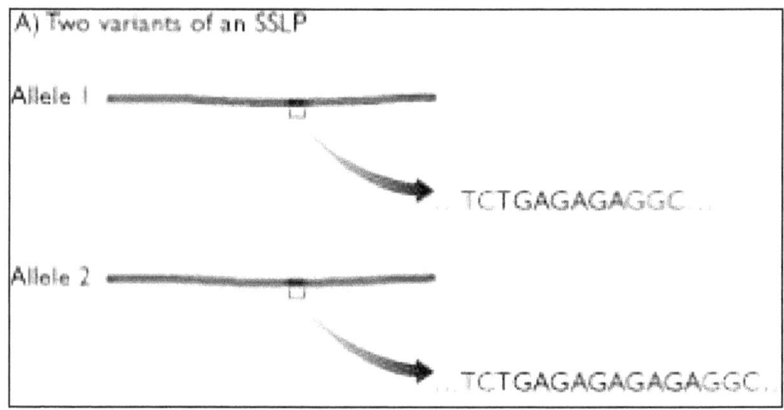

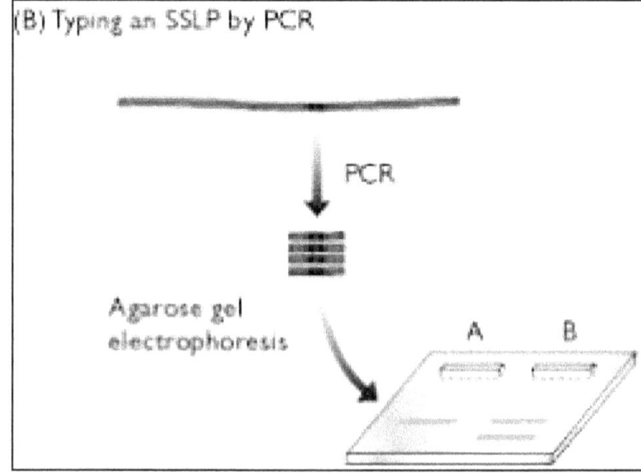

Advantages

- ✰ Transferable between populations
- ✰ Abundant and uniformly distributed in the genome
- ✰ Hyper variable (large number of alleles per locus)
- ✰ Highly reliable and robustable

Disadvantages

- ✰ Expensive and time consuming to detect SSR loci
- ✰ Not available in all crop plants

Write the difference between SSR and ISSR markers.

SSR	*ISSR*
Simple sequence repeats	Inter simple sequence repeats
Amplified region contains repeats	Contains region between repeats *i.e.* non repeat sequence
Both forward and reverse primers have different sequence	Only one primer acts as forward and reverse primers
Genetic basis of primer is only unique flanking region of repeats	Genetic basis of primer is repeat perse
Genetic basis of polymorphism is variation in number of repeats at particular locus.	Variation in inter region between repeats
Amplification profile is single locus	Amplification profile is multilocus because non repeat region may vary
Genetic nature of polymorphism is co dominant	Dominant marker

Single Nucleotide Polymorphism (SNP)

- ✰ Polymorphism observed between individuals may arise due to deletion or addition of multiple bases or due to single nucleotide substitution
- ✰ SNPs are biallelic.

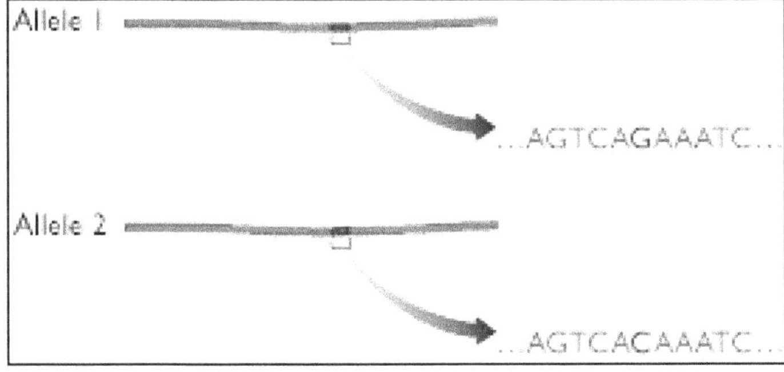

Advantages

- ✰ Co dominant marker
- ✰ Abundant in number
- ✰ Well suited for automated high through put genotyping
- ✰ Most useful to development of haplotype tags
- ✰ Serve to integrate physical and genetic maps

Disadvantages

- ✰ Requirement of sequence information
- ✰ High development or start up cost

SNPs have only Two Alleles

Any four nucleotides could be present in any position in the genome, so it might be imagined that SNP have four nucleotides. Theoretically it is possible but practically most SNPs exist just as two variants. This is because the way in which SNP arise and spread in population.

SNP arises when a point mutation occurs in a genome converting one nucleotide to another nucleotide. If this mutation occurs in reproductive cells of an individual, one or more offspring inhert this mutation and after many generations, this SNP become established in the population.

But there are two alleles *i.e.* original and mutated version. For third allele to arise a new mutation must occur at the same position in the genome in another individual and his or her offspring must reproduce in such a way that the new allele becomes established. This scenario is not impossible, but it is unlikely. Consequently most SNPs are biallelic.

SNP Detection

SNPs are detected by Oligonucleotide hybridization, DNA chip technology and Solution hybridization technique

Oligonucleotide Hybridization

- ✰ An oligonucleotide is a short, single stranded DNA molecule, usually less than 50 nucleotides in length that is synthesised in a test tube.
- ✰ If the conditions are just right, then the oligonucleotie will hybridize with another DNA molecule only if the oligonucleotide forms a completely base paired structure with the second molecule.
- ✰ If there is a single mismatch a single position within a oligonucleotide that does not form a base pair, then hybridisation does not occur. Oligonucleotide hybridisation can therefore discriminate between the two alleles of an SNP.

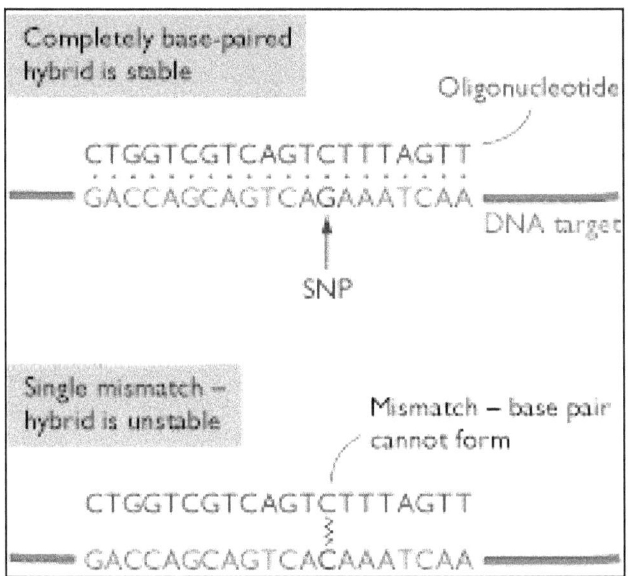

DNA chip and solution hybridisation techniques also used to detect the SNPs.

Solution hybridisation techniques are carried out in the wells of microtitre tray, each well containing a different oligonucleotide, using detection system that can discriminate between unhybridized, single stranded DNA and the double-stranded product that results when an oligonucleoide hybridizes to an test DNA.

☆ Several systems have been developed, one of which makes use of a pair of labels comprising a fluorescent dye and a compound that quenches the fluorescent signal where brought into close proximity with the dye.

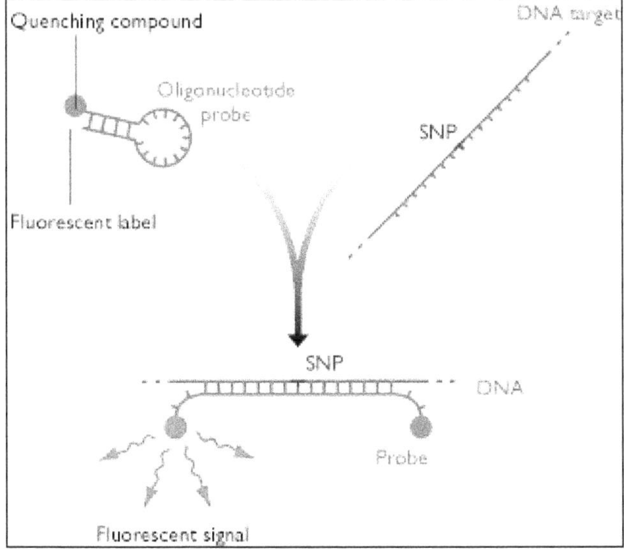

- ☆ The dye is attached to one end of an oligonucleotide and the quenching compound to the other end.
- ☆ Normally there is no fluorescence because the oligonucleotide is designed in such a way that the two ends base pair to one another placing the quencher next to the dye.
- ☆ Hybridization between oligonucleotide and the test DNA disrupts this base pairing, the quenches away from the dye and enabling the fluorescent signal to be generated.

Diversity Array Technology Marker (DArT)

Diversity Arrays Technology (DArT) is the name of a technology used in molecular genetics to develop sequence markers for genotyping and other genetic analysis.

DArT is based on microarray hybridizations that detect the presence versus absence of individual fragments in genomic representations. The technology has significant advantages over other array based Single-nucleotide polymorphism detection technologies in the analysis of polyploid plants.

Functional Markers

The molecular markers, generated by utilizing (gene) sequence data are known as 'functional markers' (FMs, Anderson and Lubberstedt, 2003). These markers are superior to random DNA markers (derived from genomic DNA) owing to complete linkage with potential trait locus alleles.

The FMs has advantages over genomic DNA markers as the later were designed from genomic DNA sequences such markers sample genetic variation in the genome more or less randomly and are sometimes referred to 'neutral' or random markers (RMs).

Explain in brief about the three specific applications of molecular markers with specific examples?

Three specific applications of molecular markers with specific examples:

1. Gene Introgression

Wild relatives of crop plants constitute a rich source of genetic variation which can be efficiently utilized for improvement of both qualitative and quantitative traits.

- ☆ *e.g.* Genes resistant to *Meloidogyne incognita* and tobacco mosaic virus have been introduced from related species *Lycopersicon peruvianum* in tomato.
- ☆ Problem of linkage drag during gene introgression by backcrossing *i.e.* undesirable linked genes is often faced by breeders. Here molecular linkage maps provide a method to reduce the problem of linkage drag by allowing selection of individual containg recombinant chromosome which breaks linkage drag.
- ☆ It has been estimated that the use of molecular markers can reduce linkage drag at least 10 fold in a fraction of the time needed by traditional breeding.

2. Gene Pyramiding

Pyramiding of Xa4, Xa5, Xa13 and Xa21 genes conferring resistance to four different races of BLB pathogen has been achieved in rice by using molecular markers at IRRI, Philippines. Two of these genes (Xa13 and Xa21) have been recently been combined with basmati quality characters in rice in IARI, New Delhi.

3. Construction of Heterotic Hybrids

In maize, through marker aided transfer of yield QTLs from inbreeds lines like TX303 to B73 and Oh43 to Mo17, Improved inbreeds were generated, which are called as enhanced lines. 15 enhanced B73 lines crossed with 18 enhanced Mo17 lines to produce 93 hybrids at North Carolina, USA. Six of hybrids exceed national check hybrids by two standard deviation and two hybrids gave 15 per cent more yield than checks.

Define MAS. Explain the advantages of marker assisted selection over conventional breeding.

Marker Assisted Selection (MAS)

Marker assisted selection involves selection of plants carrying genomic regions that are involved in the expression of traits of interest through molecular markers.

With the development and availability of an array of molecular markers and dense molecular genetic maps in crop plants, MAS has become possible for traits both governed by major genes as well as quantitative trait loci (QTLs). In general, the success of a marker based breeding system depends on three main factors:

- A genetic map with an adequate number of uniformly-spaced polymorphic markers to accurately locate desired QTLs or major gene(s)
- Close linkage between the QTL or a major gene of interest and adjacent markers
- Adequate recombination between the markers and rest of the genome and
- Ability to analyze a large number of plants in a time and cost effective manner

Advantages of Marker Assisted Selection over Conventional Breeding

- The segregants can be scored at seedling stage for traits that are expressed late in plant development. This includes traits such as grain quality, male sterility and photoperiod sensitivity.
- It is possible to screen for traits that are extremely difficult, expensive or time consuming to score or measure such as tolerance to drought, salt, mineral deficiencies and toxicity, root morphology, resistance to nematodes or to specific races or biotypes of diseases or insects.
- Selection can be practiced for several traits simultaneously which is difficult or even impossible by conventional means.

- Heterozygotes are easily identified and distinguished from the either homozygotes without resorting to progeny testing. This saves time and effort.
- Single plants can be selected using conventional screening methods for many traits, plant families or plants are grown because single plant selection is unreliable due to environmental factors. With MAS, individual plants can be selected based on their genotype.
- Markers can also be used to replacement for phenotyping, which allows selection in off season nurseries making it more cost effective to grow more generations per year.
- With MAS, total number of lines to be tested can be reduced and also selection of traits with low heritability.

Advantage of molecular markers is in terms of enabling "precise selection" and "unambiguous selection" even when the trait is unable to screened for. The inheritance is same as any other trait with the difference being one is able to see the genotype segregating rather than phenotype segregating. Thus the same approach can be adopted for.

- Pedigree breeding whether the selections can be specifically effected right from F_2 to minimize the number by picking up maximum number of homozygous loci through co dominant markers.
- Backcross breeding where both recurrent parent genome background can be assembled with minimum number of backcrosses looking for homozygous loci distributed over all chromosomes by opting for markers detectable spread over all the chromosome on each arm.

While the target gene can easily be identified for the homozygous form in the BC_1F_2 generation through its tagged marker.

- Recurrent selection where markers tagged to desire alleles or loci can be tracked for picking up the recombinants possessing most of the desired alleles with the minimum number of intermatings.
- Hybrid breeding where markers can be employed to pick up heterotic divergent parents and confirmed hybrid progenies in large number of crosses with minimum field evaluations in initial stages.

Define and explain in detail about marker assisted backcross breeding:

Marker Assisted Backcross Breeding (MABC)

Marker Assisted Backcrossing (MABC) is regarded as the simplest form of MAS and it aims to transfer one or few genes/QTLs of interest from one genetic source (donor parent) into a superior cultivar or elite breeding line (recurrent parent) to improve the target trait.

- Unlike traditional backcrossing, MABC is based on the alleles of markers associated with or linked to genes/QTLs of interest instead of phenotypic performance of target trait.
- In general backcrossing, the recovery of recurrent parent gene is delayed still BC_4 and BC_5 generations. Where the average recovery of genome is 96.87 per cent and 98.43 per cent respectively.
- It takes more time and resources for generating these populations.
- But it is observed that BC_1 itself some plants have recurrent genotype up to 90 per cent.
- These types of plants can be selected by polymorphic markers between parents. By employing background selection plants with minimum recurrent genome are identified and use for crop improvement.
- Thus the use of DNA markers in backcrossing greatly increases the efficiency of selection.

MABC is essential and advantageous when,

- Phenotyping is difficult or expensive or impossible
- Heritability of the target trait is low
- The trait is expressed in later stages of development and growth such as flowers, fruits and seeds *etc.*
- The traits controlled by genes, which require special conditions to express
- Traits are controlled by recessive genes and
- Gene pyramiding is needed for one or more traits.

Three general levels of marker-assisted backcrossing (MAB) can be described.

- In the first level, markers can be used in combination with or to replace screening for the target gene or QTL. This is referred to as 'foreground selection'. This may be particularly useful for traits that have laborious or time-consuming phenotypic screening procedures. It can also be used to select for reproductive-stage traits in the seedling stage, allowing the best plants to be identified for backcrossing. Furthermore, recessive alleles can be selected, which is difficult to do using conventional methods.
- The second level involves selecting BC progeny with the target gene and recombination events between the target loci and linked flanking markers. This is referred to as 'recombinant selection'. The purpose of recombinant selection is to reduce the size of the donor chromosome segment containing the target locus (*i.e.* size of the introgression). This is important because the rate of decrease of this donor fragment is slower than for unlinked regions and many undesirable genes that negatively affect crop performance may be linked to the target gene from the donor parent and this is referred to as 'linkage drag'.

☆ Third level involves the markers used for selection of recessive parent genome and this is referred to as 'Background selection'.

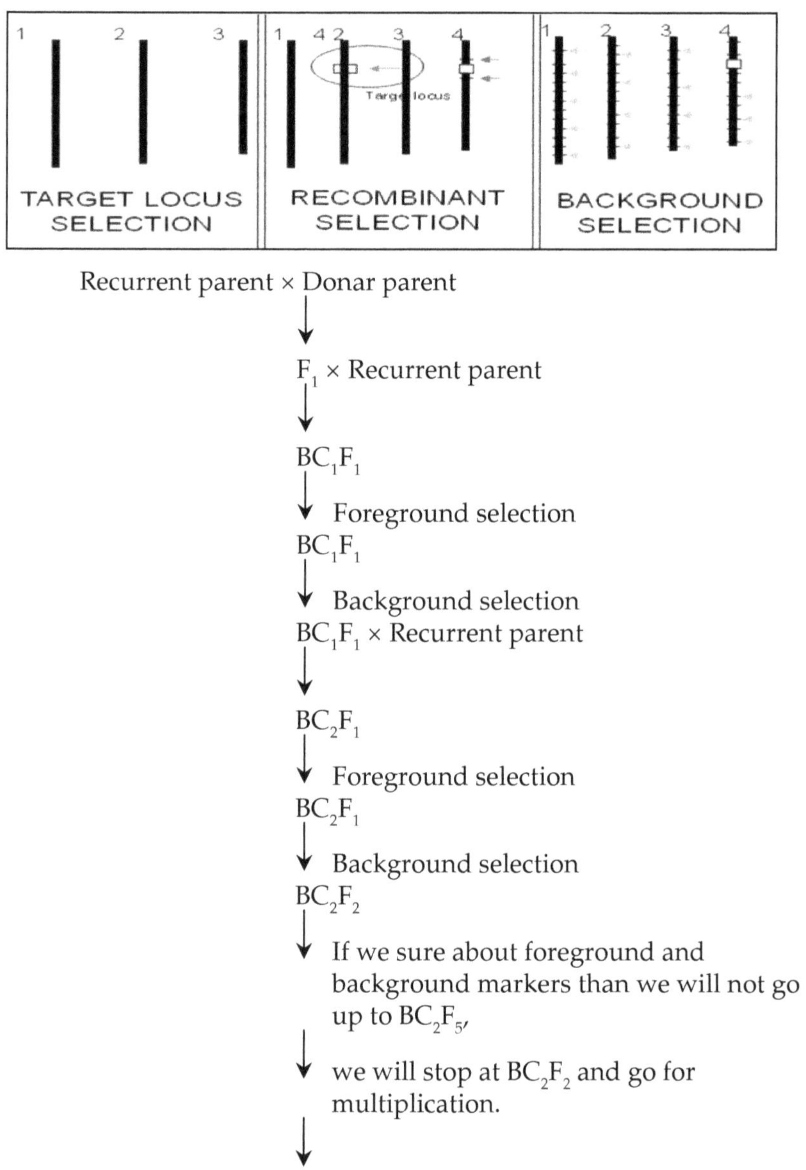

The use of background selection during marker assisted breeding to accelerate the development of a recurrent parent with an additional gene is referred to as complete conversion (Ribaut *et al.,* 2002).

Define and explain in brief about the Marker Assisted Recurrent Selection?

Marker Assisted Recurrent Selection (MARS)

It is one of the MAS method, used for identification and selection of several genomic regions (up to 20 or even more) for complex traits within a single population.

- When much variation is controlled by minor QTLs, Marker assisted backcrossing has limited applicability because estimates of QTL effects are inconsistent and pyramiding becomes increasingly difficult as the number of QTL increases.
- To avoid such difficulty, MARS particularly helps in integrating multiple favourable genes from different sources through recurrent selection based on multiple-parental population. *i.e.* it increases the frequency of favourable marker alleles in the population.
- MARS mainly involves.
  i. Defining a selection index for F_2 or F_2- derived progenies with desirable alleles at target QTLs.
  ii. Recombining selfed progenies of the selected individuals and
  iii. Repeating the procedure for a number of cycles.

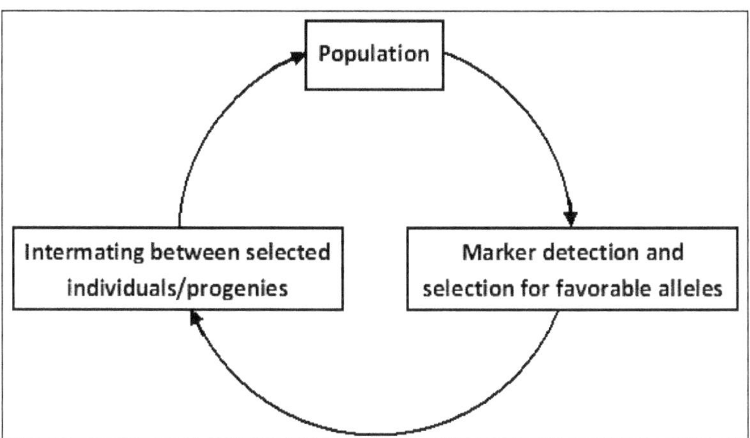

What is meant by Marker Assisted Gene Pyramiding? If you the responsibility of pyramiding four genes from four different sources into one elite line, which approach will you follow? Why?

Pyramiding is the process of combining several genes together into a single genotype.

- Pyramiding may be possible through conventional breeding but it is usually not easy to identify the plants containing more than one gene.
- But DNA markers can greatly facilitated selection because DNA marker assays are non destructive and markers for specific genes can be tested

using a single DNA sample without phenotyping. *i.e.* Marker Assisted Gene Pyramiding.

- ☆ The most widespread application for pyramiding has been for combining multiple disease resistant genes.
- ☆ The main aim of this has been the development of 'durable' or stable disease resistance since pathogens frequently overcome single gene host resistance over time due to the emergence of new plant pathogen races.
- ☆ In Marker assisted gene pyramiding, there may be three strategies or breeding schemes can be used for pyramiding the genes.
    i. Step wise transfer method
    ii. Simultaneous transfer method
    iii. Stepwise but simultaneous transfer method/Convergent backcrossing
- ☆ Suppose, recurrent parent (RP) superior in comprehensive performance but lack of trait of interest, and four different genes (A, B, C and D) contributing to the trait have been identified in four germ plasm lines (donor parents, DP).
- ☆ Three schemes for pyramiding genes can be described as follows:

Step-wise Transfer Method

- ☆ In this approach, the recurrent parent is first improved for one character
- ☆ The improved recurrent parent (IRP) is then used as recurrent parent in a fresh backcross programme for the transfer of next character.
- ☆ The advantage is that gene pyramiding is more precise and easier to implement as it involves only one gene at one time and thus the population size and genotyping amount will be small.
- ☆ The disadvantage is that it takes a longer time for transfer of more genes.

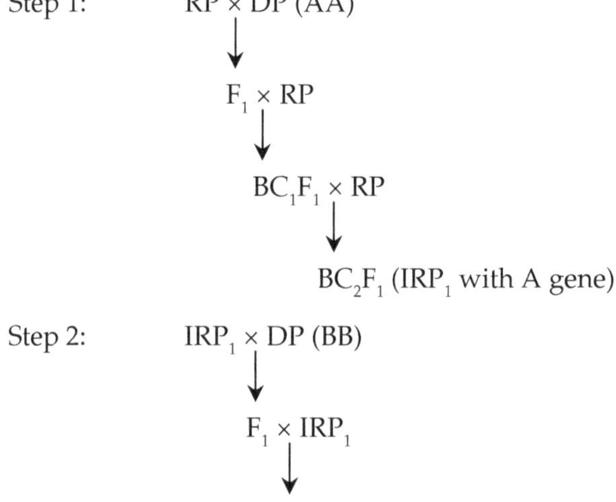

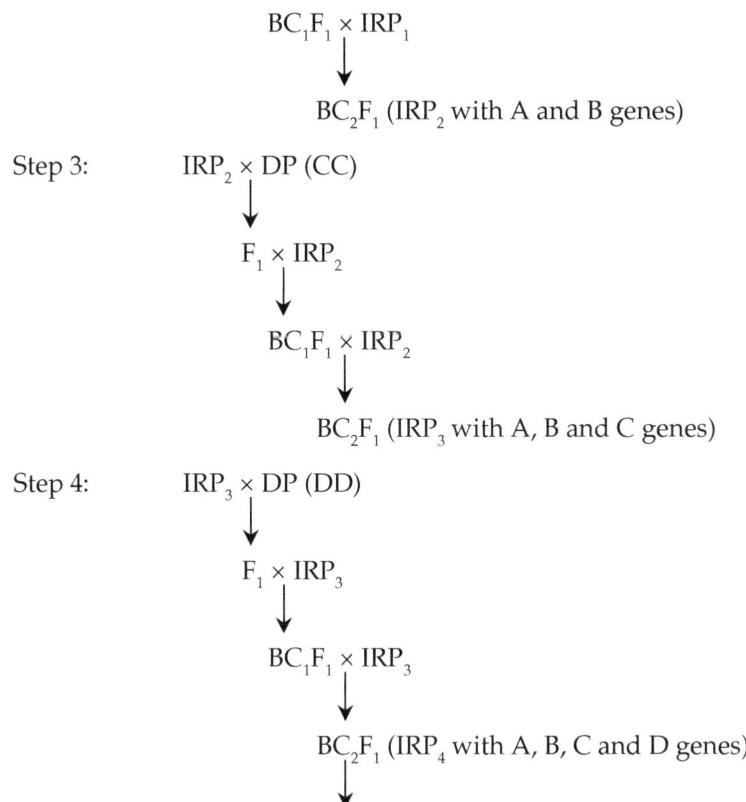

Selfing and select the segregants with all desirable genes ($RP_{AABBCCDD}$)

Simultaneous Transfer Method

- In this approach, the recurrent parent is first crossed to each of four donor parents to produce four single cross F1s.
- Two of the four single crosses F1s are crossed with each other to produce two double – cross F1s, and these two double – cross F1s are crossed again to produce a hybrid integrating all four target genes in heterozygous state.
- The hybrid or progeny with heterozygous markers for all four target genes is subsequently backcrossed to recurrent parent until a satisfactory recovery of recurrent parent genome, and finalized by one generation of selfing.
- The advantage of this method is that it takes the shortest time to complete the pyramiding of genes.
- The disadvantage is that, in backcrossing, all target genes are involved at the same time, and thus it requires a large population and more genotyping.

Crossing

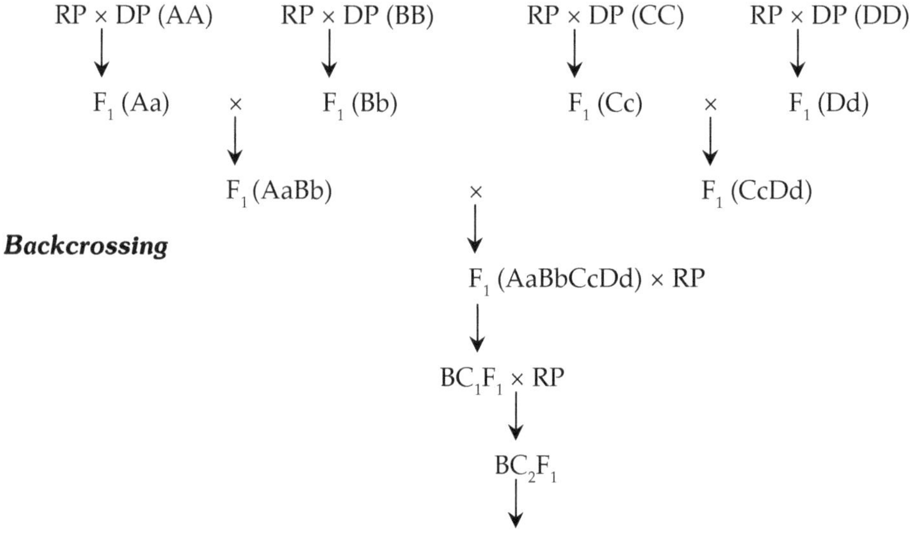

Selfing and select the segregants with all desirable genes ($RP_{AABBCCDD}$)

Simultaneous but Step-wise Transfer Method

- In this, first four target genes are transferred separately from the donors into the recurrent parent by single crossing followed by backcrossing based on markers linked to the target gene, to produce improved lines.
- Two of the improved lines are crossed with each other and the two hybrids are than intercrossed to integrate all four genes together and develop the final improved line with all four genes.
- This approach is more acceptable because in this scheme not only is time reduced (compared to step wise transfer) but gene fixation and/ or pyramiding is also more easily assured (compared to simultaneous transfer).

Backcrossing

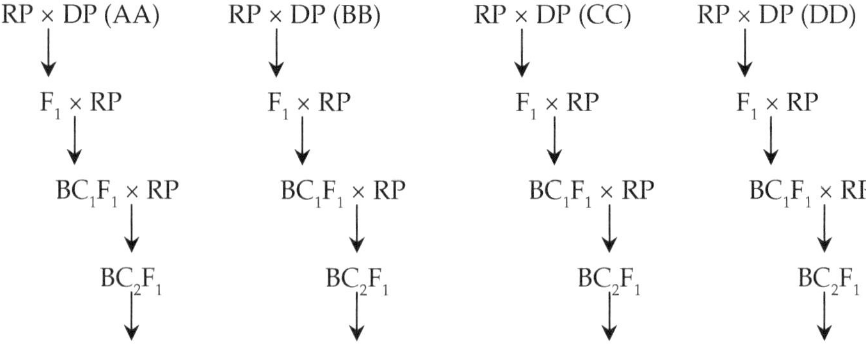

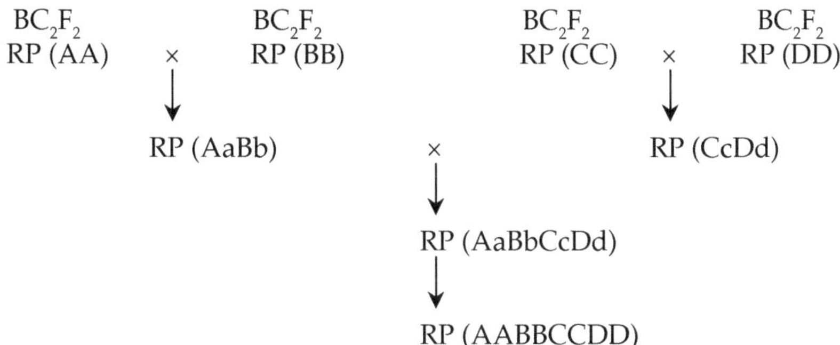

What is meant by gene mapping? Explain in brief about the steps involved in construction of genetic linkage map?

Gene Mapping

Gene mapping is the process of establishing the locations of genes on the chromosomes. Early gene maps used linkage analysis. The closer two genes are to each other on the chromosome, the more likely it is that they will be inherited together.

- Linkage maps indicate the position and relative genetic distances between markers along chromosomes.
- The most important use for linkage maps is to identify chromosomal locations containing gene and QTLs associated with traits of interest.
- Three main steps involved in construction of linkage map, which are
  - i. Production of mapping population,
  - ii. Identification of polymorphism and
  - iii. Linkage analysis of markers.

Mapping population

- The construction of linkage map requires a segregating plant population *i.e.* population derived from sexual reproduction.
- The parents selected for the mapping population will differ for one or more traits of interest.
- Population size for preliminary genetic mapping ranges from 50 to 250 individuals, however longer populations are required for high resolution mapping.
- Commonly used mapping populations include F_2, RILs, BILs and DHs populations.

Identification of Polymorphism

- It is critical that sufficient polymorphism exists between parents in order to construct a linkage map.

- Cross pollinated species possess higher levels of DNA polymorphism compared to inbreeding species generally requires the selection of parents that are distantly related.
- Once polymorphic markers have been identified, they must be screened across the entire mapping population, including the parents and F_1 hybrids. This is known as marker 'genotyping' of the population.

Linkage Analysis of Markers

- Construction of a linkage map involves coding data for each DNA marker on each individual of a population and constructing linkage analysis using computer programmes.
- Commonly used software programmes include Map-Maker, Map-Manager and Join Map *etc.*

Write the differences between genetic, cytogenetic and physical maps? And also mention the correlation between these maps?

Feature	*Genetic Maps/Linkage Maps*	*Cytogenetic Maps*	*Physical Maps*
Definition	It refers to schematic representation of places where genes reside on chromosomes in linear specific order analogous to beads on strings. These are based on recombination frequency between markers and frequency of recombinants directly propotional to distance between the genes.	It refers to representation of location of various genes in chromosome relative to microscopically visible land marks in chromosome with respect to banding pattern	It refers to genes are depicted in same order as they occurred in chromosome
Map construction	Linkage map also constructed by using molecular markers such as RFLP and SSR *etc.*	These are constructed mainly using FISH, Somatic cell hybridization and analysis of changes in polytene chromosomes in drosophila	These are constructed mainly based on Restriction mapping and STS mapping etc.
Genetic distance meassurement	These percent frequencies are used as map units for preparing linkage maps. A map unit is also called, a centi- Morghan (cM), after the name of the scientist Morghan, who first constructed the linkage map in drosophila.	Cytogenetic map is only accurate within limits of -5Mbs (1 Mb = 10^6 bp) along the chromosome; the resolution is much better in species having polytene chromosomes.	The distances between them are shown in base pairs or Kbs or Mbs

Correlation between Linkage, Cytogenetic and Physical Maps

1. *In situ* hybridization is a common method in which YAC or cosmid clones can be hybridized to intact chromosomes to determine their location. This approach allows correlation of a cytological map with physical map.
2. In case of those genes that have already been cloned, linkage map can be correlated with a cytogenetic map by using the cloned gene as a probe in insitu hybridization.
3. A linkage map can be correlated with cytogenetic map under genetically mapped gene can be localized to particular chromosomal region in human somatic cell hybrids.
4. Polymorphic STS markers can be used in both physical and linkage mapping, they provide a powerful approach to correlate location on linkage map with those of physical map.
5. *e.g.* For linkage map, distance between Sc (scute, abnormal bristle formation) and w (white eye) of drosophila is located at 1.5 map units.

   For cytogenetic map, gene Sc located on 1B3 on X – chromosome and w is located on 3C2.

   For physical map, genes sc and w were separated by 1.5×10^6 bp.

   And also 1 cM = 1000 kb in human beings, 1600 kb in mice and 1500 kb in maize. These differences arise due to differences in frequencies of recombination in different species or chromosomal regions.

Explain in detail about mapping populations with advantages and limitations.

Types of Mapping Populations

1. F_2 population
2. F_2 derived F_3 (F_2:F_3) population
3. Recombinant inbred lines (RILs)
4. Back cross populations
5. Doubled haploids (DHs)
6. Near – isogenic lines (NILs)

F_2 Population

- ☆ Best population for preliminary mapping
- ☆ Requires less time for development
- ☆ Can be developed with minimum efforts

Limitation

- ☆ It is not an immortal population
- ☆ F_2 population are of limited use for fine mapping
- ☆ Quantitative traits cannot be precisely mapped

F_2 Derived F_3 (F_2:F_3) Population

- ☆ Developed by selfing F_2 population
- ☆ Suitable for Quantitative trait loci mapping
- ☆ Mapping recessive genes

Limitation

- ☆ Not immortal population.

Recombinant Inbred Lines (RILs)

RILs are produced by continuous selfing or sib mating the progeny of individual members of an F_2 population until complete homozygosity is achieved.

Single seed descent (SSD) method is best suited for developing RILs. Bulk and Pedigree methods without selection can also be used for developing RILs.

- ☆ Once homozygosity is achieved, RILs can be propagated indefinitely without further segregation. This becomes an immortal mapping population.
- ☆ Since RILs are immortal population, they can be replicated over locations and years and therefore are of immense value in mapping QTLs.
- ☆ RILs being after several cycles of meiosis, are very useful in identifying the tightly linked markers.
- ☆ Can become a public mapping tool.

Limitation

- ☆ Requires many seasons/generations for development.
- ☆ Developing RILs is relatively difficult in crops with high inbreeding depression.

Back Cross Populations

Backcross population generated by backcrossing the F_1 hybrid with any of the parent.

- ☆ Quick and easy to generate but have poor resolution.
- ☆ Like F_2 population, the backcross populations require less time to be developed, but are not immortal population.
- ☆ The specific advantage of backcross populations is that, the populations can be further utilized for Marker assisted backcross breeding.

Limitation

- ☆ However, the recombination information in case of backcross is based on only one parent (the F_1).

Doubled Haploids (DHs)

Chromosomal doubling of anther culture derived haploid plants from F_1 generates Double haploids.

- ☆ DHs are products of one meiotic cycle, and hence comparable to F_2 in terms of recombinant information.
- ☆ DHs are permanent mapping population and hence can be replicated over years and locations and maintained without any genotypic changes.
- ☆ Useful for mapping both qualitative and quantitative characters.
- ☆ Instant production of homozygous lines and thus saving a time.

Limitation

- ☆ Recombination from the male side alone is accounted.
- ☆ Since it involves invitro techniques, relatively more technical skills are required in comparison with the development of other mapping populations.
- ☆ Often suitable culturing methods/haploid production methods are not available for number of crops and different crops differ significantly for their tissue culture response.
- ☆ Further, anther culture induced variability should be taken care of.

Near-isogenic Lines (NILs)

NILs are generated either by repeated selfing or backcrossing F_1 plants to the recurrent parent.

- ☆ NILs developed through backcrossing are similar to recurrent parent but for the gene of interest, while NILs develop through selfing are similar in pair but for the gene of interest.
- ☆ Like DHs and RILs, NILs are also immortal mapping population
- ☆ NILs are quite useful in functional genomics.

Limitation

- ☆ Require many generations for development.
- ☆ Directly useful only for molecular tagging of the gene concerned, but not for linkage mapping.
- ☆ Linkage drag is a potential problem in constructing NILs, which has to be taken care of.

Genetic Segregation Ratio at Marker Loci in different Mapping Populations

Mapping Population	*Co-dominant Loci*	*Dominant Loci*
F_2 population	1:2:1	3:1
Back cross population	1:1	1:1
RILs	1:1	1:1
DHs	1:1	1:1
NILs	1:1	1:1

- ☆ Mapping populations such as RILs and DHs equalize (neutralize) the type of marker because fixation of parental alleles at marker locus in homozygous condition. So these populations result in 1:1 segregation ratio at marker locus irrespective of genetic nature of marker.
- ☆ While F_2 population segregates in 1:2:1 for co dominant marker and 3:1 ratio for dominant marker *i.e.* F_2 population does not neutralize the marker type.
- ☆ In backcross populations, the backcross with dominant parent would segregate in a ratio of 1:0 and 1:1 for dominant and co-dominant markers, respectively. However, backcross with recessive parent or test cross would segregate in a ration of 1:1 irrespective of the nature of the marker.

Answer the following questions?

1. **What advantage RILs have over DHs in genetic mapping?**
2. **How an F_2 population can be immortalized for dissecting both additive and dominance variation?**
3. **Compare the strengths and limitations of $F_{2:3}$ and RILs mapping population in relation to QTL mapping in crop plants?**

Advantage of RILs over DHs in Genetic Mapping

- ☆ The advantage of RILs and DH populations are that they produce homozygous or true breeding lines that can be multiplied and reproduced without genetic change occurring.
- ☆ The main advantage of RILs over DHs in genetic mapping is that higher amount of recombination possible in RILs. So it is very useful in identifying linked markers.
- ☆ While DHs are products of single meiotic cycle and hence comparable to F_2 interms of recombination information.
- ☆ DH populations are quicker to generate than RILs but the production of DHs only possible for species with a well established protocol for haploid tissue culture.
- ☆ The time required for developing RILs is a major constraint in mapping studies.

Immortalized F_2 Population

- ☆ Immortalized F_2 population can be developed by paired crossing of the randomly chosen RILs derived from a cross in all possible combinations excluding reciprocals.
- ☆ The set of RILs used for crossing along with the F_1s produced, provide a true representation of all possible genotype combinations (including the heterozygotes) expected in the F_2 of the cross from which the RILs are derived.

- ☆ The RILs can be maintained by selfing and required quantity of F_1 seed can be produced at will by fresh hybridization.
- ☆ This population therefore provides an opportunity to map heterosis QTLs and interaction effects from multilocus data.

Strengths and limitations of $F_{2:3}$ and RIL mapping populations in relation to QTL mapping in crop plants

$F_{2:3}$ Population	*RIL Population*
Strengths: ☆ Ability to measure the effects of additive and dominance gene action at specific loci. ☆ Effect of heterozygosity estimated	**Strengths:** ☆ Estimates only additive gene action. So multilocation trials are possible ☆ Immortal population, so replicated over locations and years
Limitations: ☆ Non immortal population. So not possible to grow over locations	**Limitations:** ☆ Take more time for developing ☆ Inbreeding depression on cross pollinated crops ☆ Effect of heterozygosity cannot be estimated

Explain in brief about the bulk segregant analysis.

Bulk segregant analysis: It is simple strategy which is employed to quickly identify markers, which co segregate with trait of interest *i.e.* to find out whether marker is linked to gene of interest or not. This was proposed by Michelmore *et al.*, 1991.

- ☆ Here isogenic lines, two lines similar in whole genetic background except for gene of interest *i.e.* to find out cM distance between marker and gene of interest.
- ☆ In BSA, two bulked DNA samples are drawn from a segregating population originating from single cross.
- ☆ The bulks are screened for DNA polymorphisms and these differences are compared against a randomized genetic background of unlinked loci.
- ☆ A marker that differs between the two bulks is expected to be linked to the particular trait under investigation.

What is meant by QTL? Explain how they are identified and used in crop improvement programme?

QTL (Quantitative Trait Loci)

A region of genome that is associated with effect on a quantitative trait.

Concept given by Sax and term coined by Gelderman.

Types of QTLs

Based on presence and absence of epistasis.

1. Main effect QTL: Defined as single mendalian factor at which effect on a given phenotype arises from allelic substitution.
   - ✰ Detected by marker trait associations using single factor ANOVA or interval mapping models
   - ✰ Includes two groups of genes
     a. Major genes with very large effects on highly heritable traits.
        - ✰ Detected with very large LOD scores of >10.0
        - ✰ Each explains a large portion of total trait variation in mapping population
        - ✰ *e.g.*: Sd – 1 for semi dwarf nature and Xa4 for BLB resistance

     b. Represents most QTLs reported to date and have relatively small effects
2. Epistatic QTL: Defined as loci at which trait values are determined by interactions between alleles at two or more loci.
   - ✰ Detected by associations between trait values and multilocus marker genotypes using epistatic models.
   - ✰ Three types.
     a. Type I: Two main effect QTLs are involved in epistasis and affect the same phenotypes.
     b. Type II: Involves interactions between alleles at a main effect QTL and a modifying locus.
     c. Type III: Represents epistasis between two complementary loci that donot have detectable main effects.

Principle

QTL analysis is based on the principle of detecting an association between phenotype and genotype of markers. And QTL analysis is depends on "linkage disequilibrium".

Conceptual Framework

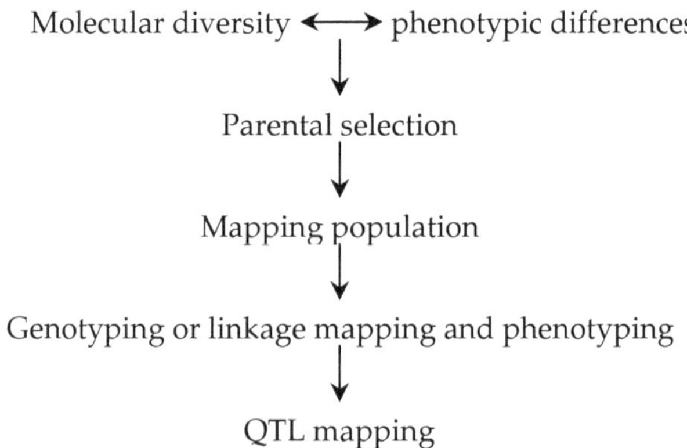

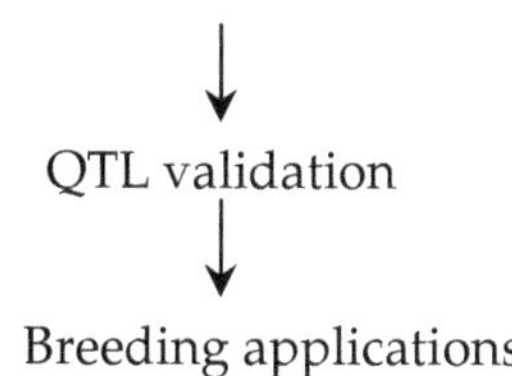

Factors Affecting QTL Mapping

- Number of genes controlling the trait
- Heritability of trait
- Distribution of genetic effects and existence of genetic interactions
- Number of genes segregating in mapping populations
- Type and size of mapping populations
- Density of linkage map
- Statistical methodology employed
- Significance level used for QTL mapping

Approaches

1. Single marker analysis
2. Simple interval mapping
3. Composite interval mapping
4. Multiple interval mapping
5. Bayesian interval mapping

Single Marker Analysis

Association of trait mean with marker locus.

- It involve comparison of different class for each marker locus
- Marker class as factor and individual values within class as error term.
- Determined by simple t-test, ANOVA, Linear regression and maximum likelihood estimation

Advantage: Does not require gene order and complete linkage map

Disadvantage: Unqualified presumption of tight linkage of QTL with marker locus

- No idea about precise QTL position
- No estimation of its effects, no recombination (r) value.

Simple Interval Mapping

Proposed by Lander and Botstein

- It involves use of two marker on chromosome, which bracket the putative QTL.

- ☆ Assuming only one QTL on chromosome.
- ☆ Use of complete linkage map.
- ☆ Provide value of r, so better estimation of QTL effect is possible, independent of its position.
- ☆ More probable location of QTL can be decided.

Advantage

Better estimation of location and effect of QTL, calculate r between marker and QTL.

Disadvantages

- ☆ Assumption of only single segregating QTL is unrealistic
- ☆ Resolution of exact number of QTLs not possible
- ☆ Exact position of QTL can't resolve when other QTLs are segregating in the population.

Complete Interval Mapping

Combine interval mapping for a single QTL with multiple regression analysis.

Advantage

Powerful, because of elimination of variance caused by other QTLs

Disadvantages

- ☆ Highly dependent on background information.
- ☆ Genetic marker incorporated a co factor in the frame work of simple interval mapping.

Multiple Interval Mapping

- ☆ Use of marker as co factor to reduce genetic background noise.
- ☆ Based on regressing the phenotype on a putative QTL in the given interval and at the same time on a number of markers located in the genome.
- ☆ Need large mapping population.
- ☆ Gives idea about number and position of QTL.
- ☆ Interaction of significant QTLs and their contribution to the genetic variance.
- ☆ Based on Cochran's model for interpreting genetic parameters.
- ☆ Maximum likelihood.

Bayesian Interval Mapping

- ☆ Provide information regarding number, position and effect of QTLs.
- ☆ Information about posterior estimate of marker QTL marker generated.
- ☆ It can estimate QTL effect and position separately.

Role in Crop Improvement

- Methods of QTL analysis mainly provide the information about the putative location and effects of QTLs influencing a quantitative trait.
- Identification of putative QTL locations and DNA markers linked to QTLs mainly useful for isolation and molecular characterization of QTLs via map-based cloning.

  e.g. Positional cloning of QTLs like *Brix 9-2-5, fw2.2, Hd1, Hd6, FRI* in tomato, rice and Arabidopsis *etc.*
- After preliminary QTL mapping, once fine mapping or high resolution mapping, QTLs can serve as useful tools for comparative genomics, functional genomics and evolutionary studies.
- Once molecular nature of the QTL was known, introgression of QTLs into elite line or germplasm and marker assisted selection (MAS) for QTLs in breeding could be undertaken easily.
- With the advancement in molecular technology, it is expected to enable greater power and precision in detection of QTL and utilization of QTL information for crop improvement.

What is LOD? Write the threshold LOD for the detection of major gene/QTLs for a target trait. What will be the output of the QTL mapping software and give their interpretations?

Linkage between markers is usually calculated using odds ratios (*i.e.* ratio of linkage versus no linkage). This ratio is more conveniently expressed as the logarithm of the ratio, and is called a logarithm of odds (LOD) value or LOD score.

$$LOD = \log_{10} [L(r)/L(r_o)]$$

where,

L(r) = likely hood calculated with most probable value of 'r' (maximum likely hood)

$L(r_o)$ = The probability of 'r' is equal to 0.5

The numerator expresses the probability of obtained results observed *i.e.* the probability of the distribution observed under the hypothesis of linkage with an estimated combination rate.

And denominator expresses the probability under hypothesis of independence. The the transformation into decimal is very large number and requires proper interpretation.

- LOD value of 2 signifies that linkage is 100 times (100:1) more probable than independence.
- LOD value of 3 signifies that linkage is 1000 times likely (*i.e.* 1000:1) than no linkage.
- LOD values of greater than 3 are typically used to construct linkage map.

☆ LOD values may be lowered in order to detect a greater level of linkage or to place additional markers within maps constructed at higher LOD values.

A typical output of linkage map is shown in figure. (Hypothetical framework of genetic map of three chromosomes represented by linkage groups and 28 markers).

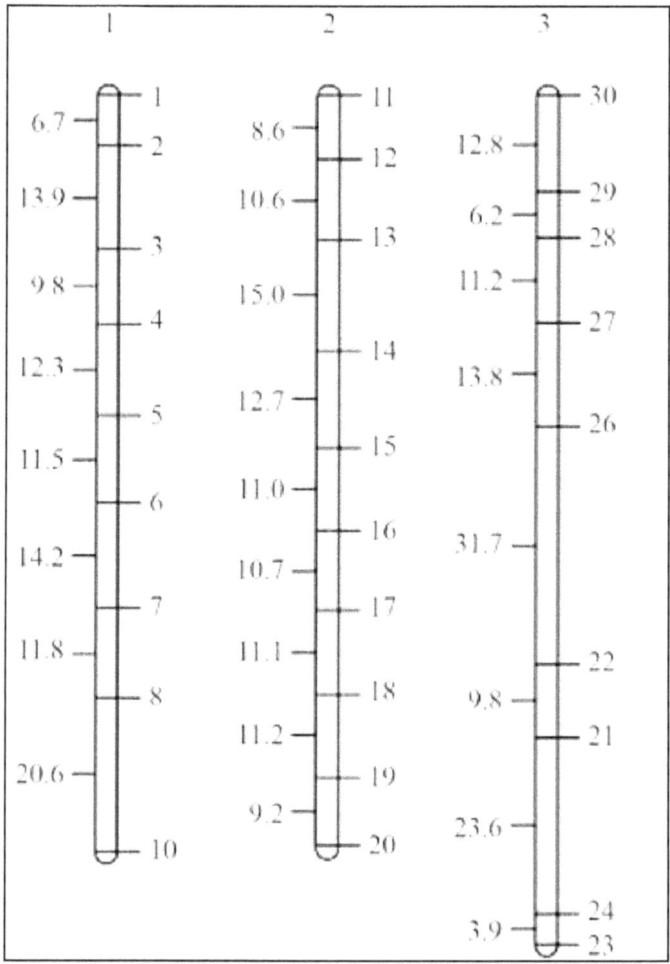

☆ Linked markers are grouped together into 'linkage groups' which represent chromosomal segments or entire chromosomes.

☆ The accuracy of measuring the genetic distance and determining marker order is directly related to the number of individuals studied in the mapping population.

☆ Ideally, mapping populations should consist of a minimum of 50 individuals for constructing linkage maps.

- The greater the distance between markers, the greater the chance of recombination occurring during meiosis.
- Distance along a linkage map is measured in terms of the frequency of recombination between genetic markers.
- Mapping functions are required to convert recombination fractions into centimorghans (cM) because recombination frequency and frequency of crossing over are not linearly related *i.e.* recombination frequency are not additive while genetic distances are additive.
- When map distances are small (<10 cM), the map distance equals to the recombination frequency. However, this relationship does not apply for map distances that are greater than 10cM.
- Two commonly used mapping functions are,
  i. Kosambi mapping function, which assumes that recombination events influence the occurrence of adjacent recombination events, *i.e.* it accounts for phenomenon of interference.
  ii. Haldane mapping function, which assumed no interference between crossover events. But more appropriate is kosambi distance.
  iii. A given distance may be valid in the genomic region and also 1cM usually corresponds to 1 per cent recombination.

Answer the following questions?

1. **Write the basic principle involved in marker assisted selection for a major-gene controlled and QTL controlled trait selection?**
2. **How many QTLs should be selected for MAS?**
3. **How many markers should be used in MAS?**

MAS for Major-Gene Controlled and QTL Controlled Trait Selection

MAS for Major Genes

- In crop plants, most of the economically important traits are controlled by major genes, which are often mono or oligogenic inheritance in nature. *e.g.* Resistance to pests/diseases and traits related to shape and colour *etc.*
- The marker loci which are tightly linked to major genes can be used for selection.
- It is more efficient than direct selection of target traits due to higher expression of the marker mRNA as marker held within a gene.
- *e.g.* Soya bean cyst nematode (*Heterodera glycines*) is an example of MAS for major genes. In this SSR marker Satt309 has been identified to be located only 1-2 cM away from the resistance gene *rgh1*, which forms the basis of many public and commercial breeding efforts.

MAS for Quantitative Traits

- ☆ In crop plants, most of the agronomic traits are polygenic or controlled by multiple QTLs.
- ☆ MAS for the improvement of such traits are complex and difficult because it is related to many genes or QTLs involved QTL × E interaction and epistasis.
- ☆ The QTL × E interaction reduces the efficiency of MAS and epistasis can result in a skewed QTL effect on the trait.
- ☆ To improve the efficiency of MAS for quantitative traits, attention should be given to repeated field tests with replications over locations for accurate characterization of effects of the QTLs and to evaluate the stability across environments.
- ☆ Composite interval mapping (CIM) allows the integration of data from different locations for joint analysis to estimate QTL- environment interaction so that stable QTLs across environments can be identified.
- ☆ *e.g.* in wheat, the effect of *Fusarium* head blight (FHB) resistance QTLs could be accumulated and resistance could be feasibly enhanced by selection of favourable marker alleles for multiple loci (Jiang *et al.*, 2007).

QTLs for MAS

- ☆ Theoretically, all the QTLs contributing to the trait of interest, but it is almost impossible to incorporate all QTLs into target individual due to the limitation of resources and facilities.
- ☆ Relative efficiency of MAS decreases as the number of QTLs increases.
- ☆ MAS will be less effective for highly complex traits governed by many genes than for a simply inherited trait controlled by a few genes.
- ☆ Typically no more than three QTLs are regarded as an appropriate and feasible choice (Ribaut and Betran, 1999).
- ☆ Lecomte *et al.*, in 2004 used five QTLs for improvement of fruit quality in tomato through marker assisted introgression.
- ☆ If SNP markers were developed, selection of more QTLs at the same time might be preferred and practicable (Kumpatla *et al.*, 2012).
- ☆ If the QTLs are selected on the basis of linked markers, limit the number of genes to three to four, and to five to six if they are known loci selected directly (Hospital, 2003).
- ☆ But priority should be given to the major QTLs because of greatest proportion of phenotypic variation and consistent detection across the range of environments and different populations.

Markers for MAS

- ☆ The more markers associated with a QTL are used, the greater opportunity of success in selecting the QTL of interest.

- For MAS efficiency and effectiveness, for a single QTL, use two markers that are tightly linked to the QTL of the interest (*i.e.* flanking markers).
- If marker is not tightly linked to a gene of interest, it leads to reduce the efficiency of MAS because recombination between the marker and gene may alternate the linkage association.
- If a marker is peak marker (located within a gene), such a marker only should be preferable because of no recombination between marker and gene.
- The efficiency of MAS decreases as the recombination frequency (genetic distance) between the marker and gene increases.
- In case of double cross overs, result in selection errors, but the frequency of double cross over is considerably rare.
- Flanking markers with an interval of 20 cM will result in higher probability (99 per cent) for recovery of the target gene than only one marker used.

Define Linkage disequilibrium and discuss the causes of Linkage disequilibrium

Linkage Disequilibrium (LD)

Non random association of alleles at different loci in gametic phase is called gametic phase disequilibrium (GPD). Gametic phase disequilibrium is more accurate term for Linkage disequilibrium.

Causes of LD: 1. Linkage 2. Multi locus selection due to epistasis 3. Random drift in small population 4. Bottle necks in population size 5. Migration 6. Mutation

- LD increase by inbreeding, small population size, genetic isolation between lineages, population subdivision, low recombination rate, population admixture, natural and artificial selection and balancing selection *etc.*
- LD decrease by out crossing, high recombination rate and high mutation rate.
- LD decaying: LD decrease/decaying with distance, it is called as LD decaying.
- Incomplete LD due to crossing over and gene conversion.
- Some genes have high LD while some genes have low LD because
    a. Possibility of recombination- Genes which are nearer to centromere and telomere have low chance of recombination.
    b. Mating system: out crossing and selfing. *e.g.* Rice has more LD than pearl millet and maize.
- Linkage disequilibrium represented by [D] and r^2 i.e [D] = 1 and $r^2 = 1$ But, in case of linkage equilibrium [D] = 0 and $r^2 = 0$.

Define association mapping. Explain the significance of association mapping in plant genetics. And write the differences between linkage analysis and association mapping?

Association Mapping

Association mapping (genetics), also known as "linkage disequilibrium mapping", is a method of mapping quantitative trait loci (QTLs) that takes advantage of historic linkage disequilibrium to link phenotypes (observable characteristics) to genotypes (the genetic constitution of organisms). *i.e.* Genotyping of naturally available genotypes.

- ☆ In association mapping the genetic markers usually lie within candidate genes and association mapping relies on linkage disequilibrium (LD) between the candidate gene markers and the causal polymorphisms in the gene.
- ☆ So association mapping is suitable for identifying the polymorphism within genes that are responsible for phenotypic differences.
- ☆ The main feature of association mapping are
  a. No production of mapping population
  b. No development of linkage map
  c. Just genotype the naturally available genotypes

Significance of Association Mapping in Plant Genetics

In traditional QTL mapping approach generates linkage disequilibrium between genetic markers and QTL through crossing of different genotypes and creation of segregant population. But central problem with the conventional mapping approaches is limited number of meiosis have been occurred. In association mapping, recombination's over many generations has broken up the linkage disequilibrium that exists initially between marker and novel QTL allele.

The current major uses of association mapping in plants are:

- ☆ The detection of marker-trait association in natural populations and subsequent marker assisted selection and
- ☆ Studies of genetic diversity in natural populations and studies of population genetics.

Limitation

- ☆ Difficulties can arise in association mapping due to population structure. Genetically heterogeneous populations can cause spurious associations, thus constraining the use of association studies.
- ☆ Trait of interest is segregating over generations, so it is not possible to identify and integration of allele from wild species to cultivated species. But it is achieved by Advance Back cross-QTL analysis (AB-QTL). Linkage disequilibrium mapping is best when there is strong selection pressure on trait of interest.

Sl.No.	Linkage Analysis	Association Mapping
1.	Family mapping	Population mapping
2.	Structured population	Unstructured population
3.	Based on two diverse mapping populations	Based on diverse collections
4.	Biparental mapping	Multiparental mapping
5.	Only two alleles *i.e.* from both parents	Multiple allelic variation
6.	Few cycles of recombination *i.e.* one which leads to shuffling of genome	Several cycles of recombination *i.e.* extensive shuffling of genome because of more meiotic events
7.	Less resolution	More resolution
8.	Cannot go much nearer to the gene	Can go more nearer to genome
9.	High power and moderate marker density	Low power and high marker density
10.	Genome scan	Candidate gene approach

Write short notes on the following?

1. **AB QTL analysis**
2. **Nested association mapping**
3. **Candidate gene approach**
4. **Breeding by design**

AB QTL Analysis

AB-QTL strands for Advanced Backcross QTL analysis, proposed by Tanksley and Nelson (1996), in which the QTL mapping is delayed to BC_2 or BC_3 so that the contribution of epistatic interactions is reduced as compared with selfing generations.

- ✰ AB-QTL analysis is a method for the simultaneous discovery and transfer of valuable QTLs from unadapted germplasm into elite breeding lines. It is a method of QTL mapping.
- ✰ The lines derived from backcross generations will be more close to the recurrent parent as a result of which QTL-NILs can be obtained with less additional efforts.

Steps in AB-QTL Analysis

- ✰ In the AB-QTL analysis the elite inbred lines of cultivated species are crossed to exotic stocks and the resulting F_1s again crossed to IL to generate nearly 100 BC_1 progeny.
- ✰ After rejecting plants with obviously undesirable features of the wild stocks, the others are again backcrossed to elite line to produce a population of BC_2 generations of about 200 individuals.
- ✰ The BC_2 plants are assayed for the polymorphic DNA markers.

- ☆ The selfed seed of these plants *i.e.* BC_2S_1 and BC_2S_2 can be evaluated and compared with the parental elite line in replicated trail to identify BC_2 individuals with desirable QTLs from the wild donar stock.
- ☆ The molecular data from BC_2 plants can be used to search for QTL association markers with performance in BC_2S_1 or BC_2S_2 families to identify useful QTL.
- ☆ In this way the alleles from wild species can be evaluated for their agronomic worth as well as obsociation with markers to be used as index of such alleles for their transfer to elite line.

Nested Association Mapping (NAM)

It is nothing but structured sub population nested within unstructured population.

- ☆ It is mainly useful to combine the advantages and eliminating the disadvantages of two traditional methods for identifying the QTLs *i.e.* linkage analysis and association mapping.
  - ❒ Linkage analysis depends mainly on recombination between two different plant lines to identify the gene of interest. But it has low mapping resolution and low allele richness.
  - ❒ While association mapping takes the advantage of historic recombination and has high mapping resolution and high allelic richness.
- ☆ NAM takes advantage of both historic and recombination between two different lines. It has
  - ❒ Moderate marker density
  - ❒ High map resolution
  - ❒ High statistical power
  - ❒ High allelic richness
  - ❒ Analysis of many alleles

Candidate Gene Approach

Candidate genes either genes with molecular polymorphism genetically linked to major loci or QTLs or genes with molecular polymorphism statistically associated with variation of the trait being studied.

- ☆ CG approach is based on the hypothesis that known functional genes could correspond to the loci controlling trait of interest.
- ☆ CGs refers 1. Either to cloned genes presumed to effect the given trait (Functional CGs) 2. or to genes suggested by their close proximity on linkage maps to loci controlling the trait (Positional CGs).

Advantage of candidate gene approach over other approaches: Main objective of molecular genetics is to identify and isolate a gene; methods such as positional cloning and insertional mutagenesis have been used with success to identify major genes. However, these methods are limited by genome size and/or by lack of transposons in the species being studied.

In this regard CG approach is an alternative strategy.

- ✫ CG approach provides markers of gene itself.
- ✫ Validated CGs for QTLs provides very efficient molecular marker because recombination between marker and QTL would be absent.

Breeding by Design

The application of markers in breeding can not only improve existing selection process but can aid in creating novel varieties bearing new characteristics of agronomic importance. Understanding of genetic basis of all agronomically important characters and the allelic variation at those loci, would enable the breeder to design superior breeding lines '*in silico*'. This concept is referred to as 'Breeding by design' and uses two types of mapping methods to generate the knowledge required.

a. Mapping traits using population which simplify complex characters, such as introgression libraries.
b. Mapping of traits and allelic variation by linkage disequilibrium (LD) mapping on germplasm.

The knowledge of the map positions of all loci of agronomic interest, allelic variation at those loci, and their contribution to the phenotype will enable the breeder to design superior genotypes consisting of a combination of favorable alleles at all loci.

Define reverse genetics. Enlist the different approaches of reverse genetics.

Reverse Genetics

Reverse genetics is an approach to discover the function of a gene by analyzing the phenotypic effects of specific gene sequences obtained by DNA sequencing. It is needed to elucidate gene and protein function in context of whole organism.

The reverse genetic strategies broadly fall in to two general categories depends on whether mutagenesis is targeted specifically to the locus of interest or its performed throughout the genome.

1. Targeting specific loci: It is attractive when only a few genes of interest exist. For this post transcriptional gene silencing (PTGS) increasingly popular. Here strategies include are:
   - i. Antisense RNA suppression
   - ii. RNA_i based PTGS
   - iii. Homologous recombination
   - iv. Chimeric RNA/DNA oligonucleotides
2. Genome wide mutagenesis and screening: These are inherently suitable for reverse genetics on a large scale. These are:

i. Insertional mutagenesis (T-DNA and Transposable elements)
ii. Fast neutron mutagenesis and size screening
iii. Tilling

Write the short notes on the following

1. **Anti sense RNA technology**
2. **Flavr savr tomato**

Anti Sense RNA Technology

Anti sense RNA technology is one of the approaches to reduce endogenous gene expression.

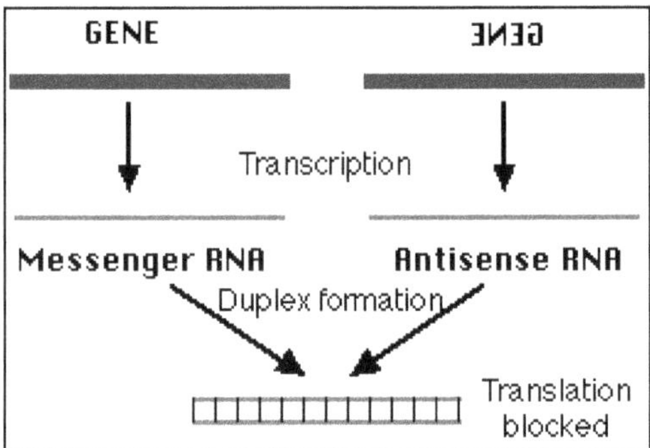

An anti sense RNA is produced by reversing the orientation of protein encoding region of gene in relation to its promoter. The RNA produced by this gene is known as antisense RNA. When an antisense RNA is present in same nucleus along with normal endogenous gene, it causes pairing of antisense RNA and sense RNA finally double stranded RNA formed. The implication of double stranded RNA as follows:

- ☆ mRNA unavailable for translation
- ☆ dS RNA degraded by dS specific RNAse
- ☆ Methylation of promoter and coding region leads to silencing of endogenous gene. So, Antisense RNA technology used for gene silencing and used for various purposes. Like

1. Slow Fruit Softening Tomato

a. Flavr savr-Polygalacturonase gene used. Improved flavor and, shelf life.
b. Endless summer – ethylene synthesis gene used (ACC synthase) ethylene <1 per cent

2. Change in Fattyacid Composition of Brassica Oil

Antisense gene construct of Brassica compestris stearyl-ACP desaturase gene.

a. Increase in level of steric acid (up to 40 per cent)
b. Decrease in oleic acid content

3. *Delayed Senescense in Carnation*

a. Gene *aco* suppressed using this thechnology.

b. Vase life increases by 200 per cent.

4. *Male Sterility*

a. Flavonoid is essential for normal pollen development. Chalcone synthase (CHS) key enzyme of flavonoid synthesis. Use of antisense construct of CHS gene used to create male sterility.

b. Apply flavonol during pollination of CHS antisense MS line to obtain 100 per cent male sterile progeny.

c. Also used to restore fertility-MS induced by role gene of *A. rhizogenes.*

'Flavr Savr' Tomato

- The first approval for commercial scale of food product was a transgenic tomato 'Favr Savr' with delayed ripening, developed by (calgene USA in 1994).
- Calgene used antisense RNA against polygalacturonase (PG) enzyme encoding gene.
- On the basic of PG gene sequence, an antisense PG gene was constructed and tomato plants were transformed.
- The tomato plants produced both sense and antisense mRNA for the PG gene resulting in RNA-RNA pairing this would result in non production of PG gene product, and thus preventing the attack of PG gene upon pectin in the cell wall of ripening fruit there by preventing softening of fruit.
- Calgene has given the brand name Mac Gregor to its transgenic tomato; it can stay on the market shelf for approximately two weeks longer without softening.

Explain the steps involved in RNAi induced silencing mechanism.

RNA$_i$ Based PTGS

RNA molecules involved here are two types of 1. Short interfering RNAs (Si RNAs) 2. Micro RNAs (mi RNAs). Steps involved in RNA interference mechanism are

- A large double stranded RNA molecule is diced in to small, double stranded interfering 21-28 bp long fragments by dicer enzyme.
- The small interfering RNAs and proteins assemble in to ribonucleoprotein particles.
- The small interfering RNA in ribonucleoprotein particle is unwound to produce an RNA- induced silencing complex (RISC).
- RISC targets a sequence in a messenger RNA that is complementary to the interfering RNA.

- ☆ The RISC's interfering RNA basepairs with its target in the mRNA.
- ☆ If perfectly base paired, mRNA is cleaved. It leads to degradation of mRNA. The RISC associated with this RNA interference is called Short interfering RNAs.
- ☆ If imperfectly base paired, translation of the mRNA is arrested *i.e.* polypeptide synthesis from the mRNA is repressed. The RISC associated with this RNA interference is called Micro RNAs.

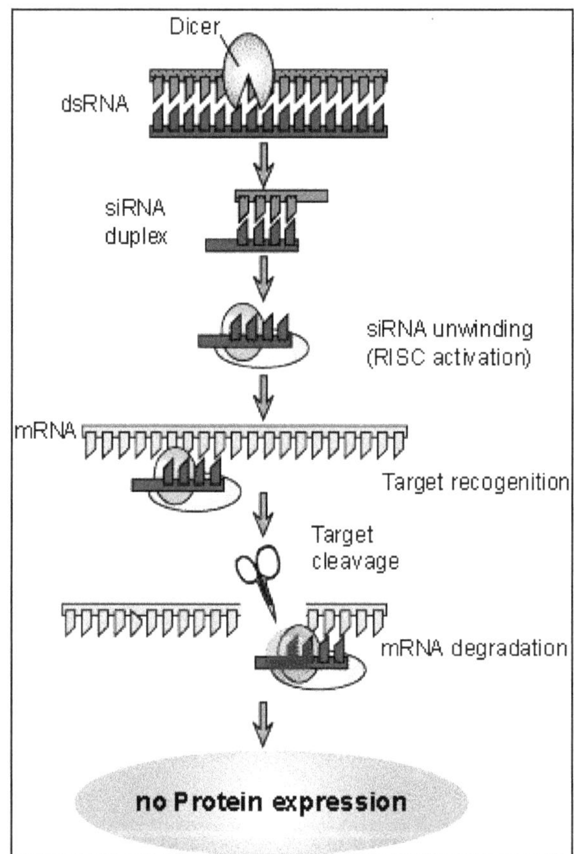

Define TILLING. Write the steps involved in TILLING process?

TILLING (Targeting Induced Local Lesions in Genomics)

(Discuss the process of TILLING and its usefulness in identification of candidate genes.)

TILLING is s new reverse genetic strategy that combines the high density of point mutations provided by traditional conventional mutagenesis with rapid mutational screening to discover induced lesions.

TILLING has the potential to become a standard reverse genetic strategy for plant functional genomics. Facile and efficient tilling depends on the availability of two resources. 1. A well mutanized population and 2. Genomic information.

Procedure

1. Mutaginized seeds obtained from treatment with EMS (Ethyl methane sulfonate).
2. Self fertilization of resulting M_1 plants.
3. Preparation of DNA samples for mutational screening in M_2 generation.
4. DNA samples are pooled, arrayed on microtiter plates and subjected to gene-specific PCR.
5. Amplification products are incubated with an endonuclease such as CELI, a member of the S1 nuclease family of single strand-specific nucleases.
6. CELI cleaves the 3′ side of mismatched DNA where the heteroduplex between the wild-type and the mutant strands of DNA loops out; homoduplexes are left intact.
7. Cleavage products are electrophoresed using an automated sequencing gel apparatus, and gel images are analyzed by examining the gel readout with the aid of a standard commercial image-processing program.
8. Differential double end labeling of amplification products allows for rapid visual confirmation because mutations are detected on complementary strands and can be easily distinguished from amplification artifacts.
9. Upon detection of a mutation in a pool, the individual DNA samples are similarly screened to identify the plant carrying the mutation.
10. This rapid screening procedure determines the location of a mutation to within ±10 bp for PCR products that are 1-kb in size.

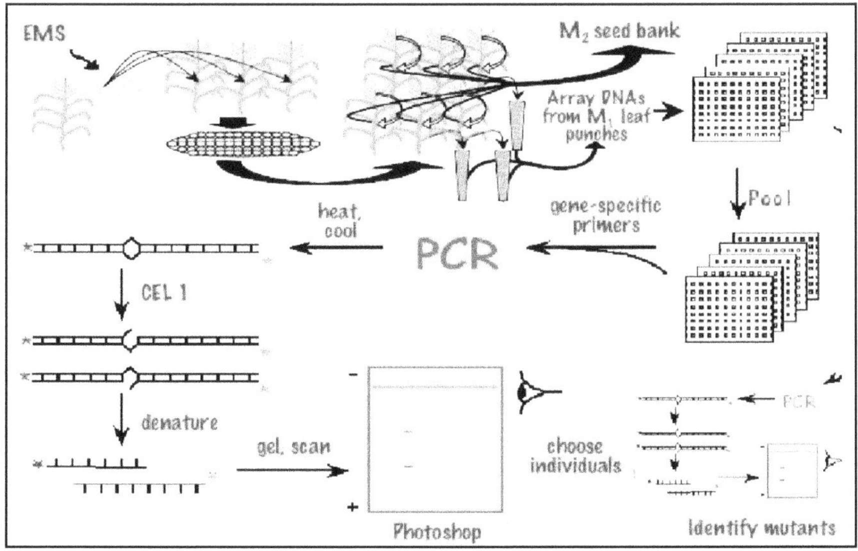

Advantages of TILLING

- ☆ The likelihood of recovering deleterious mutations can be calculated in advance.
- ☆ It minimizes efforts required to find mutations.
- ☆ Appropriate for both small and large scale screening.
- ☆ It is high through put. Most steps of tilling are suitable for automation.
- ☆ It can be applied even if less information about selected targeted genes.
- ☆ As chemical mutagenesis is widely applicable and mutation duration for tilling dependent only on sufficient yield of PCR products.

Eco-TILLING

Use of high through put tilling technology for the detection of natural variation in DNA *i.e.* SNP discovery. It can be practical for SNP discovery and genotyping

- ☆ Screen many individuals at one locus
- ☆ Discover rare natural polymorphism
- ☆ Good in heterozygous populations.

What is meant by allele mining? Write the approaches of allele mining along with its applications?

Allele Mining

Allele mining aims to identify the allelic variation present in genetic resource collection. It helps in the tracing the evolution of alleles, identification of new haplotypes and development of allele specific markers for use in Marker assisted selection.

There are two approaches available for the identification of sequence polymorphism for a given gene in the naturally occurring population such as Eco-TILLING and Sequence-based allele mining.

Eco-TILLING

It is a cost effective approach for discovery of haplotypes and SNP variation, this technique needs more sophistication and involves several steps starting from making DNA pools of reference and test genotypes, conditions for effective cleavage by nuclease, detection in polyacrylamide gels using Li-cor genotyper and confirming through sequencing.

Sequence-based Allele Mining

This technique involves amplification of alleles in diverse genotypes through PCR followed by identification of nucleotide variation by DNA sequencing. Sequence based allele mining would help to analyze individuals for haplotype structure and diversity to infer genetic studies in plants.

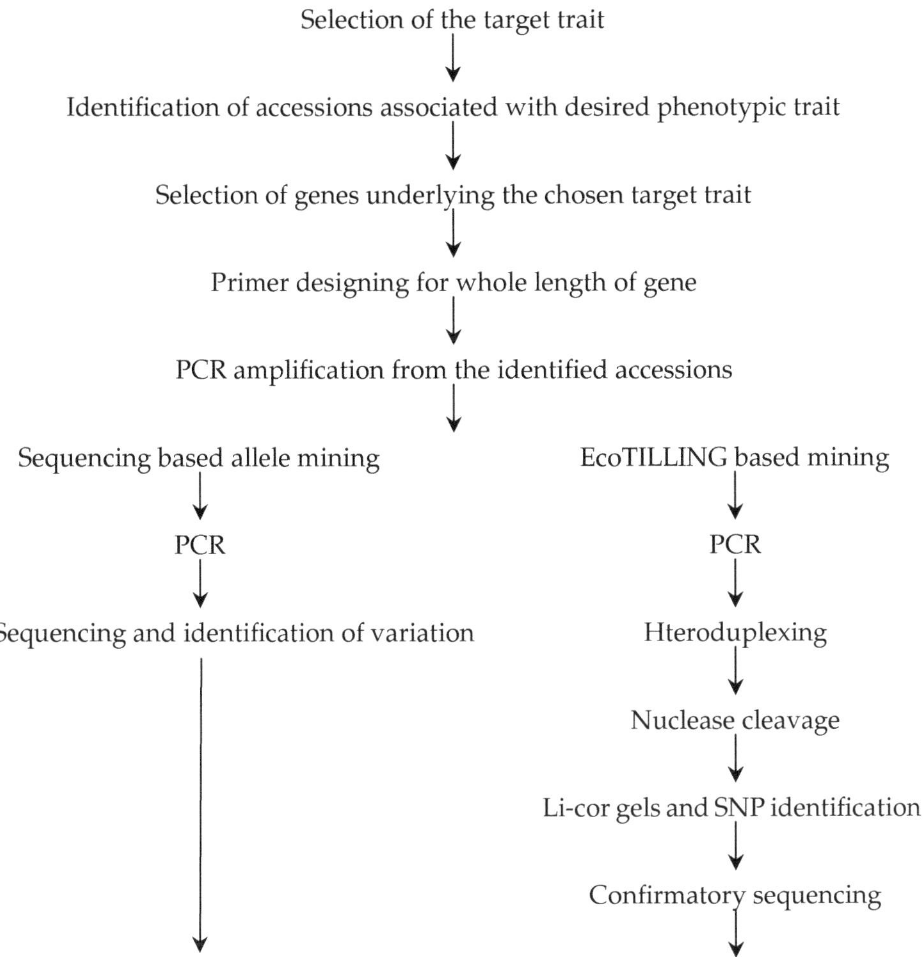

Applications

1. Allele mining can be effectively used for discovery of superior alleles through mining the gene of interest from diverse genetic resources.
2. It can also provide insights into molecular bases of novel trait variations and identify the nucleotide sequence changes associated with superior alleles.
3. In addition, the rate of evolution of alleles, allele similarity/dissimilarity at a candidate gene and allelic synteny with other members of the family can also be studied.
4. It may also pave way for molecular discrimination among related species, development of allele-specific molecular markers, facilitating introgression of novel alleles through MAS or deployment through genetic engineering.

What is meant by genomics? And write its applications?

Genomics

Genomics is the study of genes and their function in relation to the environment. In contrast to genetics, which focuses on genes and inheritance, the goal of genomics is to understand genes, their products and how, when, and why these products are synthesized.

Applications of Genomics

- Genomics will be able to help in both the diagnosis and treatment of the diseases.

  e.g. Many diseases, such as sickle cell anemia, cystic fibrosis and Huntington's disease, are caused by abnormalities in the sequence of DNA that code for a specific protein or proteins.
- Genomics will hopefully lead to an increase in the number of drug targets used in pharmaceuticals.
- It may also provide information on the genetic basis for side effects and the effectiveness of treatments that can be used to tailor prescriptions for individuals.
- Two specific types of gene therapies have been advanced.
  1. Somatic cell therapy involves the insertion of therapeutic genes into specific cells in the body. This will hopefully allow those cells to synthesize proteins that they are unable to produce or to turn off genes that are over expressed.
  2. Germ line therapy involves the insertion of normal genes into an egg cell, with the hope that the normal gene will be incorporated in to the genome of the offspring and that a genetic disease will not be inherited.
- In addition to their importance in medicine, bacteria, viruses and fungi play key roles in agriculture. Because their genomes are small, the genomes of at least 40 species of microorganisms have been sequenced. Understanding the genomics of these organisms has the potential to improve crop yields, decrease damage done by pest species and increase the nutritional value of food.
- As part of their metabolism, some microorganisms have the ability to break down harmful products and to produce energy as a product. Understanding the gene products involved in these transformations may lead to industrial uses, with the potential for solving different types of environmental problems and providing new energy sources.
- Genomics should also provide new techniques for identifying biological agents on the battlefield. One of the most promising technologies is the biochip or DNA chip, which is a microarray of molecular probes on a silicon chip that specifically bind to the DNA of biological threats.

What is meant by structural genomics? And explain various approaches used for structural genomics?

Structural genomics seeks to describe the 3-dimensional structure of every protein encoded by a given genome. This genome-based approach allows for a high-throughput method of structure determination by a combination of experimental methods using genomic sequences and modeling approaches based on sequence or structural homology to a protein of known structure or based on chemical and physical principles for a protein with no homology to any known structure.

- The principal difference between structural genomics and traditional structural prediction is that structural genomics attempts to determine the structure of every protein encoded by the genome, rather than focusing on one particular protein.
- One advantage of structural genomics, such as the Protein Structure Initiative, is that the scientific community gets immediate access to new structures, as well as to reagents such as clones and protein. A disadvantage is that many of these structures are of proteins of unknown function and do not have corresponding publications.

Approaches for Structural Genomics

- Structural genomics takes advantage of completed genome sequences in several ways in order to determine protein structures.
- The gene sequence of the target protein can also be compared to a known sequence and structural information can then be inferred from the known protein's structure.
- Structural genomics can also take modeling-based approach that relies on homology between the unknown protein and a solved protein structure.

de novo Methods

- Completed genome sequences allow every open reading frame (ORF), the part of a gene that is likely to contain the sequence for the mRNA and protein, to be cloned and expressed as protein.
- These proteins are then purified and crystallized, and then subjected to one of two types of structure determination: X-ray crystallography and Nuclear Magnetic Resonance (NMR).
- The whole genome sequence allows for the design of every primer required in order to amplify all of the ORFs, clone them into bacteria, and then express them.
- By using a whole-genome approach to this traditional method of protein structure determination, all of the proteins encoded by the genome can be expressed at once.This approach allows for the structural determination of every protein that is encoded by the genome.

Modelling-based Methods

1. *Ab Initio Modeling*

- ☆ This approach uses protein sequence data and the chemical and physical interactions of the encoded amino acids to predict the 3-D structures of proteins with no homology to solved protein structures.
- ☆ One highly successful method for *ab initio* modeling is the Rosetta program, which divides the protein into short segments and arranges short polypeptide chain into a low-energy local conformation.
- ☆ Rosetta is available for commercial use and for non-commercial use through its public program, Robetta.

2. *Sequence-based Modeling*

- ☆ This modeling technique compares the gene sequence of an unknown protein with sequences of proteins with known structures.
- ☆ Depending on the degree of similarity between the sequences, the structure of the known protein can be used as a model for solving the structure of the unknown protein.
- ☆ Highly accurate modeling is considered to require at least 50 per cent amino acid sequence identity between the unknown protein and the solved structure. 30-50 per cent sequence identity gives a model of intermediate-accuracy, and sequence identity below 30 per cent gives low-accuracy models.
- ☆ It has been predicted that at least 16,000 protein structures will need to be determined in order for all structural motifs to be represented at least once and thus allowing the structure of any unknown protein to be solved accurately through modeling.
- ☆ One disadvantage of this method, however, is that structure is more conserved than sequence and thus sequence-based modeling may not be the most accurate way to predict protein structures.

3. *Threading*

Threading bases structural modeling on fold similarities rather than sequence identity. This method may help identify distantly related proteins and can be used to infer molecular functions.

What is meant by functional genomics? And explain the various techniques used for functional genomics?

Functional genomics is a field of molecular biology that attempts to make use of the vast wealth of data produced by genomic projects (such as genome sequencing projects) to describe gene (and protein) functions and interactions.

- ☆ Unlike genomics, functional genomics focuses on the dynamic aspects such as gene transcription, translation, and protein–protein interactions, as opposed to the static aspects of the genomic information such as DNA sequence or structures.

- ☆ A key characteristic of functional genomics studies is their genome-wide approach to these questions, generally involving high-throughput methods rather than a more traditional "gene-by-gene" approach.

Techniques for Functional Genomics

- ☆ Functional genomics includes function-related aspects of the genome itself such as mutation and polymorphism (such as single nucleotide polymorphism (SNP) analysis), as well as measurement of molecular activities.
- ☆ Functional genomics uses mostly multiplex techniques to measure the abundance of many or all gene products such as mRNAs or proteins within a biological sample. Together these measurement modalities endeavor to quantitate the various biological processes and improve our understanding of gene and protein functions and interactions.

I. At the DNA Level

1. **Genetic interaction mapping:** Systematic pairwise deletion of genes or inhibition of gene expression can be used to identify genes with related function, even if they do not interact physically.
2. **The ENCODE project:** The ENCODE (Encyclopedia of DNA elements) project is an in-depth analysis of the human genome whose goal is to identify all the functional elements of genomic DNA, in both coding and noncoding regions.

II. At the RNA Level: Transcriptome Profiling

1. **Microarrays:** Microarrays allow for identification of candidate genes involved in a given process based on variation between transcript levels for different conditions and shared expression patterns with genes of known function.
2. **SAGE** (Serial analysis of gene expression)**:** SAGE is an alternate method of gene expression analysis based on RNA sequencing rather than hybridization.
   - ☆ SAGE relies on the sequencing of 10–17 base pair tags which are unique to each gene. These tags are produced from poly-A mRNA and ligated end-to-end before sequencing.

III. At the Protein Level: Protein–Protein Interactions

1. Yeast Two-hybrid System

- ☆ Yeast two-hybrid (Y2H) screen tests a "bait" protein against many potential interacting proteins to identify physical protein–protein interactions.

- ✰ This system is based on a transcription factor, originally GAL4, whose separate DNA-binding and transcription activation domains are both required in order for the protein to cause transcription of a reporter gene.

2. *Affinity Purification and Mass Spectrometry (AP/MS)*

- ✰ Affinity purification and mass spectrometry (AP/MS) is able to identify proteins that interact with one another in complexes. Complexes of proteins are allowed to form around a particular "bait" protein.
- ✰ The bait protein is identified using an antibody or a recombinant tag which allows it to be extracted along with any proteins that have formed a complex with it.
- ✰ The proteins are then digested into short peptide fragments and mass spectrometry is used to identify the proteins based on the mass-to-charge ratios of those fragments.

IV. *Loss-of-Function Techniques*

1. **Mutagenesis:** Gene function can be investigated by systematically "knocking out" genes one by one. This is done by either deletion or disruption of function (such as by insertional mutagenesis) and the resulting organisms are screened for phenotypes that provide clues to the function of the disrupted gene.
2. **RNAi:** RNA interference (RNAi) methods can be used to transiently silence or knock down gene expression using ~20 base-pair double-stranded RNA typically delivered by transfection of synthetic ~20-mer short-interfering RNA molecules (siRNAs) or by virally encoded short-hairpin RNAs (shRNAs).
   - ✰ RNAi screens, typically performed in cell culture-based assays or experimental organisms (such as *Caenorabdities elegans*) can be used to systematically disrupt nearly every gene in a genome or subsets of genes; possible functions of disrupted genes can be assigned based on observed phenotypes.

V. *Functional Annotations for Genes*

1. **Genome annotation:** Putative genes can be identified by scanning a genome for regions likely to encode proteins, based on characteristics such as long open reading frames, transcriptional initiation sequences, and polyadenylation sites.
   - ✰ A sequence identified as a putative gene must be confirmed by further evidence, such as similarity to cDNA or EST sequences from the same organism, similarity of the predicted protein sequence to known proteins, association with promoter sequences, or evidence that mutating the sequence produces an observable phenotype.

2. **Rosetta stone approach:** The Rosetta stone approach is a computation method of de novo protein function prediction, based on the hypothesis that some proteins involved in a given physiological process may exist as two separate genes in one organism and as a single gene in another.
   - ✰ Genomes are scanned for sequences that are independent in one organism and in a single open reading frame in another. If two genes have fused, it is predicted that they have similar biological functions that make such coregulation advantageous.

What is meant by comparative genomics? And discuss the various computational tools for analyzing genomic sequences?

Comparative Genomics

Comparative genomics is a field of biological research in which the genomic features of different organisms are compared. The genomic features may include the DNA sequence, genes, gene order, regulatory sequences, and other genomic structural landmarks.

- ✰ In this branch of genomics, whole or large parts of genomes resulting from genome projects are compared to study basic biological similarities and differences as well as evolutionary relationships between organisms.
- ✰ The major principle of comparative genomics is that common features of two organisms will often be encoded within the DNA that is evolutionary conserved between them.
- ✰ Therefore, comparative genomic approaches start with making some form of alignment of genome sequences and looking for orthologous sequences (sequences that share a common ancestry) in the aligned genomes and checking to what extent those sequences are conserved.

Computational tools for analyzing sequences and complete genomes are developed quickly due to the availability of large amount of genomic data.

- ✰ Visualization of sequence conservation is a tough task of comparative sequence analysis.
- ✰ As we already know, it is highly inefficient to examine the alignment of long genomic regions manually.
- ✰ Internet-based genome browsers provide many useful tools for investigating genomic sequences due to integrating all sequence-based biological information on genomic regions. When we extract large amount of relevant biological data, they can be very easy to use and less time-consuming.
  1. **UCSC Browser**: This site contains the reference sequence and working draft assemblies for a large collection of genomes.
  2. **Ensembl**: The Ensembl project produces genome databases for vertebrates and other eukaryotic species, and makes this information freely available online.

3. **MapView**: The Map Viewer provides a wide variety of genome mapping and sequencing data.
4. **VISTA** is a comprehensive suite of programs and databases for comparative analysis of genomic sequences. It was built to visualize the results of comparative analysis based on DNA alignments. The presentation of comparative data generated by VISTA can easily suit both small and large scale of data.

An advantage of using online tools is that these websites are being developed and updated constantly. There are many new settings and content can be used online to improve efficiency.

What is meant by gene library? Explain in brief about the cDNA library and genomic library?

Gene Library

A gene library is a random collection of cloned fragments in a suitable vector that ideally includes all the genetic information of that species. It is also called as Shotgun collection. It is of two types *i.e.* cDNA library and Genomic library.

Genomic Library

For construction of a genomic library, total DNA of the cell is isolated. It is also of great use because cDNA libraries lack introns and so not complete genome coverage is possible.

- ✰ Genes isolated form genome library contain both exon as well as intronic sequences.
- ✰ For construction of a genomic library, total of DNA of the cell is isolated.
- ✰ The DNA is then cut into small pieces with a restriction enzyme.
- ✰ Each piece of DNA is then joined to vectors. Vectors derived from λ phage are commonly used for preparation of genomic libraries.
- ✰ The recombinant DNA is then introduced into bacterial cells such as *E. coli.*
- ✰ The bacterial cells are plated on agar plates containing suitable medium on which they grow and form colonies or plaques.
- ✰ Each colony/plaque represents a recombinant clone carrying a different piece of genomic DNA.
- ✰ A collection of all these clones is called a genomic library.
- ✰ To isolate genes, genomic library is screened with the help of probes.
- ✰ Genomic libraries are useful for study of structure of genes, to identify regulatory regions *i.e.* DNA sequences needed for correct expression of the gene.

cDNA Library

cDNA is DNA derived from mRNA. The mRNA in this case is processed transcript. It is used when it is known that gene of interest is expressed in a particular tissue or cell type.

Steps

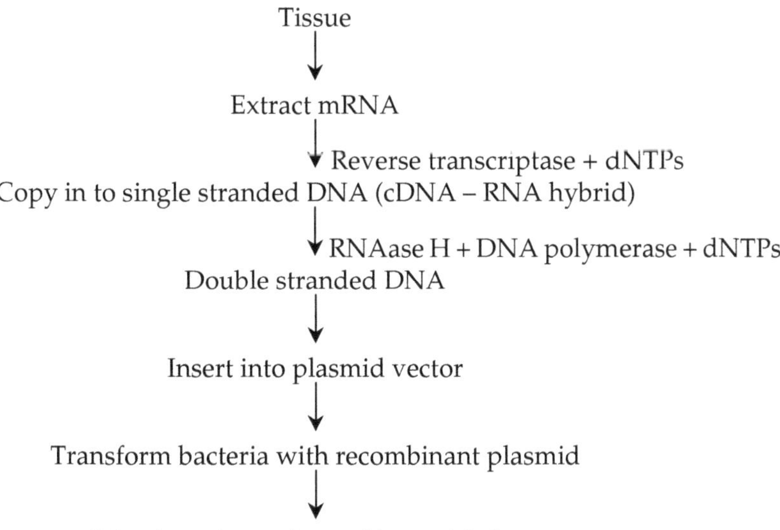

What is meant by gene targeting and gene tagging? Explain in brief about the strategies for gene tagging?

Gene Targeting

It is a genetic technique that uses homologous recombination to change an endogenous gene.

- It is mainly used to delete a gene, remove exons, add a gene, and introduce point mutations.

Gene Tagging

Gene tagging is the use of an insertional mutagen to mark interrupted genes with a unique DNA sequence.

- This DNA sequence subsequently can be used as a target for hybridization or as an annealing site for PCR primers, allowing flanking sequences to be isolated.
- The main strategies for gene tagging involves T-DNA tagging, transposon tagging, activation tagging and entrapment tagging.

T-DNA Tagging

- Plant transformations by *A. tumefaciens* involve the transfer of a small segment of DNA, known as T-DNA from the plasmid harbored by tumefaciens.
- It means T-DNA acts as an insertional mutagen and can have been used for genome wide mutagenesis programs.
- As T-DNA is not a transposon, it has no ability to 'jump' following integration. Therefore, having advantage of generating stable insertions.
- In this, Gene insertion is essentially random. And also it affects only the gene, where it incorporated (within the gene space).
- The main disadvantage is the tendency to generate complex, multicopy integration patterns and sometimes deletions and gene rearrangements of surrounding genes.

Transposon Tagging

- It mainly permits the cloning of the genes whose function is not known.
- The first step in this procedure is to identify a plant stock that is mutant for a specific trait because a transposable element has been inserted into and inactivated the gene.
- Creation of a genomic library of the plant stock screening with a clone for the transposable element.
- One strategy is use of Ty element as an insertional mutagen and used as "genetic foot print".
- Transposon-induced alleles give a different spectrum of mutant phenotypes than other mutagens and are useful in generating allelic series.
- Transposons also cause mutable or epigenetically regulated alleles that are useful for generating chimeric plants.

Activation Tagging

- In this technique instead loss of function in response to insertion element, gain of function takes place.
- The insertion element carries a strong outward-facing promoter and if, it integrates adjacent to an endogenous gene, the gene will be activated by the promoter.
- This type of tagging causes over expression or ectopic expression of endogenous gene.

Entrapment Tagging

- Difficulties of RNA or antibody in-situ experiments are tedious and time consuming, they are unsuitable for large-scale analyses of many genes and these can be overcome by entrapment tagging.

- In this system, T-DNA or transposons are engineered to carry a promoterless reporter gene, such as b-glucuronidase (GUS) or green fluorescent protein (GFP), next to the insert border.
- Entrapment systems can be divided into three groups: enhancer trap, promoter trap, and gene trap.
- In the enhancer trap system, the reporter gene is fused to a minimal promoter, which is unable to drive reporter gene expression but can be activated by an enhancer element of the neighboring genes.
- In the promoter trap system, reporter gene expression can occur when the reporter is inserted into an exon and forms a translational fusion with the endogenous gene.
- In contrast, gene trap constructs have an intron with multiple splice donor and acceptor sequences in front of a reporter gene. Therefore, the reporter gene can be expressed regardless of insert position (exon, intron, or UTR).

8

Tissue Culture and Plant Genetic Engineering

Define plant tissue culture? And write the applications of plant tissue culture in crop improvement?

The term 'plant tissue culture culture' broadly refers to the *in vitro* cultivation of plants, seeds, plant parts on nutrient media under aseptic conditions.

- ☆ In 1902, Gottlieb Haberlandt, a German plant physiologist, attempted to cultivate plant tissue culture cell *in vitro.* He is regarded as the father of plant tissue culture.

Applications of plant tissue culture in crop improvement

- ☆ Micro propagation helps in mass multiplication of plants which are difficult to propagate through conventional methods.
- ☆ The vegetative propagation like grafting, budding are tedious and time consuming. In such crops micro propagation helps in rapid multiplication.
- ☆ Rapid multiplication of rare and elite genotypes such as Aromatic and Medicinal plants.
- ☆ Isolation of *in vitro* mutants for a large number of desirable character *e.g.* Isolation of biochemical mutants and mutants resistant to biotic (pest and disease) abiotic (salt and drought, cold, herbicide *etc.*) stresses through the use of somaclonal variation
- ☆ Screening of large number of cells in small space.

- ☆ Development of genetically uniform plants in cross pollinated crops is possible through tissue culture. *e.g.* Cross pollinated crops like cordamum, Eucalyptus, coconut, oil palm do not give true to type plants, when multiplied through seed.
- ☆ In case of certain horticultural crops orchids etc seed will not germinate under natural conditions, such seed can be made to germinate *in vitro* by providing suitable environment.
- ☆ Induction of flowering in some trees that do not flower or delay in flowering. *e.g.* Bamboo flowers only once in its life time of 50 years.
- ☆ Virus free plants can be produced through meristem culture.
- ☆ Large amount of germplasm can be stored within a small space and lesser cost for prolonged periods under *in vitro* condition at low temperature. The preservation of cells tissues, organs in liquid Nitrogen at – 196°C is called cryopreservation.
- ☆ Production of secondary metabolites. *e.g.* Caffine from *Coffea arabica,* Nicotine from *Nicotiana rustica.*
- ☆ Plant tissue culture can also be used for studying the biochemical pathways and gene regulation.
- ☆ Anther and pollen culture can be used for production of halploids and by doubling the chromosome number of haploids using cholchicine homogygous diploids can be produced. They are called dihaploids.
- ☆ In case of certain fruit crops and vegetative propagated plants where seed is not of much economic impor tant, triploids can be produced through endosperm culture.
- ☆ Inter specific and inter generic hybrids can be produced through embryo rescue technique which is not possible through conventional method. In such crosses *in vitro* fertilization helps to overcome pre-fertilization barrier while the embryo rescue technique helps to overcome post fertilization barrier.
- ☆ Somatic hybrids and cybrids can be produced through protoplast fusion (or) somatic hybridization.
- ☆ Ovary culture is helpful to know the physiology of fruit development.
- ☆ Development of transgenic plants.

What are the requirements for establishing a plant tissue culture laboratory? And write a note on the composition and preparation methods of plant tissue culture medium for transgenic crops?

Plant Tissue Culture Laboratory Requirements

The following facilities should be there in a biotechnology laboratory.

1. Store room: for storing chemicals and glassware
2. Cleaning and washing room: for general cleaning purposes

3. General laboratory: for routine laboratory experiments
4. Specialized rooms: preparation and sterilization room, laminar flow and sterile storage and culture rooms.
5. General instrumentation room: for PCR machine, gel documentation system, electrophoresis unit, centrifuge, pH meter, balance, laminar flow, freezers, ice machine *etc.*
6. Proper disposal of media, cultures *etc.*

Plant Tissue Culture Laboratory Organization

An ideal tissue culture laboratory should have at least two big rooms and a small room. One big room is for general laboratory work such as preparation of media, autoclaving, and distillation of water *etc.* the other big room is for keeping cultures under controlled light, temperature and humidity. The small room is for aseptic work and for keeping autoclaved articles.

Thus a tissue culture facility should have the following features.

1. Washing Facility/Area

- ☆ An area with large sinks (some lead lined to resist acids and alkalis) and draining area is necessary.
- ☆ The conventional method for cleaning laboratory glassware involves chromic acid sulphuric acid soak followed by thorough washing with tap water and subsequent rinsing with distilled water.
- ☆ However due to corrosive nature of acids, it is recommended that for routine procedures glassware should be soaked in a 2 per cent detergent cleaner for 16 h followed by washing with 60-70°C hot tap water, and finally distilled water.
- ☆ The cleaned glassware should then be dried at 150-200°C in a convection-drying oven for 1-4 h. However, with the availability of a wide range of reusable plastic ware, it is recommended that these should be washed with mild detergents followed by a rinse with tap water and then distilled water.

Essential requirements in washing area are:

- ☆ Sink with running tap water connection
- ☆ Soaps and disinfectants
- ☆ Sterilants

2. General Laboratory and Media Preparation Area

- ☆ This part is a central section of the laboratory where most of the activities are performed.
- ☆ The general laboratory section includes the area for media preparation and autoclaving of media.

- It also includes many of the activities that relate to the handling of tissue culture materials.
- The area to be set-aside for media preparation should have ample storage and bench space for chemicals, glassware, culture vessels, closures and other items needed to prepare media.

Essential equipments required in media preparation area are:

- Analytical loading single pan balance
- Autoclave
- Electric hot-air oven
- Laminar air flow hood
- Refrigerator and freezer
- Electronic balances
- Digital pH meter
- Water distillation unit
- Hot plate and magnetic stirrer
- Dissecting microscope
- Gyratory shaker

3. *Inoculation/Aseptic Transfer Area*

- Tissue culture techniques can be successfully- carried out in a very clean laboratory, dry atmosphere with some protection against air-borne microorganisms.
- The laminar airflow cabinet is the most common accessory used for aseptic manipulations.
- The cabinet should be designed with horizontal airflow from the back to the front, and should be equipped with gas cocky if gas burners are to be used.
- In the airflow cabinets, air is forced into the cabinet through small motor into the unit first through a coarse filter, where large dust particles are separated and then subsequently passes through a bacterial 0.3 μm HEPA (high efficiency particulate air) filter, it flows outwards (forward) over the working bench at a uniform rate.
- The air coming out of the fine filter is ultra clean (free from fungal and bacterial contamination) and its velocity (27 ± 3 m/min) prevents the micro contamination of the working area by a worker sitting in front of the cabinet.
- The roof of the cabinet is generally provided with an ultraviolet (UV) germicidal light often used to sterilize the interior of chamber.
- These are turned on about 30 min prior to using the chamber but must be turned off during operations.

- Though the cabinet is presterilized with UV light, before starting of the work, it is desirable to sterilize the inside of cabinet and the table top with the cotton swab dipped in 70 per cent alcohol.

Essential requirements in aseptic transfer area are:

- Laminar airflow cabinets
- Micro dissecting scissors
- Scalpel handles with blades
- Forceps of different sizes
- Needles and inoculating loops

4. *Environmentally Controlled Incubation Facilities*

- Plant tissue cultures should be incubated under conditions of well-controlled temperature, illumination, photoperiod, humidity and air circulation.
- These facilities can also be constructed by developing a room with proper air condition, perforated shelves to support culture vessels, fluorescent tubes and a timing device to set light dark regimes (photoperiods) which are the 3 standard accessories.
- For cultures requiring continuous darkness a room closed off with thick black curtains is necessary.
- The culture room can also have a variety of liquid culture apparatus including gyratory and reciprocating shakers and batch/continuous bioreactors.
- For most culture conditions the temperature control should be adequate to stay within ±1 °C in a range from 10-32 °C, even though in general we use 25±1°C for normal cultures. Lighting is adjusted in terms of quantity and photoperiod duration by using automatic clocks. Provision of humidity range of 50-70 per cent with ±3 per cent should be made.

Essential requirements in incubation room are:

- Racks with light arrangement on timers and controlled temperature
- Incubators
- Rotary shakers
- Lux meter to measure intensity of light in culture room
- Shaker
- Humidifier

5. *Acclimatization Facility*

- This facility is required as a transitional step of taking plaint materials from culture containers present in the controlled room to the field which is commonly called as Hardening.

- ✰ The hardening chamber needs higher illumination (4000-10000 lux) and high humidity (90-100 per cent) through mist and fog systems.
- ✰ Humidity is required for conditioning tissue culture plants after taken out from rooting media and transfer to pots under green house.
- ✰ Thus in the green house, plants are acclimatized and hardened before being transferred to the field conditions.

Safety Requirements

First and foremost the following general guidelines for safety in the laboratory are very essential for any plant tissue culture laboratory.

General guidelines for safety in the laboratory:

1. Do not eat, drink or smoke in the laboratory.
2. Wear a laboratory coat and gloves all the time.
3. Regard all microorganisms as potentially pathogenic and treat them accordingly.
4. Discard all the waste organisms *i.e.,* agar plate cultures, contaminated tissue culture plates and tubes *etc.* Use only after autoclaving.
5. Wash your hands and clean your working area with disinfectant before starting and after ending your work.
6. A pipettor should always be used for pipetting any solution.
7. Does not mouth pipette dangerous chemicals. Ethidium bromide is a powerful mutagen. Avoid contact with hand and skin.
8. All solutions used should be either autoclaved or filter sterile.
9. Toxic chemicals should be handled with appropriate precautions and should be discarded into separately labelled containers.
10. First-aids kits should be placed in every laboratory and the staff should know its location and how to use its contents.
11. Fire extinguishers should be provided in each laboratory.

Plant Tissue Culture Medium

The basic nutritional requirements of cultured plant cells as well as plants are very similar. However, the critical composition varies according to the type of cultures and also with respect to particular plant species. Plant tissues in culture can be in the form of callus cultures, suspension cultures or organ cultures.

Callus Cultures

Callus is defined as a mass of unorganized cells derived from an explant (stem, root, cotyledon, leaf *etc.*) cultured *in vitro* on a suitable medium. The initiation and growth of callus depends on the composition of the medium and the culture conditions.

Suspension Cultures

Continuous agitation of a piece of callus in a liquid medium gives rise to suspension cultures, where in the cells divide and multiply under optimal nutritional conditions. Suspension cultures can be conveniently initiated from soft and friable callus tissues.

Organ Cultures

Whole plant organs like embryo, ovule, anthers, endosperm, meristems *etc.* can be successfully cultured and are called organ cultures.

The appropriate composition of the medium largely determines the success of cultures. Plant materials do vary in their nutritional requirements and therefore it is often necessary to modify the medium to suit to a particular tissue.

In general the medium contains

a. Inorganic salts
b. Vitamins
c. Growth regulators
d. Carbon source
e. Organic supplements

Inorganic salts: These are divided into two groups.

Major salts: The salts of potassium, nitrogen, calcium, magnesium, phosphorous and sulphur constitute the major salts. Nitrogen is generally used as nitrate or ammonium salts, sulphur as sulphates and phosphorous as phosphates.

Minor salts: The salts of iron, zinc, manganese, boron, copper, cobalt, molybdenum, iodine etc make up the minor salts. These salts are essential for growth of tissues and are required in trace quantities.

Vitamins: The B-vitamins play an important role in the growth of tissues. Thiamine, nicotinic acid and pyridoxine are generally incorporated in all media although pantothenic acid, folic acid, biotin and riboflavin *etc.*, have also been used.

Growth regulators: Growth as well as differentiation of tissues *in vitro* is controlled by various growth regulators (auxins, cytokinins, gibberellins, ethylene, abscissic acid *etc.*)

Auxins: These auxins are used for cell division and elongation. Indole Acetic Acid (IAA), Indole Butyric Acid (IBA), Naphthalene Acetic Acid (NAA) and 2,4-dichloro phenoxy acetic acid (2,4-D) are frequently used at 0.1 to 10 mg/lit concentration in plant tissue culture media.

- ☆ NAA and 2,4-D are thermo stable and do not lose their activity on autoclaving where as indole acetic acid thermo-labile and loses most of its activity upon autoclaving.
- ☆ Hence, it is sterilized by filtration. Auxins have a clear rhizogenic action, *i.e.* induction of adventitious roots.

☆ At low auxin concentration, adventitious root formation predominates, whereas at high auxin concentrations, root formation fails and callus formation takes place.

Cytokinins: Cytokinins have a profound effect on cell division and cell differentiation. The commonly used cytokinins are kinetin, Zeatin and 6-benzyl aminopurine used in 0.1-10 mg/lit concentration.

Others like Gibberellic acid, ethylene releasing compounds and abscissic acid are incorporated in the media.

Carbon source: Plant tissues in culture can utilize a variety of carbohydrates - sucrose, glucose; fructose, starch and maltose. Sucrose at 2-5 per cent concentrations in the nutrient media, remains as the most widely used carbohydrate source.

Organic supplements: Complex substances such as yeast extract, malt extract and casein hydrolysate are also added (0.1-1 per cent w/v). Among various plant extracts, liquid endosperm of immature coconut (coconut water) is widely used (5-20 per cent v/v).

Preparation of Tissue Culture Media

Preparation of Stock Solutions

For the preparation of stock solutions the chemicals are dissolved in distilled or high purity demineralised water. Adding one compound at a time will usually avoid precipitation. Dissolving the inorganic nitrogen sources of the major salts first will avoid precipitation between phosphate and calcium sources when added subsequently which can occur when the pH approaches 6.0. Dissolving the calcium salt separately before adding it will also help to avoid precipitation.

When the salts and other ingredients have been added and dissolved, the pH is adjusted by using 0.5 N HCl or 0.2 M KOH. The kinds and quantities of stock solution being made vary with size of operation and preferences. Usually the stock solutions are prepared in 10X or 100X concentrations. The stocks can consist of groups of chemicals or nearly complete media. For example the inorganic salts and vitamins perhaps including sucrose can be combined and prepared in 10X concentrations. After the stock ingredients are dissolved, the solution is distributed in plastic bags with zipper seals. The bags are stored frozen. A bag with 100 ml could then be used to prepare 1 litre of medium.

The procedures for stock solutions and media preparation for MS and B5 media are mentioned below. This generalized approach can be adapted to any medium.-

Notes on solutions

1 molar = the molecular weight in g/l

1 mM = the molecular weight in mg/l

ppm = parts per million = mg/l

Dilutions

$$\frac{\text{Required concentration x medium volume}}{\text{Concentration of stock solution}} = \text{Volume of stock required}$$

General Methodology for Media Preparation

Stock solutions of major salts, minor salts and vitamins are prepared to be used in the preparation of media and are stored in a refrigerator. For preparing one litre of the medium

1. Transfer appropriate amounts of stock solution of salts to a one litre flask.
2. Add vitamins, auxins, cytokinins and organic supplements.
3. Add a carbohydrate source such as sucrose or fructose (2-5 per cent).
4. Adjust the pH to 5.6-6.0 using a pH meter.

   Note: Wash the electrode of the pH meter with double distilled water. Take 10-15 ml of standard buffer pH 7.0 in a 25 ml beaker. Dip the electrode in this standard buffer adjusts to pH 7. Wash the electrode and check with another buffer solution of known pH. Now dip the electrode in the medium and read the pH of the medium. If the pH is less than 5.5, add 1 N NaOH drop wise, and if it is more than 6.0 add 1 N HCl to adjust the pH at 5.6-6.0. Wash the electrode thoroughly with double distilled water.
5. Make the volume to 1 lit with double distilled water.
6. Add powdered agar 8 g for 1000 ml for making the medium semi-solid.
7. Cover the flasks with paper or aluminium foil and keep in a steamer or water bath at 100°C for dissolving the agar.
8. Shake the flask well for the uniform distribution of agar.
9. Dispense the medium into culture tubes or flasks.
10. Autoclave tubes or flasks containing medium at 121°C for 20 min.

Nutritional Components of some Plant Tissue Culture Media (Amount in mg/l)

Chemicals	*White's*	*MS*	*B5*	*N6*
Macronutrients				
$MgSO_4.7H_2O$	750	370	250	185
KNO_3	80	1900	2500	2830
$NaH_2PO_4.\ H_2O$	19	—	150	—
KH_2PO_4	—	170	—	400
NH_4NO_3	—	1650	—	—
$(NH_4)_2SO_4$	—	—	134	463
$CaCl_2\ 2H_2O$	—	440	150	166

Contd...

Contd...

Chemicals	White's	MS	B5	N6
Micronutrients				
EDTA Na ferric salt	—	—	43 mg	—
H_3BO_3	1.5 mg	6.2 mg	3.0 mg	1.6 mg
$MnSO_4$. $4H_2O$	5mg	22.3 mg	—	4.4 mg
$ZnSO_4$. $7H_2O$	3 mg	8.6 mg	2.0mg	1.5 mg
KI	0.75 mg	0.83 mg	0.75 mg	0.83 mg
Na_2MoO_4. $2H_2O$	—	0.25 mg	0.25 mg	0.25 mg
$CuSO_4$. $5H_2O$	0.01 mg	0.025 mg	0-025 mg	0,025 mg
$CoCl_2$ $6H_2O$	—	0.025 mg	0.025 mg	0.025 mg
$FeSO_4$. $7H_2O$	—	27.8 mg	—	27.8mg
Organic supplements				
Vitamins				
Nicotinic acid	0.05 mg	0.5 mg	1.0 mg	0.50 mg
Thiamine HCl	0.01 mg	0.5 mg	10 mg	1 mg
Pyridoxine HCl	0.01 mg	0.5 mg	1 mg	0.5 mg
Myoinositol	—	100 mg	100 mg	—
Others				
Glycine	3.0 mg	2.0 mg	—	2.0mg
Sucrose	20.0 g	30-0 g	20.0 g	50.0 g
GA_3 (Gibberllic acid)	0.1 mg	—	—	—
NAA	—	—	—	0.05 mg/l
Biotin	—	—	—	0.05
Agar (Difco Bacto)	8.0 g	8.0 g	8.0 g	8.0 g
pH	5.8	5.8	5.5	5.8

Answer the following questions.

1. **Define somaclonal variations. And enlist the causes of somaclonal variations?**
2. **How somaclonal variations are induced and selections made? And write the significance of somaclonal variations in crop improvement?**

Somaclonal Variations

Somaclonal variations is the variation among the tissues or plant derived from the in vitro somatic cell culture *i.e.,* callus or suspension cultures.

1. Somaclonal variations may be genetic or epigenetic out of which the genetic variation is heritable whereas the epigenetic variation caused by cultural conditions is not heritable and hence of no significance in the sexually propagated plants.

2. Genetic variation may result from the following causes:
   - ✰ Chromosomal changes: *e.g.* Numerical and structural chromosomal aberrations
     - ❐ Aneuploidy observed in oats, ryegrass, triticale and potato.
     - ❐ Deletions, inversions, translocations and duplications observed in barley, wheat, potato and maize.
   - ✰ Mitotic crossing over: M.C.O may account for some of the genetic variation that leads to recovery of homozygous recessive single gene mutations in some regenerated plants.
3. Apparent 'point' mutations
4. Cytoplasmic genetic changes

   e.g. Sensitivity to host specific toxin of *Drechlera maydis* race T, the causal agent of southern corn leaf blight is associated with all genotypes containing Texas male sterile (CMS-T) cytoplasm.
5. De-amplification and amplifications:
   - ✰ Deficiencies in ribosomal DNA *i.e.*, rDNA deamplification observed at molecular level in flax, triticale and potato.
   - ✰ Gene amplifications observed in Nicotiana and tomato.
6. Transposable element activation: *e.g.* Nicotiana, alflfa and maize.
7. Virus elimination: *e.g.* Prior infection with Barley yellow dwarf virus causes susceptibility to powdery mildew in oats.
8. Methylation/deamination of DNA: Genome activity *i.e.* transcription, replication, rearrangements and the structural organization of chromatin seems to be related to DNA methylation.
9. Altered expression of multigene families.

   e.g. Heritable somaclonal variation has been obtain for *gliadina* storage protein and β-amylases in wheat.

Induction and Selection of Somaclonal Variation

- ✰ Callus cultures are established from suitable explants and multiplied through periodic sub culturing.
- ✰ Plants are regenerated usually from long-term maintained callus/cell suspension cultures and transferred to soil and screened for variation in glass house or field.
- ✰ In vitro selection at cellular level can be carried out for some traits by growing cells from cell suspensions and calli on a medium supplemented with elevated levels of various biotic and abiotic stress factors.
- ✰ Moreover, it reduces the chances of diplonitic selection but it requires high level of correspondence between the trait selected *in vitro* and expressed *in vivo*.

Significance in Crop Management

- Somaclonal variation appears to be an important alternative for creation of genetic variability in crops where tissue culture plant regeneration systems have been established.
- Recovery of novel variants which either do not exist or are rare in the natural gene pool such as atrazine resistance in maize, glyphosphate resistance in tobacco, improved lysine and methionine content in cereals, increased seedling vigour in lettuce are of much significance.
- Genetic and cytogenetic evidences have now been provided for increased recombinant frequency through cell culture.
- This will be important for breaking undesirable linkages and obtaining introgression from alien sources through tissue culture of wide hybrids.
- New varieties have been developed through somaclonal variation in tomato, sugarcane, celery, brassica and sorghum.
- This is simple and cost effective technique possess greater potential for the improvement of apomictic and vegetatively propagated crops and for seed propagated crops with narrow genetic base.
- In India, a somaclonal variant of a medicinal plant citronella java has been released as a commercial variety 'Bio 13' which gives higher yield and oil content.
- 'Pusa jai kisan' is another variety of *B. juncea* released as somaclonal variant of 'Varuma' variety.

Define totipotency? And discuss the significance of regenerating plants from single cell or protoplast under asceptic conditions?

Totipotency

Totipotency is the ability of plant cell to perform all the functions of development which are characteristic of zygote *i.e.* Ability of single cell to develop into complete plant.

- Haberlandt first demonstrated it in 1902 by culturing palisade cells in sucrose media. He only showed the accumulation of starch but failed to show division finally it was demonstrated by R.J. Gauthert and White.
- This character is too much vital because whole tissue culture is based on it. Later on embryo culture, pollen culture, protoplast cultures have been demonstrated.

Getting Whole Plant through it has several Advantages

1. Single cells culture or protoplast culture offer the possibility for large scale clonal propogation. *e.g.* Banana or orchids.
2. Somatic embryos from cell suspension can prove useful for long term

storage in germplasm bank. *e.g.* Jack fruit and mango *etc.*

3. Secondary metabolite production – secondary metabolite production through cell culture have certain advantages over normal cultivation of whole plant. *e.g.* Nicotine –*Nicotiana tobaccum*. Vinoristrin – *Vinca rosea.* The reserpine – *Rawalphia serpentine.*
4. Pollen culture and ovule culture are also a specialized cell culture and very useful in haploid production and too much helpful in development of hybrid as well as mapping of gene.
5. One can harness somaclonal variation and by this one can produce so many novel germplasm.
6. Chemical mutagen can be effectively applied through cell culture.
7. Protoplast culture is the first step in production of somatic hybrids. This is especially useful in vegetatively propagated crop and crops where sexual hybrid production is difficult.

   e.g. CMS cytoplasm transfer
8. Production of artificial seed *i.e.* somatic embryo encapsulated by calcium alginate coating.
9. In genetic transformation the transformation of cell through transfer of foreign gene into cell. By this way transgenic plants can be produced.

Give in detail the steps involved in micro-propagation of plants with the help of flow diagram? And also mention the applications and problems of micro propagation?

Micro-propagation

The multiplication of genetically identical copies of a cultivar by asexual means is called clonal propagation.

☆ A genetically uniform assembly of individual, derived originally from single individual by asexual propagation constitutes a clone.

☆ *In vitro* clonal propagation is called micro propagation

Techniques of Micro-propagation

The process of micro propagation involves 4 distinct stages

1. Selection of suitable explants, their sterilization and transfer to nutrient medium for establishment/initiation of a sterile culture explant.
2. Proliferation of shoots from the explant on medium
3. Transfer of shoots to a rooting medium
4. Transfer of plants to soil/normal environment

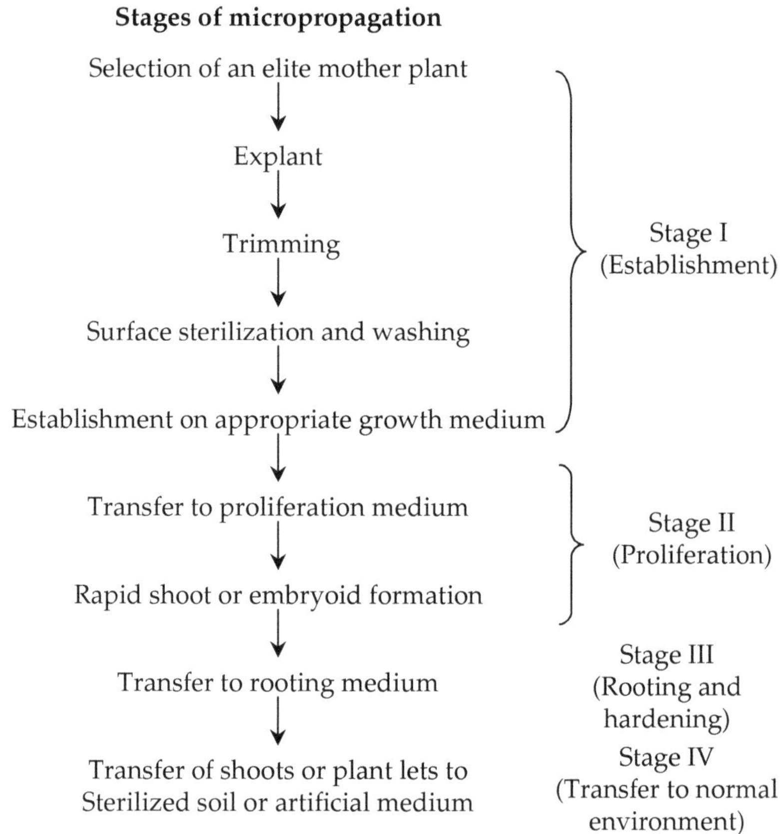

Applications of Micro-propagation

1. Micro-propagation of a hybrid has the greatest multiplication advantage since it can be result in large number of elite plants from a very small tissue clump taken from the hybrid plant.
2. Maintenance of inbred lines for producing F_1 hybrids
3. Maintenance of male sterile genotypes of wheat and onion are useful in hybridization.
4. Selective propagation of dioecious plants *e.g.* female plants of papaya, male plants of Asparagus
5. Multiplication of particular heterozygous superior genotype with increased productivity *e.g.* Oil palm
6. Shoot cultures of some species are maintained as slow growth culture for germ plasm conservation
7. Rapid production of disease free material
8. Tissue culture can be used to minimise the growing space in commercial nurseries for maintenance of shoot plant

Problems of Micropropagation

- ☆ Microbial contamination
- ☆ Callusing
- ☆ Tissue culture induced variation
- ☆ Browning of the medium
- ☆ Vitrification/Hyper hydration
- ☆ Vulnerability of micro-propagation plants to transplantation shocks

Define and write the applications of the following?

1. **Meristem culture**
2. **Anther culture**
3. **Embryo culture**
4. **Ovule culture**
5. **Ovary culture**
6. **Endosperm culture**
7. **Somatic embryogenesis**

Meristem Culture

Cultivation of axillary or apical shoot meristems, particularly of shoot apical meristem, is known as meristem culture. The first application of meristem culture was to obtain virus-free plants of dahlias.

Applications

- ☆ Virus elimination generally improves the yield by 20-90 per cent over infected controls.
- ☆ Virus-free plants serve as excellent experimental materials for evaluating the detrimental effects of infections by various viruses.
- ☆ The virus free bulbs grew more rapidly, plants were more vigorous, and they produced a greater number of larger flowers that had richer colour than the virus infected stock.
- ☆ The virus -free plants are deliberately infected by known viruses, and effects of the infection on performance of the host are assayed.
- ☆ Meristem culture can also help eliminate other pathogens, *e.g.*, Mycoplasmas, bacteria and fungi. Bacteria and fungi present in explants show up when they are cultured *in vitro* since tissue culture media provide excellent nutrition for the microbes.
- ☆ Meristem culture has been used to eliminate systemic bacteria form Diffenbachia and Pelargonium, and *Fusareum roseus* from carnations.

Anther/Pollen Culture

Anther culture is a means of obtaining haploids from the pollen grains. Haploid production through anther culture is known as "Androgenesis".

Applications

- Dihaploid production through bulbosum technique is successful in barley.
- Through the induction of genetic variability (gametoclonal variation) haploids (or) plants with various ploidy levels and mutants are obtained.
- Selection of mutants resistant to disease. Haploids provides relatively easiest system for the induction of mutations. Therefore they can be employed in rapid selection of mutants having traits for disease resistance. Eg:- Tobacco mutants resistant to Black shank disease and wheat lines resistant to scar. *Fusarium graminorium.*
- Developing asexual lines of trees (or) perennial species.
- Pollen derived rubber tree is taller by 6m which could be multiplied by asexual propagation.
- Transfer of desired alien genes/genetic transformation.
- Chromosomal instability in haploids makes them potential tool for introduction of alien chromosomes (or) genes during wide crossing programme. Haploids are used for the development of substitutional and additional lines.
- Double haploids are used for QTL mapping and construction of genetic linkage maps *etc.*

Embryo Culture

The term embryo culture means excision of embryos regardless of age, size and developmental stage from their natural environment and growing them under artificial environmental conditions.

Applications

1. Production of rare hybrids from intergeneric and interspecific crosses.

   Cereals: *Hordeum vulgare x secale cereale*

   Triticum durrum x secale cereale

   Legumes: *Arachis hypogaea x Arachis monticola*

   Phaseolus vulgaris x Phaseolus vitensis
2. Development of disease resistant plants
3. Embryo culture has been useful in evolving disease resistant plants.

   e.g. Tomato resistance to virus, fungi and nematode. *Lycopersican esculentum x L peruvianum*
4. Production of haploids: Chromosome elimination technique for production of haploids in cereals/Bulbosum technique (barley and wheat)

5. Overcoming seed dormancy: Embryo culture technique is applied to break seed dormancy which can be caused by numerous factors including endogenous inhibitors, specific light requirements, low temperature, dry storage requirements and embryo immaturity.
6. Shortening of Breeding cycle
7. Propagation of rare plants
8. Propagation of orchids
9. Prevention of embryo abortion with early ripening stone fruits
10. Clonal micro propogation
11. Rapid seed viability testing

Ovule Culture

It is the technique by which ovules are aseptically isolated from ovary and are grown on chemically defined nutrient medium under controlled conditions.

Maheshwari (1958) isolated ovules of *Papaver somniferum* and cultured them on nutrient medium.

Applications

1. To obtain interspecific and intergeneric hybrids *e.g.* Cotton, the hybrid embryo between tetraploid and diploid was rescued and obtained plants. *Gossypium barbadense* × *G. arborium* and *Gossypium hirsutum* × *G. herbaceum.*
2. Orchid seed germinate only in association with fungus, but the cultured fertilized ovules germinate even in the absence of fungus.
3. Test tube pollination and fertilization is possible through ovule culture.
4. It helps in the development of several embryos in Citrus and other crop plants.
5. Culture of unfertilized ovules helps in the formation of haploid callus.

Ovary Culture

Culture of unfertilized ovaries to obtain haploid plants from egg cells or other haploid cell of the embryo sac is called ovary culture and this process of haploid production is termed as gynogenesis.

Applications

1. Interspecific hybrids have been successfully obtained in several genera including Brassica species (*B. compestries* × *B. oleracea*) seeds with well developed embryos have been obtained when ovaries are cultured 4 days after pollination in white's medium contain casein hydrolysate. For inter specific and inter generic crosses ovaries are excised at zygote stage (or) at two celled pro embryo stage.
2. It is used for the study of understanding of physiology of fruit development.

3. The effect of phytoharmones on the development of parthenogenic fruit or haploids from the cultures of un pollinated flower can be studied.
4. It is successful in inducing poly embryony.

Endosperm Culture

Production of triploid plants from culturing the endosperm cells is called endosperm culture.

Applications of Endosperm Culture in Crop Improvement

1. Triploid cells of endosperm have totipotency. So endosperm culture technique is applied to economically important cultivars for raising superior triploid plants
2. Triploid plants are seed sterile and undersirable for plants where seeds are commercially important. *e.g.* Apple, Banana, Mulberry, Sugarbeet, Peach, Watermelon *etc.*, which are commercially important for their edible parts.
3. In some cases triploids are superior in quality than diploid. *e.g.* Triploids of populus have better quality pulpwood.
4. To exploit in the biosynthesis of some natural products *e.g.* Cultured endosperm of coffee synthesizes caffeine the level of this alkaloid in callus is synthesized by three times after two weeks by and 6 times after 4-5 weeks
5. Various trisomics developed from triploids may also be useful in gene mapping for cytogenetic studies
6. Endosperm can be used as a nurse tissue for raising hybrid embryos. *e.g.* Using *Hardeum* endosperm as a nurse tissue the young embryos of hybrid between *Hardeum* × *Triticum*, *Hardeum* × *Secale* and *Hardeum* × *Agropyron*
7. These can be induced to germinate and form normal hybrid plants
8. The multiplication of triploids should not be a serious problem with the available technique of micro propagation

Somatic Embryo genesis

Where an embryo is obtained from somatic cell rather than from zygote. Such somatic embryos can be encapsulated in a suitable matrix along with nutrients, growth regulators, antibiotics needed for development of complete plant to make what are known as synthetic or artificial seed

Synthetic Seed

The concept of synthetic seed was first conceived by Murashiqe (1978). However Kitto and Janick (1982) used it for first time in carrot

- ☆ Synthetic seed involved the production of tissue culture derived somatic embryos encapsed in protective coating
- ☆ It may be encapsulated somatic embryos, shoot buds, cell aggregates and any other tissue that can be used for sowing as a seed and that possess the

ability to convert into plant under invivo or invitro condition and returns this potential after storage also.

- ☆ Somatic embryo is bipolar structure with both apical and meristamatic regions, which are capable of forming, shoot and root respectively. Somatic embryos are structurally same as zygotic embryos found in seed and possess many of their useful features including the ability to grow into a complete plant.
- ☆ Artificial seeds may be produced by one of the two following ways 1) Dessicated and 2) Hydrated systems
- ☆ In the dessicated system the somatic embryos are first hardened to withstand dessication and then are encapsulated in a suitable coating material to yield desiccated artificial seeds.
- ☆ In the Hydrated systems, somatic embryos are enclosed in gels which remain hydrated. Of the many gels evaluated "calcium alginate" is the most suitable.

What is meant by protoplast culture? And explain in brief about somatic hybridization technique along with applications?

Protoplast Culture

Protoplast can be isolated either by mechanical (or) enzymatic methods.

1. *Mechanical Method*

The cells are kept in a suitable plasmolyticum and protoplasts are isolated by cutting the plasmolysed tissues with a sharp blade so that protoplasts are released from cells through the cell wall when the tissue is again deplasmolysed. This method is suitable for isolation of protoplast from highly vacuolated cells of storage tissues,

e.g. Onion bulb, scales, radish roots *etc.*

2. *Enzymatic Method*

Cocking (1960) demonstrated the possibility of enzymatic isolation of protoplast from higher plants. He used concentrated solution of cellulase to degrade the cell walls.

The enzymatic isolation of protoplasts can be performed in different ways

Two step (or) Sequential method: The tissue is first treated with macerozyme or pectinase enzyme which separates the cell by degrading the middle lamella. These free cells are then treated with cellulase which releases the protoplasts, cellulase enzyme digest the cellulose in plant cell walls while pectinase enzyme breakdown the pectin holding cells together

One Step (or) Simultaneous Method

The tissue is subjected to a mixture of enzyme in first step reaction which includes both macerozyme and cellulose. The one step method is generally used because it is less labour intensive while the yield of protoplast isolated by two step method are better for culture studies.

Somatic Hybridization

The technique of hybrid production through the fusion of protoplast from different genetic backgrounds is known as somatic hybridization or parasexual hybridization and protoplast fusion.

The technique of somatic hybridization is of special significance for the improvement of vegetatively propagated plants such as banana, cassava, potato, sweet potato, and sugarcane.

The production of somatic hybridization involve a number of steps

1. Collection of explants
2. Protoplast isolation
3. Protoplast fusion
4. Selection of hybrid cells
5. Regeneration of plants from hybrid tissue
6. Characterization of hybrid and cybrid plants

Applications of Somatic Hybridization

1. Production of novel interspecific and intergeneric crosses between the plants that are difficult (or) impossible to hybridize conventionally.

   e.g. Tomato x Potato = Pomato
2. Transfer of desirable genes for disease resistance. *e.g.* Potato protoplast have been fused with those of *Solanum brevidens* and *Solanum phuraja* and *Lycopersicon* the resulting somatic hybrids are resistant to potato leaf roll virus, potato virus and *Erwinia* soft rot. Somatic hybrids resistant to *Phomalingam* (Black log disease) were produced by the fusion of protoplast of *Brassica napus* with those of *Brassica nigra* (resistant to pomalingam).
3. Transfer of desirable genes for Abiotic stress resistance, Atrazine (resistance) has been transferred from wild species of tomato *Lycopersicon peruvianum* into cultivated species (*Lycopersicon esculentum*).
4. Transfer of desirable genes for quality characters: Somatic hybrids produced between *Brassica napus* and *Eruca sativa* were fertile and had low concentration of Erucic acid.
5. Transfer of cytoplasmic male sterility: Male sterility can be induced by alloplasmic association. It results in interaction between nucleus of one species and cytoplasm of another species. Male sterile lines were developed by fusing protoplast of *N. tabacum* with X-ray irradiated protoplasm of *N. africana.*
6. Identification of crop cultivar, somatic hybrids and hybrids in IPR issue and protection.

Answer the following questions.

1. **Define haploids? And explain in brief about the methods of production of haplodis?**
2. **Discuss the use of haploids in plant breeding and gene mapping in comparision to conventional methods?**

Haploid is nothing but organisms having the same number of sets of chromosomes as a germ cell, or half the diploid number of a somatic cell. The haploid number (23 in humans) is the normal chromosome complement of germ cells. And it is described by 'n'.

Haploidy Production System

Generally these are two methods haploidy production, in vivo and *in vitro*

In vivo: Parthenogenesis and apogamy chromosome elimination.

In vitro: Cultural method.

Examples

1. Spontaneous: Maize and Linum
2. Artificial introduction by external stimulus
   a. Irradiation- Crepis and Wheat
   b. Wounding/injury- Oenothera, Tobacco and Maize
   c. Temperature- Datura and Secale
   d. Chemical- Populus and Capsicum
3. Delayed pollination- Maize
4. Wide hybridization- S. *tuberosum*
5. Alien cytoplasm- Wheat
6. Haploid inducing genes
   a. Independent gametophyte - Maize
   b Haploid initiator gene (hap)-Barley
7. Semigamy – Cotton and Arabidopsis
8. Pseudogamy – Potato and Strawberry
9. Chromosome elimination:
   a. Somatic reduction *e.g*. Sorghum
      Chlorophenicol and paraflurophenyl alanine induction of haploidy
   b. Barley haploidy by crossing with *H.bulbosum.*
      Also other Hordeum species wheat (Barley)
   c. Wheat haploid from wheat X maize cross.

   The four important methods for invitro production of haploids are:
   i. Anther culture

ii. Isolated microspore culture
iii. Unpollinated ovary culture
iv. Embryo rescue from wide crosses
   a. Bulbosum method
   b. Wheat haploids from wheat × maize crosses

Role of Haploidy in Crop Improvement

Applications

1. Production of homozygous lines:- Used directly as cultivar (or) in breeding programme *e.g.* marglobe tomato (doubled haploid) Chase used-DH lines as inbred line in hybrid maize programme.
2. Hybrid sorting:- Refers to selection of recombinant superior gametes.

   e.g. F-211 (tobacco), Haayul, 2(Rice) Yuhna 1,2 (wheat), *Mais haptona* (*Brassica napus*)
3. Short breeding cycle:- Only 4-5 years as compare to 10 years in conventional method.
4. Analytical breeding:- Selection at haploid level and improved dihaploids used for chromosome doubling eg:-potato
5. Generation of exclusive male Asparagus (commercially male are preferred)
6. Mutation research:- *e.g.* Tobacco-cell line resistance to methionine sulphoxamide and wild fire (*Pseudomonas tabaci*)
7. Cytogenetic research:- determining basic chromosome number and polyploidy and production of alien addition and substitution line (wheat)
8. Evolutionary studies.
9. Genetic studies:- Hybrid development.
10. Indirect resistance- Rice- Hwacheongbye

Comparison between DHs and purelines produced through conventional breeding:

- ✰ In self pollinated crops, long period is required to recombine desirable gene combinations from different sources in homozygous form. Generally, it takes 8-10 years to develop stable, homozygous and ready to use material from a fresh cross of two or more lines.
- ✰ Due to inbreeding depression in cross pollinated crops, it is difficult to obtain vigorous inbreds for hybrid seed production programmes.
- ✰ Haploids possessing gametic chromosome number are very useful for producing instant homozygous true breeding lines *i.e.* by colchicines treatment.

- Haploids constitute an important material for induction and selection of mutants particularly for recessive genes.
- Period required to develop homozygous true breeding line is less through haploid breeding as compared to conventional breeding.
- In conventional breeding, the early segregating generation population involves variation due to both additive as well as nonadditive effects whereas DH lines exhibit variation only due to additive gene effects along with additive × additive type of epistasis which are easily fixable through one cycle of selection.
- Elimination of dominant effects and availability of sufficient seed of each doubled haploid line for replicated testing leads to high heritability in such populations.
- In contrast to relatively large sized segregating populations in conventional methods, less number of DH lines are required for the purpose of selection of desired recombinants through the production of haploids.
- The DH populations derived through anther culture from highly inbred parents show variability for quantitative as well as reduced vigour.
- DH approach is increasingly being used for rapid development of populations for QTL mapping and construction of genetic linkage map for traits of interest.

Write short notes on the following?

1. **Secondary metabolite production in tissue culture**
2. **Cryopreservation**

Secondary Metabolite Production

Cultured plant cells often produce reduced quantities and different profiles of secondary metabolites when compared with the intact plant and these quantitative and qualitative features may change with time.

- This method of production called as phytoproduction.
- Routian and Nickell obtained the first patent for the production of substances by plant tissue culture in 1956.
- This extraction of metabolites can be possible from both callus and suspension cultures.

  e.g. Suspension cultures of *Thalictrum minus* produced the stomachic and antibacterial berberine. And callus cultures of *Catharanthus roseus* produced the antihypertensive ajmalicine.
- There are a number of examples of cultured cells producing metabolites observed in the plant.

Compound	Plant Species	Culture Type
Shikonin	*Lithospermum erythrorhizon*	Suspension
Ginsenoside	*Panax ginseng*	Callus
Anthraquinones	*Morinda citrifolia*	Suspension
Ajmalicine	*Catharanthus roseus*	Suspension
Rosmarinic acid	*Coleus blumeii*	Suspension
Ubiquinone-10	*Nicotiana tabacum*	Suspension
Diosgenin	*Dioscorea deltoides*	Suspension
Benzylisoquinoline Alkaloids	*Coptis japonica*	Suspension
Berberine	*Thalictrum minor*	Suspension
Berberine	*Coptis japonica*	Suspension
Anthraquinones	*Galium verum*	Suspension
Anthraquinones	*Galium aparine*	Suspension
Nicotine	*Nicotiana tabacum*	Callus
Bisoclaurine	*Stephania cepharantha*	Suspension
Tripdiolide	*Tripteryqium wilfordii*	Suspension

Cryopreservation

Cryopreservation is the use of very low temperatures to preserve structurally intact living cells and tissues. It is also termed as cryoonservation.

- ☆ Cryonic storage at very cold temperatures is presumed to provide an indefinite longevity to cells, and it can be achieved at –196 °C (77 K; –321 °F), the boiling point of liquid nitrogen.
- ☆ Cryopreservation methods seek to reach low temperatures without causing additional damage caused by the formation of ice during freezing.
- ☆ Traditional cryopreservation has relied on coating the material to be frozen with a class of molecules termed cryoprotectants.
- ☆ Many compounds have such properties, including glycerol, dimethyl sulfoxide, ethanediol, and propanediol.
- ☆ Ice can be avoided by vitrification—the production of a glassy state that is defined by the viscosity reaching a sufficiently high value to behave like a solid, but without any crystallization.
- ☆ Toxicity is the major problem in the use of vitrification methods.
- ☆ Whether freezing is permitted (conventional cryopreservation) or prevented (vitrification), the cryoprotectant has to gain access to all parts of the system.

Answer the following questions.

1. **Define genetic engineering and write the applications of genetic engineering in crop improvement?**

2. **Explain in brief about the methods of transformation of foreign genes into plant cells?**
3. **What is T-DNA? And discuss the structural organization of the Ti plasmid and T-DNA?**

Genetic Engineering

Genetic Engineering is a term used to refer the manipulation of existing genes with the new or foreign genes isolated from other organisms.

- ☆ Plant Genetic Engineering is defined as Isolation, introduction, expression of foreign DNA in plant.
- ☆ Genetic engineering enables the transfer of genes across taxonomic boundaries unlike conventional breeding where it is possible to transfer genes from closely related species only.
- ☆ It also offers new avenues of plant improvement in shorter period compared to conventional breeding and new possibility of incorporating new genes without problems incompatibility.

The various applications of genetic engineering in crop improvement includes

1. Herbicide resistance: Herbicides normally affect processes like photosynthesis or biosynthesis of essential amino acids.
   - ☆ Transformation of cereal crops with Glyphosate resistant gene (Glyphosate = herbicide).
   - ☆ Herbicide tolerant (HT) soybean and canola are released for commercial cultivation.
2. Insect resistance: The genes which responsible for the production of delta-endotoxine in *Bacillus thuringiensis* is used as biological insecticide.
   - ☆ The transgene Cry 1AC has been transferred to many crops for example boll worm resistance in cotton, looper resistance in soybean, pod borer resistance in groundnut, head borer resistance in sunflower, semi-looper resistance in castor *etc.*
   - ☆ Snowdrop lectin gene from snow drop (*Galanthus nivalis*) was transferred to brassica and safflower for aphid resistant.
3. Resistance against viral infection: Coat protein gene from Tobacco Mosaic Virus (TMV) was transferred to develop resistant varieties of crop plants.
   - ☆ The resistant varieties developed in crop plants like soybean for resistant to yellow mosaic virus, groundnut for resistant to bud and stem necrosis, clump and stripe virus resistance, whereas in sunflower, resistance developed for bud necrosis.
4. Resistance against bacterial and fungal pathogens:
   - ☆ Chitinase genes was transferred to crops like Brassica, Soybean, Sunflower, Sesame etc for alternaria leaf spot disease, where as in case

of groundnut which was introduced against leaf spot and alternaria blight and in castor for Botrytis resistance.

- ✰ Acetyl transferase gene was transferred for wildfire disease of tobacco caused by pseudomonas syringae.

5. Improvement of the nutritional qualities in crop plants:
   - ✰ The carotene gene has been transferred from daphoddils to rice grains (Golden Rice) for increasing Beta-carotene content in grains and for solving the blindness in childerns.
   - ✰ Antisense Fae 1 gene transferred to Brassica napus and Brassica juncea for low erucic acid content and also for low linolenic acid content in case of linseed.
   - ✰ Antisense ricin gene transferred to castor for reduction of ricin content and RCA endosperm in castor seeds. Antisense sterol desaturase/+ ac1 inserted into sunflower for developing high oleic acid containing types.
6. Improvement of crop plants against abiotic stresses:
   - ✰ Transcription factor genes, structural genes, regulatory genes were introduced into the groundnut, soybean, Brassica juncea, B. napus to develop drought and salinity tolerant types.
7. Development of transgenic male sterile lines:
   - ✰ Transgenic male sterile lines of safflower Brassica juncea were developed through the transfer of Barnase gene from Bacteria (Bacillus amyloliquefaciens).
8. A long term goal in agriculture is to introduce the genes (Nif genes) for nitrogen fixation in crop plants.

Genetic engineering yielded genetically modified (GM) crops having novel genes with favourable characteristics like higher yields, herbicide resistant, insect and disease resistant, drought resistant, salinity resistant, *etc.*

Depending upon their preference, there are four methods for transformation of the novel genes are as follows:

1. Biological method
2. Mechanical method
3. Chemical method
4. Genotype independent methods

Biological Method

This method involves the use of biological systems for transformation of the desired gene or genes.

i. **Agrobacterium mediated transformation:** Agrobacterium remains on the preferred mechanism to introduce exogenous genes into plant cells.

- The main reasons for this is the wide spectrum of plants that are susceptible to transformation by this bacterium *i.e.* bacterium had the ability to transfer its own genome in to the plant cell.
- This method works especially well for dicotyledonous plants like potatoes, tomatoes, and tobacco. Agrobacterium infection is less successful in crops like wheat and maize.

ii. **A modified T- DNA** region of the Ti (Tumor inducing) plasmid in which the genes responsible for tumor formation are removed and inserted the foreign novel gene by genetic engineering.

iii. **Viruses**: Phages (Viruses that infect bacteria) and Tobacco Mosaic virus can be used as vectors for transformation of genes into plant cells.

Mechanical Method

It includes the use of equipments for transformation of gene is as follows;

i. **Electrophoration:** Is a process whereby electric pulses of high field strength are used to reversibly permiabilise cell membranes to facilitate uptake of DNA.

- Electrophoration has been successfully used for transforming plants in which efficient regeneration of plants from protoplast is possible.

ii. **Microinjection:** In this case needles used for injecting DNA are with a diameter greater than cell diameter.

- DNA (0.3 ml) solution is injected with conventional syringe into region of plant which will develop into floral tillers fourteen days before meiosis.

iii. **Biolistic or Microprojectile method:** This method involves bombardment of particles carrying DNA or RNA of interest into target cells using high velocity transfer mechanism.

- The DNA separates from the metal and is integrated into plant DNA inside the nucleus.
- This method can be used for any plant cells, leaves, root sections, embryo seeds and pollens.
- This method is successful in many cultivated crops, especially monocots like wheat or maize for which Agrobacterium method has been less successful
- The major disadvantage of this procedure is that serious damage can be done to the cellular tissues.

Chemical Method

This method using polyethylene glycol and calcium phosphate for gene transfer.

Genotype independent transformation methods includes

i. Shoot apical meristem

ii. Pollen tube pathway

iii. Flower dip method

T-DNA

The transfer DNA (abbreviated T-DNA) is the transferred DNA of the tumor-inducing (Ti) plasmid of some species of bacteria such as *Agrobacterium tumefaciens* and *Agrobacterium rhizogenes.*

☆ It derives its name from the fact that the bacterium transfers this DNA fragment into the host plant's nuclear DNA genome.

Ti Plasmid

This plasmid contains all the genes which required for tumor formation.

The basic elements of the vectors designed for *Agrobacterium* mediated transformation includes

i. The T-DNA border sequences,

☆ The T-DNA region is flanked at both ends by 24 bp direct repeat border sequence called T-DNA borders.

☆ The left T-DNA left border is not essential, but the right border is indispensable for T-DNA transfer *i.e.* it initiates the integration of the T-DNA region into the plant genome.

ii. The *vir* genes, present in virulence region are required for transfer of T-DNA region into the plant.

iii. Oncogenic genes responsible for production of auxins and cytokinens resulting in uncontrolled cell growth and crown gall formation.

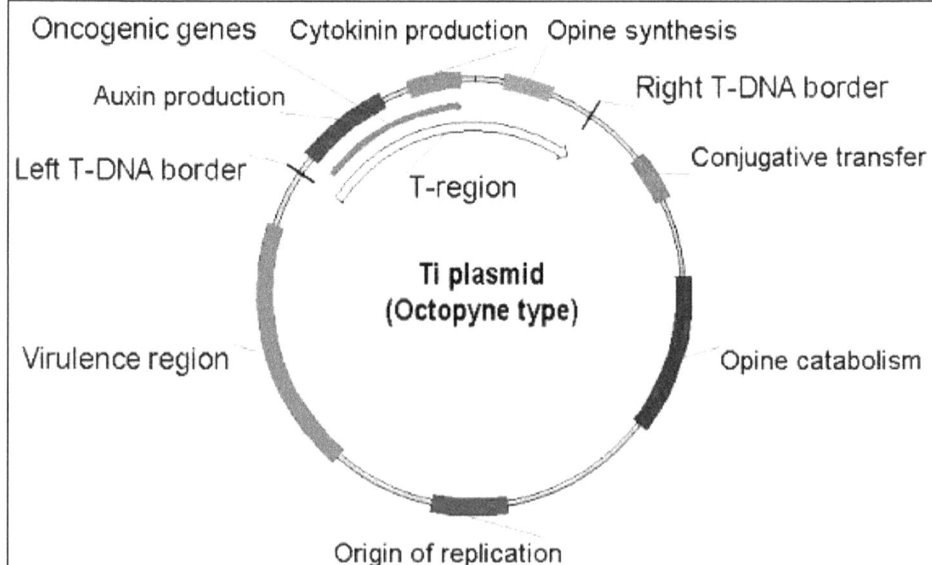

iv. OS region responsible for unnatural aminoacids opines and napolines as a source of carbon and sucrose.

A modified T-DNA region of the Ti plasmid, in which the genes responsible for tumor formation are removed by genetic engineering and replaced by foreign genes of diverse origin *e.g.* from plants, bacteria and virus.

What are the genetic engineering strategies to create the following traits in transgenic crops?

1. **Abiotic stress**
2. **Virus resistance**

Genetic Engineering Strategies for Abiotic Stress Resistance

Plant engineering strategies for abiotic stress tolerance rely on the expression of genes that are involved in signaling and regulatory pathways or genes that encode proteins conferring stress tolerance or enzymes present in pathways leading to the synthesis of functional and structural metabolites.

Stress-related Genes and Proteins

These can be grouped in to major categories like

1. **Signaling cascades and transcriptional control:** Genes involved in signaling cascades and in transcriptional control mainly are mitogen-activated protein (MAP) and salt overly sensitive (SOS) kinases, phoapholipases and transcription factors (Heat shock factors) and the C-repeat binding factor/dehydration-responsive element binding protein (CBF/DREB) and ABA-responsive element binding factor/ABA-responsive elements (ABF/ABRE).

   e.g. Constitutive expression of the tobacco mitogen-activated protein kinase Kinase Kinase/Nicotiana protein kinase 1 (MAPKKK/NPK1) in maize activates an oxidative signal cascade and leads to cold, heat and salinity tolerance in the transgenic plants.

2. **Heat-shock proteins and chaperons:** Heat-shock proteins (Hsps) and molecular chaperones, as well as late embryogenesis abundant (LEA) protein families are mainly involved in plant abiotic stress tolerance.

   - Hsps and LEA proteins help to protect against stress by controlling the proper folding and conforming of both structural (*i.e.* cell membrane) and functional (*i.e.* enzymes) proteins.
   - Overexpression of HAP101 from *Arabidopsis* in rice plants resulted in a significant improvement of growth performance during recovery from heat stress.
   - Overexpression of HVA1, a group 3 LEA protein isolated from barley conferring dehydration tolerance to transgenic rice plants.

3. **Reactive oxygen species:** Stress-induced production of reactive oxygen species (ROS) is another aspect of environmental stress in plants.

- ☆ Alleviation of oxidative damage by the use of different antioxidants and ROS scavangers can enhance plant resistance to salt and drought.
- ☆ Transgenic tobacco plants overexpressing *Chlamydomonas* glutathione peroxidase in the cytosol and in the chloroplast displayed increased tolerance to oxidative stress.
- ☆ Overexpression of the aldehyde dehydrogenase *AtALDH3* gene in *Arabidopsis* conferred tolerance to drought and salt stress.

Stress Associated Changes in Metabolites and Metabolomics

Severe osmotic stress causes detrimental changes in cellular components. A wide range of metabolites that can prevent these detrimental changes have been identified, including amino acids (*e.g.* proline), amines (*e.g.* glycine-betaine and ployamines) and a variety of sugars and sugar alcohols (*e.g.* Mannitol and trehalose) *etc.*

- ☆ Two general strategies for the metabolic engineering of abitotic stress tolerance are increased production of specific desired compounds or reduction in the levels of unwanted (toxic) compounds.

Amino Acids

Proline accumulation was correlated with improved plant performance under salt stress.

- ☆ *e.g. Arabidopsis* transformation with proline dehydrogensae antisense or a knockout of this enzyme resulted in increased free proline accumulation and better growth performance under salt stress.

Amines

Betaine aldehyde decarboxylase from the haplophyte *Suaeda liaotungensis* was introduced in to tobacco plants and the *in vitro* plantlets were significantly resistant to salt conditions.

Sugars and Sugar Alcohols

Rice tolerance to multiple abiotic stresses through engineering trehalose overexpression was reported.

Metabolic Profiling

Overall metabolic profiling of plants under stress is an important tool to study stress-induced changes in metabolites.

- ☆ *Arabidopsis* metabolic profiling revealed that plants subject to a combination of drought and heat stress accumulate sucrose and other sugars such as maltose and glucose.
- ☆ Heat stress was found to reduce the toxicity of proline, suggesting that during the more severe, combined stress, sucrose replaces proline in plants as the major osmoprotectant.

Therefore, comprehensive profiling of stress-associated metabolites, combined with stress metabolomics of major crop plants will be a key factor in molecular breeding of tolerance.

Genetic Engineering Strategies for Virus Resistance

Genetic engineering offers a means of incorporating new vrirus resistance traits into existing desirable plant cultivars.

There are mainly two approaches for developing genetically engineered resistance depending on the source of the genes used.

1. Pathogen derived resistance (PDR), a part or a complete viral gene is introduced into the plant, which subsequently interferes with one or more essential steps in the life cycle of the virus.
2. Non-pathogen derived resistance, is based on utilizing host resistance genes and other genes responsible for adaptive host processes, elicited in response to pathogen attack, to obtain transgenics resistant to the virus.

Transgenics with Pathogen-Derived Resistance

This can be achieved by either viral coat protein or replicase gene encoding sequencesor movement proteins or *Satellite* RNAs or defective interfering viral nucleic acids.

Coat Protein Mediated Resistance (CPMR)

The expression of viral coat protein gene in transgenic plants was shown to induce protective effects similar to classical cross protection, and was therefore distinguished as "coat protein mediated protection".

- ✰ The level of protection conferred by CP genes in transgenic plants varies from immunity to delay and attenuation of symptoms.
- ✰ Powell-Abel *et al* in 1986 first reported resistance against TMV in transgenic tobacco expressing the TMV CP gene.
- ✰ The success of CPMR has prompted the production of transgenic plants expressing multiple CP genes from more than one virus.

  e.g. Transgenic tobacco containing the CP gene of three viruses ahs been shown to develop resistance to all of them, namely tomato spotted wilt virus (TSMV), tomato chlootic spot virus and groundnut ring spot virus.
- ✰ CPMR confers additional advantage of resistance to vector inoculation *i.e.* CP plays a major role vector transmission.

  e.g. Potato expresses PVX and PVY CP nad tobacco, tomato and cucumber expressing CMV CP were seen to be highly resistant to aphid transmissions.

Replicase Mediated Resistance

Engineering virus resistance by using genes encoding viral RNA- dependent RNA-polymerases (RdRps).

- ✰ Replicase (Rep) protein mediated resistance against a virus in transgenic plants was firat shown in tobacco against TMV in plants containing the 54 KDa putative *Rep* gene.

Movement Protein Mediated Resistance

Movement proteins (MP) are essential for cell to cell movement of plant viruses.

- ✰ These proteins have been shown to modify the gating function of plasmodesmata, thereby allowing the virus particles or their nucleoprotein derivatives to spread to adjacent cells.
- ✰ This phenomenon was first used to engineer resistance against TMV in tobacco by producing modified MP which are partially active as a transgene.

Satellite RNA Mediated Resistance

the presence of sat-RNA modulates the symptoms induced by the helper virus and often depresses helper virus accumulation in different host species.

- ✰ CMV sat RNA depends on its helper virus CMV for replication, movement within the plant, encapsulation and transmission.
- ✰ Transgenic tobacco plants expressing multiple or partial copies of CMV sat-RNA showed attenuated symptoms when challenged with CMV. In addition, tobacco plants transformed with anti-sense sat-RNA showed delayed symptom development with the cognate virus.

Defective Interfering Viral Nucleic Acids

In several viruses, truncated genomic components are often detectable in infected tissues, which interfere with the replication of the genomic components. These species of DNA are also called defective interfering (DI) DNA.

- ✰ The expression of delayed disease symptoms and recovery, coupled with increased resistance upon repeated inoculation have been observed in plants engineered with DI DNA.
- ✰ Incorporation of subgenomic DNA B that conferes with the replication of full length genomic DNA A and B confers resistance to ACMV in *Nicotiana benthamiana*.

Transgenics with Non-Pathogen-Derived Resistance

Strategies for non-pathogenic derived resistance mainly includes utilizing genes derived from either the host plant or any other non-pathogenic source.

- ✰ The other non-pathogenic derived strategies are the utilization of plant disease resistance genes, the ribosome –inactivating proteins, plant proteinase inhibitors, human interferon-like systems, antiviral antibodies expressed in plants, systemic acquired resistance and secondary metabolic engineering.

- ☆ A new phenomenon called post-transcriptional gene silencing (PTGS) has recently been shown to be responsible for the inherent ability of many plants to specifically degrade nucleic acids in a sequence specific manner.

Post-Transcriptional Gene Silencing (PTGS)

It is a specific RNA degradation mechanism of any organism that that take care of aberrant, unwanted excess or foreign RNA intra-cellularly in a homology-dependent manner.

- ☆ This activity could be present constitutively to help normal development or induced in response to cellular defense against pathogens.
- ☆ In this mechanism, the elicitor double-stranded RNA (ds RNA) commonly produced during viral infection, is degraded to 21-25 nucleotides, termed as small interfering RNA (siRNA).
- ☆ When viral RNA is either the elicitor or target of PTGS, the degradation mechanism is known as virus induced gene silencing (VIGS).
- ☆ VIGS comes into play when plants recover from initial viral infection or plants resist superinfection of viruses with genomes bearing homology with those of the viruses used as primary inoculums.

  e.g. If tobacco rattle virus (TRV) infects *Nicotiana benthamiana*, the plant develops initial symptoms of viral infection at the inoculated region. But the plant shows signs of recovery later and newly emerged leaves are free of TRV.

Plant Disease Resistant Genes

Disease resistance genes mainly encode products which respond to viral signals (avirulence (*avr*) gene products) culminating in a number of resistance responses in the plant.

- ☆ R genes in palnts are defined by the classical gene for gene hypothesis, which states that for every incompatible host pathogen interaction, there exist matching R genes in the host and *avr* genes in the pathogen.
- ☆ Resistance reaction against pathogen results generally by direct interaction between the products of R and *avr* genes.
- ☆ The resistance reaction due to this interaction is called as hypersensitive reaction (HR).
- ☆ All known R genes encode products having two basic functions: to act as sensors for the corresponding *avr* factors/elicitors and to initiate signaling cascades for the expression of defense related genes.
- ☆ A number of structural features are conserved across several R gene products includes leucine-rich repeat (LRR), nucleotide-binding site (NBS), serine-threonine kinase, leucine zipper and toll-interleukin region (TIR) *etc.*

e.g. N gene of tobacco provides resistance against TMV. And the N gene product has a prominent TIR signaling domain at the amino-terminus and a LRR a recognition domain at the carboxyl-terminus of the polypeptide.

Ribosomal Inactivating Proteins

Ribosomal inactivating proteins (RIPs) inhibit the translocation step of translation by catalytically removing a specific adenine base from 28S ribosomal RNA.

- ✰ They are synthesized either as pre or pre-pro proteins and targeted to vacuoles.
- ✰ Because of their specific intracellular localization, RIPs donot affect the endogenous 28S RNA.
- ✰ RIPs enter in to the cells together with the viruses and exert the damage to the host ribosome or possibly viral RNA.
- ✰ When RIPs are mixed with viruses and applied on plants, virus multiplication and symptom development are dramatically suppressed.
- ✰ RIPs not only inhibit local virus multiplication in RIP-treated leaves but also block viral multiplication systematically.

  e.g. Transgenic *Nicotiana benthamiana* plants expressing PAP (Pokeweed) have been shown to offer broad-spectrum virus resistance to both mechanical and aphid transmission.

Protease Inhibitors from Plants

Many viruses, namely poty-, tymo-, nepo-, como-, and closteroviruses need cysteine protease activity to process their own polyproteins for their replication and propagation.

e.g. Tobacco lines expressing the rice cysteine protease inhibitor gene were examined for resistance against tobacco etch virus (TEV) and PVY infection.

Interferon like Systems

Higher vertebrates resist virus infections in part by catalysis of RNA decay using the interferon regulated 2-5A system.

- ✰ The 2-5A synthetase that makes 5′ phophorylated, 2′-5′- linked oligoadenylates (2-5A) in response to double stranded DNA, and the 2-5A dependent RNAse L.
- ✰ In plants, homologues of this system are not yet known but the inducers, *i.e.* interferon-like molecules have been reported.

  e.g. The transgenic tobacco produced low-level but functional 2-5A synthetase and activated RNAse L. these transgenic lines were tested positive for their proficiencies to resist at least three different types of viruses: TEV, TMV and AMV.

Anti-viral Plantbodies

Expression of specific anti-viral antibodies in plants to control plant viruses are commonly known as plantbodies.

- ☆ A panel of monoclonal antibodies was raised against ACMV and the gene for the most reactive of the above panel was cloned and expressed in *Nicotiana benthamiana*.

Systemic Acquired Resistance

Plants develop active resistance first at active localized site only, but spreads systemically in due course. This resistance is called systemic acquired resistance (SAR).

- ☆ SAR is characteristically associated with accumulation of salicylic acid (SA), enhanced expression of pathogen related (PR) proteins leading to the synthesis of higher phenolic compounds, increase of active oxygen species and reinforcement of cell wall by the deposition of lignin and suberin.

  e.g. Involvement of SA in TMV resistance has been shown by expressing the bacterial salicylate hydroxylase (NahG) gene in tobacco palnt, thus decreasing its endogenous salicylic acid, and causing susceptibility to TMV infection.

Secondary Metabolite Pathways

Metabolic pathways are important in viral pathogenesis as they are key targets for intervention against viral infection.

- ☆ S-adenosyl homocysteine hydrolase (SAHH), is key enzyme in trans-methylation reaction.
- ☆ The antisense RNA for tobacco SAHH was expressed in transgenic tobacco plants. Though 50 per cent of the plants showed stunting, they were resistant to infection by various plant viruses.

9

Transgenics, Biosafety and Regulations

Answer the following questions.

1. **Define transgenic crops? And write the role of transgenics in crop improvement along with its limitations?**
2. **Give the detail account of the current status of transgenic crops at global level?**
3. **Area under commercial transgenic crops has increases continuously since the 2000. Still a large number of NGOs are opposing commercialization of transgenics.**
   a. **What is the controversy behind the GM crops?**
   b. **What are the major concerns for the opposition?**
   c. **What is the India's stand on GM crops?**

Transgenic plants are plants that have been genetically engineered, a breeding approach that uses recombinant DNA techniques to create plants with new characteristics. They are identified as a class of genetically modified organism (GMO).

☆ The first genetically modified crop plant was produced in 1982, an antibiotic-resistant tobacco plant. The first field trials occurred in France and the USA in 1986, when tobacco plants were engineered for herbicide resistance.

☆ The People's Republic of China was the first country to allow commercialized transgenic plants, introducing a virus-resistant tobacco in 1992, which was withdrawn in 1997.

☆ The first genetically modified crop approved for sale in the U.S., in 1994, was the *FlavrSavr* tomato. It had a longer shelf life, because it took longer to soften after ripening.

Role of Transgenic Plants in Crop Improvement

Transgenic plants or genetically modified plants (GMP) have both basic and applied uses which are briefly summerised below:

1. Transgenes will be important in increasing the efficiency of crop production systems. For instance tragenic plants resistant to herbicides, insects, viruses and other biotic and abiotic stress have already been produced.
2. Transgenic breeding is an effective means of inducing male sterility in crop plants. *e.g.* Barnase and barstar systems in *Brassica napus.*
3. Transgenic plants which are stable for food processing have also been produced *e.g.* Bruise resistant and delay ripening in tomato.
4. Several gene transfers have been aimed in improving the produce quality. *e.g.* Protein and lipid quality: improved quality may be achieved by either rganizatio or over production of endogenous genes.
5. Transgenic plants are aimed to produce novel biochemicals like interferon, Insulin, Immunoglobin etc useful biopolymers like poly rganiza butyrate which are not produced by normal plants. These compounds are extracted from plants can be used as pharmaceutical (or) industrial substrates.
6. The utilization of transgenic plants as rganizatio (or) factories for the production of special chemical and pharmaceuticals is known as Molecular forming.
7. Transgenic plants have been produced that express a gene encoding antigenic protein from a pathogen. Use of transgenic plants as vaccines for immunization against pathogens fast emerging as an important objective.
8. Transgenic plants have proved to be extremely valuable tools in studies on plant molecular biology. Regulation of gene action, identification of regulatory, promoters sequences *etc.*

Limitations

1. Transgenic plants sometimes exhibit instable performance for character under consideration.
2. The transformants had the undesirable side effects of trans genes
3. The position and integration of foreign gene in host genome effects the expression of transgene in the transformant.
4. Inability to transfer polygenic traits

5. Transgenic breeding is a very expensive method of crop improvement and requires high technical skills.
6. Transgenic cells are recovered at a very low frequency in cell culture, more over regene ration of transgenic cells into whole plants is also difficult and time consuming task.
7. Transgenic breeding acts against natural evolution.
8. There are chances of developing new weed species through transgenic breeding
9. Sometimes the foreign genes have adverse effects on genome of the recipient parent, in such cases it may give rise useless gene combinants which may become a problem.

Current Global Status of Transgenic Crops

In recent years, development of transgenic crops expressing a variety of novel traits such as insect resistance, disease resistance, herbicide tolerance, hybrid production, improved oil quality etc have led to large scale cultivation of transgenic crops.

- ✰ In 2013, GM crops were planted in 27 countries; 19 were developing countries and 8 were developed countries. 2013 was the second year in which developing countries grew a majority (54 per cent) of the total GM harvest.

Country	*2013–GM Planted Area (million hectares)*	*Biotech Crops*
USA	70.1	Maize, Soybean, Cotton, Canola, Sugarbeet, Alfalfa, Papaya, Squash
Brazil	40.3	Soybean, Maize, Cotton
Argentina	24.4	Soybean, Maize, Cotton
India	11.0	Cotton
Canada	10.8	Canola, Maize, Soybean, Sugarbeet
Total	175.2	—

- ✰ According to the 2013 International Service for the Acquisition of Agri-biotech Applications (ISAAA), a total of 36 countries (35 + EU-28) have granted regulatory approvals for biotech crops for food and/or feed use and for environmental release or planting.
- ✰ Japan has the largest number (198), followed by the U.S.A. (165, not including "stacked" events), Canada (146), Mexico (131), South Korea (103), Australia (93), New Zealand (83), European Union (71 including approvals that have expired or under renewal process), Philippines (68), Taiwan (65), Colombia (59), China (55) and South Africa (52).

- Maize has the largest number (130 events in 27 countries), followed by cotton (49 events in 22 countries), potato (31 events in 10 countries), canola (30 events in 12 countries) and soybean (27 events in 26 countries).
- In November 2014, the USDA approved a GM potato that prevents bruising.
- In February 2015 Arctic Apples were approved by the USDA, becoming the first genetically modified apple approved for US sale. Gene silencing is used to reduce the expression of polyphenol oxidase (PPO), thus preventing enzymatic browning of the exposed fruit after it has been sliced open.

To Date GM Crops

Crop	*Traits*
Alfalfa	Tolerance of glyphosate orglufosinate
Apples	Delayed browning
Canola/Rapeseed	Tolerance of glyphosate or glufosinate High laurate canola, https://en.wikipedia.org/wiki/Genetically_modified_crops - cite_note-128Oleic acid canola
Corn	Herbicides glyphosateglufosinate, and 2,4-D. Insect resistance. Added enzyme, alpha amylase, which converts starch into sugar to facilitate ethanol production.
Cotton (cottonseed oil)	Insect resistance
Eggplant	Insect resistance
Papaya (Hawaiian)	Resistance to the papaya ringspot virus.
Potato (food)	Resistance to Colorado beetleResistance to potato leaf roll virusand Potato virus YReduced acrylamide when fried and reduced bruising
Potato (starch)	Antibiotic resistance gene, used for selectionBetter starch production
Rice	Enriched with beta-carotene (a source of vitamin A)
Soybeans	Tolerance of glyphosate orglufosinateReduced saturated fats (high oleic acid);Kills susceptible insect pests
Squash (Zucchini/ Courgette)	Resistance to watermelon, cucumber and zucchini/courgette yellow mosaic viruses
Sugar beet	Tolerance of glyphosate, glufosinate
Sugarcane	Pesticide tolerance and High sucrose content.
Sweet peppers	Resistance to cucumber mosaic virus
Tomatoes	Suppression of the enzymepolygalacturonase (PG), retarding fruit softening after harvesting, while at the same time retaining both the natural colour and flavor of the fruit

Transgenic Research Status in India

Bt cotton was the first crop approved for commercial cultivation in India. The Bt cotton (Bollgard) was approved in 2002.

- In the year 2002, first transgenic crop planting of Bt cotton was cultivated in about 50,000 ha in India.

- ☆ Presently, about 11 million hectare (2013) area is estimated to be under approved *Bt* cotton hybrids.
- ☆ But in February 2010, the government imposed a moratorium on the commercial cultivation of Bt brinjal EE-I, the first biotech food crop, due to concerns aired by green activists.
- ☆ The institutions under the ICAR, DST, DBT and CSIR have established a large number of facilities where most advanced transgenic research are being done.

Transgenic Research on other Crops in India

Several public and private companies in India are engaged in doing research on some 17 crops for various transgenic traits.

Traits	*Crops*
Insect resistance	Blackgram, brinjal, cabbage, cauliflower, chickpea, cotton, maize, pigeon pea, potato, rice, tobacco, tomato and wheat
Disease resistance	Black gram, brinjal, chickpea, coffee, rice and tomato
Herbicide resistance	Blackgram, cotton, mustard and rice
Stress tolerance	Mustard and rice
Edible vaccines	Muskmelon and tomato
Fruit ripening	Banana, potato and tomato
Nutrition enhancement	Mustard and potato

Controversy behind the GM Crops

GM foods are controversial and the subject of protests, vandalism, referenda, legislation, court action and scientific disputes. The controversies involve consumers, biotechnology companies, governmental regulators, non-governmental organizations and scientists. The key areas are whether GM food should be labeled, the role of government regulators, the effect of GM crops on health and the environment, the effects of pesticide use and resistance, the impact on farmers, and their roles in feeding the world and energy production.

Broad scientific consensus states that currently marketed GM food poses no greater risk than conventionally produced food. No reports of ill effects have been documented in the human population from GM food. Although GMO labeling is required in many countries, the United States Food and Drug Administration does not require labeling, nor does it recognize a distinction between approved GMO and non- GMO foods. Advocacy groups such as Greenpeace and the World Wildlife Fund claim that risks related to GM food have not been adequately examined and managed, and have questioned the objectivity of regulatory authorities and scientific bodies.

Opposition

Like many rganizations around the world, SJM (Swadeshi Jagran Manch) and BKS (Bhartiya Kisan Sangh) also have raised many concerns. Their main concerns are food and environment safety. They have also raised questions as to whether *Bt* Cotton with one gene is a failure.

Some Concerns Raised by SJM and BKS

1. If GM crops are successful, why are two genes stacked in the new *Bt* Cotton?
2. If India's biosafety system is robust, how did Monsanto conduct illegal trials for its genetically modified Okra (ladyfinger)?
3. Can India produce its own genes so that the country is not dependent on MNCs for GM technology?
4. Why are GM crops not increasing yields?
5. Can there be a way by which farmers can keep GM seeds with them and use it again and again?
6. Are genetically modified foods safe to consume?
7. How will GM crops affect biodiversity?

India's Stand on GM Crops Now

"Since the BJP is in power at the Centre and in most states now, implementation won't be an issue as the government is technology friendly. Besides, the government is pumping more money into GM crops. They are also setting up universities and projects, which are indications that it accepts the technology and wants India to use it."

The pro-technology, Modi government is desperate to introduce genetically modified (GM) crops. It is now holding closed-door meetings with the right wing-affiliated groups – the Swadeshi Jagran Manch (SJM) and the Bhartiya Kisan Sangh (BKS) – to end their opposition to the crops.

"The aim is to dispel fears about GM crops, so that the country can join the US, China and Canada. GM crops are very important for India's agriculture growth," says a senior agriculture ministry official.

Explain in detail about the bio-safety guidelines and regulation of GM crops in India?

Biosafety or the prevention of potential risks from products derived from genetic engineering has emerged as an important scientific and political concern worldwide.

- ☆ The Cartagena protocol on biosafety, 2003, negotiated under the United Nations Convention on Biological Diversity (CBD) is the first legally binding international agreement that purports to safeguard biodiversity concerns by regulating GMOs intended for use in agriculture.

- The Cartagena protocol on Biosafety to the convention on Biological Diversity states that that the protocol aims to ensure an adequate level of protection in the field of the safe transfer, handling and use of living modified organisms resulting from modern biotechnology that may have adverse effect on the conservation and sustainable use of biological diversity, taking into account the risks to human health and specifically focusing on transboundary movements.
- In India, the Ministry of Environment and Forest (MoEF) promulgated in December, 1989 the rules and procedures for the manufacture, import, use, research and release of GMOs as well as products made from such organisms; this was done under the provisions of the Environment Protection Act, 1986 (EPA).
- In addition, The Indian Recombinant DNA safety Guidelines and Regulations have been prepared (initially in 1990; revised in 1994) by and are available on request from Recombinant DNA Advisory Committee (RDAC), Department of Biotechnology (DBT), New Delhi.
- The DBT implements the research and development experiments utilizing GMOs and recombinant DNA products, while MOEF implements the large scale commercial use of these.
- The Indian regulatory structure comprises of six committees to regulate GMOs.

1. *Recombinant DNA Advisory Committee (RDAC)*

- Monitors development in GMOs at the national and international levels.
- Gives recommendations on safety regulations of GMOs and its products.
- Prepared the first Indian Recombinant DNA Biosafety Guidelines in 1990.

2. *Institutional Biosafety Committee (IBSC)*

- All institutes that conduct research in rDNA and different aspects of biosafety are to be set up Institutional Biosafety Committees (IBSCs). And institutions are awarded official recognition by the department of biotechnology.
- Representatives from department of biotechnology and a medical officer are to oversee the safety aspect.
- All research using GMOs is to be inspected by an investigator who reports the status and result of experiments to the IBSC.
- The progress of such experiments is to be reported to the Review Committee on Genetic Manipulation of the DBT at least once in every six months.

3. *Review Committee on Genetic Manipulation (RCGM)*

- Constituted under the department of biotechnology with members from Indian Council of Medical Research, Indian Council of Agricultural Research, Council of Scientific and Industrial Research, *etc.*

☆ Monitors and regulates all safety aspects of research on genetic manipulation, including small experimental field trials.

☆ Clears import of genetic material meant for research purposes, including vectors used for producing or cloning genetically modified micro-organisms, animals and plants.

☆ Publishes manuals of guidelines for the regulatory process of genetically engineered organisms in research and applications.

☆ Approves applications for the generation of information on transgenic organisms and plants.

☆ A monitoring-cum-evaluation committee (MEC) comprising of scientists, agricultural experts, and other officials nominated by relevant ministries is formed under RCGM.

☆ The MEC conducts on the spot visits to field trial sites and advises the RCGM on the kind of initiatives to be taken, on the basis of their assessments.

☆ Assists the RCGM in conducting trials and collecting data. Collects information on comparative agronomic advantages of transgenic plants.

4. *Genetic Engineering Approval Committee (GEAC)*

☆ Constituted under the ministry of environment and forests, with members from DBT, Indian Council of Agricultural Research, Indian Council of Medical Research, Council of Scientific and Industrial Research, Directorate of Plant Protection, Central Pollution Control Board, *etc.*

☆ Authorises clearance for large scale field trials, industrial application, and the commercial cultivation of GMOs, mainly from the viewpoint of environmental safety.

☆ Can approve or prohibit GMOs used for import, export, transfer, manufacture, processing, use or sale of GMOs. Approved commercial cultivation of Bt cotton in March 2002.

5. *State Biotechnology Coordination Committee (SBCC)*

☆ Members include, chief secretary of the state government, secretaries, department of environment, health, agriculture, commerce, forests, public works, public health, chairman, state pollution control board, microbiologists and pathologists from the state.

☆ States that conduct research in GMO are to constitute a SBCCs for effective enforcement of biosafety regulations.

☆ Assists and coordinates with the GEAC. Can inspect sites, and take punitive actions when necessary.

☆ Coordinates with the central ministries and nominates state government officials as representatives to conduct field inspections on GMOs.

6. *District Level Committee (DLC)*

- Members include the district collector, factory inspector, representative of the pollution control board, chief medical officer, district agricultural officer, representative from the public health department, district microbiologists or pathologists, municipal corporation commissioner, *etc.*
- Nodal agency at the district level to assess damage from GMO release. Reports infringement of regulation to the SBCC or the GEAC.
- Monitors safety regulation and ensures smooth functioning and compliance with the rDNA guidelines and procedures.

The salient features of these guidelines and regulations are as follows (Ghosh, 1997).

1. Every organization involved in research and development using recombinant DNA technology is required to set up an Institutional Biosafety Committee (IBC), which has a DBT nominee.
   - IBC is the nodal point for interaction of the organization with the Government.
   - It is the competent authority to which experiments that are likely to have bid-hazard potential and genetic manipulation of plants under containment have to be reported.
   - It also authorizes interstate exchange of etiologic agents, diagnostic specimens, *etc.* IBC will provide half-yearly reports on the on-going projects to Review Committee for Genetic Manipulation (RCGM).
2. The DBT has a RCGM, which reviews all the approvals of ongoing projects on GMOs and several other issues related to recombinant DNA research and development. Based on the recommendations of RCGM, trial permits are issued by DBT.
3. Each state has, in addition, a State Biotechnology Coordination Committee, and a District Level Committee, which are also involved, along with RCGM, in the inspection and monitoring of the experiments at the field sites.
4. The MOEF has an inter-ministerial committee, called The Genetic Engineering Approval Committee (GEAC), which has subject specialists as members, and is the competent authority to decide on the large-scale use of GMOs.
   - As in the case of GEAC, state and district level committees also participate, along with GEAC, in the watch on the field applications of GMOs.
   - GEAC would also involve the user Ministries and the other regulatory authorities, *e.g.* Directorate General of Health Services in the case of drugs and pharmaceuticals, Ministry of Agriculture for transgenic plants and other organisms related to agriculture, *etc.*

5. The guidelines recognize four levels of risk in case of experiments with microorganisms, based on pathogenicity of the microorganisms, local prevalence of the concerned disease and of epidemic causing strains in India.
6. Experiments with microorganisms, plants and animals are grouped into the following three categories:
   i. Exempt category (for self-cloning experiments),
   ii. Category requiring intimation of initiation to competent authority (*e.g.* experiments involving non-pathogenic DNA vector systems), and
   iii. Category requiring review and approval by the competent authority (*e.g.* cloning of genes for toxins, antibiotic resistance, *etc.*).
7. Four different biosafety levels are recognised and containment facilities for each level are recommended for necessary safeguards.
8. Physical containment envisages to limit the spread of dangerous microorganisms by good laboratory practices, safety equipment and laboratory design and facilities.
9. Biological containment consists of the use of vectors and hosts in such a way so that it can limit the infectivity of vector to specific hosts and, control host-vector survival in the environment.
10. Experiments with plants are to be done under two types of special environment that may be achieved by using glass-house containment.
   i. Glass-house containment A is used when no plant pathogens are used, and where non-pathogenic vector systems and regeneration from single cells are used.
   ii. Glass-house containment B is recommended for experiments involving genetically manipulated plant pathogens, including plant viruses, and the growth of plants regenerated from cells transformed by genetically manipulated pathogen vector system, which still contain the pathogen.
11. Application for recognition of research facility to carry out genetic manipulation should be made to the Department of Environment before the commencement of work.
12. All products obtained using recombinant DNA technology shall be subject to the general regulations normally applicable for such products.
13. The controlled release of GMOs should be done under appropriate containment facilities to ensure safety and to prevent unwanted release in the environment.
14. Prerelease tests of GMOs in agriculture should include elucidation of requirements for vegetative growth and persistence and stability in small plots and experimental fields.

15. Planned field experiments with transgenic plants are permitted only after a step-wise (laboratory to growth chamber and greenhouse) evaluation, either in India or elsewhere, to generate data on the following:
    - ✰ Promoter sequence,
    - ✰ Target gene sequence,
    - ✰ Regulatory mechanism utilized in the expression cassette,
    - ✰ Diagram of the expression Cassette to describe fully the marker genes used,
    - ✰ Cell lines used for shuttling and amplification of the cassette and several other details of the procedure, and the consequences thereof, used for production of the transgenic plants.

In addition, the following is also insisted upon:

- ✰ Laboratory data to show that protein products of transgenes are safe to the environment and human beings.
- ✰ Isolation distance applicable to foundation seed of the crop be provided all round the field having transgenic crops.
- ✰ A few rows of the same crop as the transgenic one should be planted beyond the isolation distance to act as pollen trap.
- ✰ Non-transgenic plants should be grown within the isolation distance at 1 or 5 m intervals to determine the distance of pollen escape.
- ✰ All the vegetative plants and left-over seeds must be destroyed by burning after the experiments are concluded.
- ✰ After the experiment, the land may be left fallow and all plants that emerge must be destroyed.
- ✰ The experimental field may be visited by the company authorized personnel only, and all records of visits are to be maintained.
- ✰ Full account of transgenic seeds produced is to be kept and no part of this seed can be transacted or further propagated without authorization.

Define and explain in brief about the Bioprospecting?

Bioprospecting

Bioprospecting, also known as biodiversity prospecting is the exploration of biological material for commercially valuable genetic and biochemical properties.

- ✰ It involves mainly systemic search for and development of new sources of chemical compounds, genes, micro-organisms, macro-organisms and other valuable products from nature *i.e.* looking for ways to commercialize biodiversity.
- ✰ Biodiversity refers to all living things, including plants, animals, insects and marine life.

- ✰ When biodiversity or related knowledge is collected without permission from the owners of these resources and then patented, it is known as biopiracy.
- ✰ The main aim of bioprospecting is to find new resources and products from nature that can be used by humans.
- ✰ Bioprospecting plays a dominant role in discovering leads for drug development, since existing/known compounds for developing drugs for human use are limited.
- ✰ Bioprospecting can be devided into three phases: collecting, analysis and commercialization.
- ✰ The most significant change and the starting point of international regulation and legislation took place at the Rio Earth Summit in 1992, when participating countries discussed bioprospecting and signed the Convention of Biological Diversity (CBD).
- ✰ The CBD recognizes the need for conservation, sustainable use and equitable benefit sharing as cornerstones, and for the first time, acknowledged the sovereign rights of countries over their natural resources.

10

Biomolecules: Carbohydrates and Proteins

Define carbohydrates? And explain in detail about the structure, functions of carbohydrates?

Carbohydrates

Carbohydrates are polyhydroxy aldehydes or ketones or substances that yield such compounds on hydrolysis. Many but not all carbohydrates have the empirical formula $(CH_2O)_n$.

- ☆ Some carbohydrates also contain nitrogen, phosphorous or sulfur.
- ☆ So, the carbohydrates are defined as polyhydroxy aldehydes or ketones or their condensation products or derived products.
- ☆ The basic condensing bond is glycosidic bond.
- ☆ A sugar that bears an aldehyde group is called an aldose whereas a sugar with a ketone group is a ketone

Classification

Carbohydrates vary greatly in size ranging from smaller glyceraldehydes molecule to larger starch molecule. Large carbohydrate molecules are polymers of small molecules. The carbohydrates have been classified based on the degree of polymerization into:

1. *Monosaccharides*

These are the simple sugars which cannot be further broken down *i.e.* they cannot be further hydrolysed under normal conditions of hydrolysis.

- ☆ They are further classified based on the number of carbon atoms present in the monosaccharide into trioses (3C), tetroses (4C), pentoses (5C), hexoses (6C) and heptoses (7C).
- ☆ Based on the functional group, they are classified as aldoses and ketoses depending on whether they have aldehyde (CHO) or ketone (>C=O) as functional group.
- ☆ The smallest carbohydrates are trioseses, thus glyceraldehydes is a triose that has an aldehyde group (aldose). Thus it can also be called an aldotriose. Similarly, dihydroxy acetone is a ketotriose.

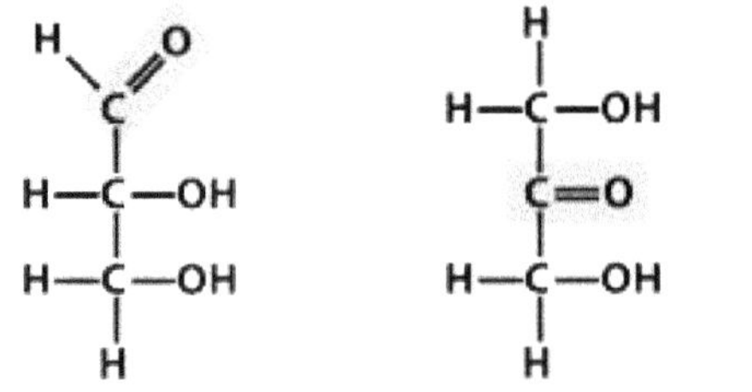

Glyceraldehyde, an aldotriose   Dihydroxyacetone, a ketotriose

Classification of Monosaccharides

Monosaccharides	*Aldoses*	*Ketoses*	*Occurrence*
Triose	D-Glycerose	Dihydroxyacetone	Intermediary metabolites in glucose metabolism
Tetrose	D-Erythrose	D-Erythrulose	–
Pentose	D-Ribose	D-Ribulose	Ribose is a constituent of nucleic acid
	L-Arabinose	–	Occurs in oligosaccharides
	D-Xylose	D-Xylulose	Gum arabic, cherry gums, wood gums, proteoglycans
Hexose	D-Glucose	D-Fructose	Fruit juices and cane sugar
	D-Galactose	–	Lactose, constituent of lipids
	D-Mannose	–	Plant mannosans and glycoproteins
Heptose	–	D- Sedoheptulose	Intermediate in carbohydrate metabolism

Isomerism

Isomerism is the phenomenon where by certain compounds, with the same molecular formula, exist in different forms owing to their different organizations of atoms.

i. **Stereoisomerism:** Stereoisomers are isomeric molecules that have the same molecular formula and differ only in how their atoms are arranged in the three dimensional orientation. And this phenomenon is called stereoisomerism.

- ☆ The number of stereoisomers possible is equal to 2^n where 'n' is number of asymmetric carbon atoms.
- ☆ Glyceraldehyde has a single asymmetric carbon atom and so two stereoisomers are possible, that is two forms of glyceraldehyde (denoted as D- and L- glyceraldehyde), which are mirror images of each other.

H–C=O
H—C—OH
CH_2OH

H–C=O
OH—C—H
CH_2OH

D-Glyceraldehyde L-Glyceraldehyde

- ☆ Monosaccharides belong to D or L series depending on the position of OH group on the penultimate carbon atom. If OH is towards right side of the penultimate carbon it is called as D sugar and if OH is towards left side of the penultimate carbon atom is called as L- sugar.
- ☆ In nature, D- sugars are more widely distributed than L- sugars.
- ☆ Enantiomers: D and L forms of sugars which are non super impossible mirror images of each other are called enantiomers. *e.g.* D and L threoses.

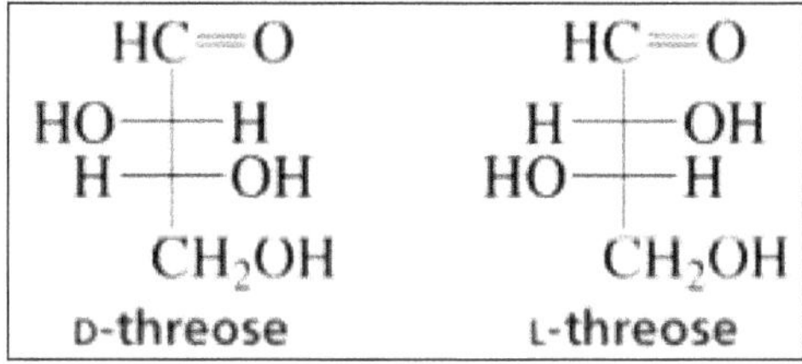

- ☆ Diastereoisomers: The stereoisomers which are not mirror images of each other are called diastereoisomers. *e.g.* D- glucose, D- mannose and D- galactose.
- ☆ Among the diastereoisomers, those which differ in configuration at a single carbon atom are called epimers. *e.g.* Mannose is an epimer of glucose at 2nd carbon atom whereas galactose is an epimer of glucose

D-Mannose (epimer at C-2)	D-Glucose	D-Galactose (epimer at C-4)
¹CHO	¹CHO	¹CHO
HO—²C—H	H—²C—OH	H—²C—OH
HO—³C—H	HO—³C—H	HO—³C—H
H—⁴C—OH	H—⁴C—OH	HO—⁴C—H
H—⁵C—OH	H—⁵C—OH	H—⁵C—OH
⁶CH₂OH	⁶CH₂OH	⁶CH₂OH

at 4th carbon atom whereas galactose and glucose bear no epimeric relationship.

ii. **Structural isomerism:** Compounds having same molecular formula but different structural formulae are called structural isomers. *e.g.* Glucose, galactose and mannose have same molecular formula but different structures.

iii. **Functional isomerism:** Compounds having the same molecular formulae but different functional groups. *e.g.* Glucose and fructose have same molecular formula but glucose is an aldose while fructose is a ketose.

iv. **Optical isomerism:** Compounds having same molecular and structural formula but differ in their behavior towards plane polarized light.

☆ An optical isomer rotating the plane of polarized light toward right is called dextrorotatory 'd' (+) while one rotating the plane of polarized light toward left is called levorotatory 'l' (-). *e.g.* D glucose and L glucose are optical isomers.

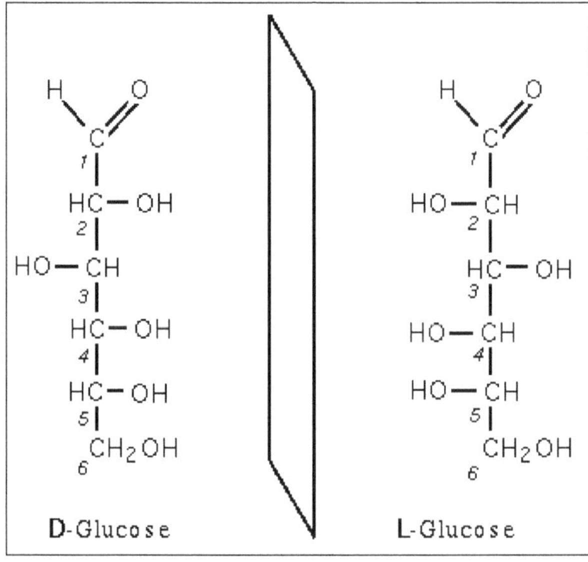

v. Ring – straight chain isomerism

- The aldehyde or ketone group of straight chain monosaccharide will react reversibly with a hydroxyl group on a different carbon atom to form a hemiacetyl or hemiketal, forming a heterocyclic ring with an oxygen bridge between two carbon atoms.
- Rings with five and six atoms are called furanose and pyranose forms, respectively, and exist in equilibrium with the straight chain form.
- During the conversion from straight chain from to the cyclic form, the carbon atom containing the carboxyl oxygen, called the anomeric carbon, becomes a stereoisomeric centre with two possible configurations.
- The oxygen atom may take a position either above or below the plane of the ring. The resulting possible pair of stereoisomers is called anomers.
  - In α anomer, the –OH substituent on the anomeric carbon rests on the opposite side (trans) of the ring from the CH_2OH side branch.
  - β anomer, the alternative form in which the CH_2OH substituent and the anomeric hydroxyl are on the same side (cis) of the plane of the ring.

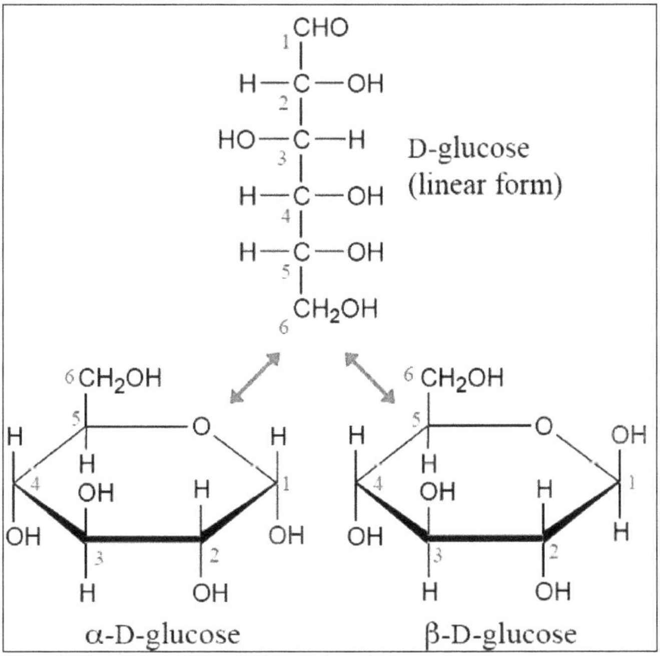
CHO
H—C—OH
HO—C—H
H—C—OH
H—C—OH
CH2OH
D-glucose
(linear form)
CH2OH
O
H
OH
α-D-glucose
β-D-glucose

2. *Oligosaccharides*

The monosaccharides condense with each other through glycosidic linkage to form oligosaccharides. Depending upon the number of monosaccharides presence they can be classified as

i. Disaccharides: The hydroxyl group on the anomeric carbon atom of one monosaccharide can react with the hydroxyl group of a second monosaccharide to form a disaccharide.

   a. Sucrose: Produced from alpha glucose beta fructose by α-1,2 glycosidic linkage. It is a non reducing sugar.

   b. Maltose: Consists of two units of glucose linked together by α-1,4 linkage. It is reducing sugar found in germination seeds largely.

   c. Lactose: Consists of one molecule of β- D glucose and one molecule of β-D galactose linked together by β-1,4 linkage.

   d. Cellobiose: Consists of two units of glucose but the bond involved is β-1,4 linkage. It is a reducing sugar.

ii. Trisaccharides: Three monosaccharide units condense with each other to form trisaccharides. Raffinose is formed by condensation of Galactose, Glucose and Fructose

iii. Tetrasaccharides: It consists of four monosaccharide units. Stachyose is tetrasaccharide consisting of one glucose, one fructose and two galactoses.

3. *Polysaccharides*

They are composed of several monosaccharide units. They are classified depending on the function, nature of branching and repeating unit.

i. Functional Classification

a. Structural polysaccharide: Polysaccharides belonging to this class help in maintaining the cell structure.

   e.g. Cellulose, Chitin, Hemicellulose and Pectin.

b. Storage polysaccharide: Polysaccharides belonging to this class help in storing carbohydrate material in the cell.

   e.g. Starch, Glycogen and Inulin.

ii. Nature of Branching

a. Linear: Polysaccharides belonging to this class have a linear glycosidic bonding only.

   e.g. Cellulose, Chitin and Amylose

b. Branched: Polysaccharides belonging to this class have a branched glycosidic bonding.

   e.g. Starch, Amylopectin and Glycogen.

iii. Repeating Unit

a. Homopolysaccharide: Polysaccharides belonging to this class contain the same basic repeating monosaccharide unit.

   e.g. Starch, Glycogen, Chitin and Inulin

b. Heteropolysaccharides: Polysaccharides belonging to this class contain more than one basic repeating unit.

   e.g. Hemicellulose and Pectin.

- ☆ Cellulose is an unbranched polysaccharide of glucose units linked by β-1,4 glycosidic linkage.
- ☆ Hemicellulose is a branched polysaccharide consists of glucose, xylose, mannose, galactose, rhamnose and arabinose units linked by xylose - β-1,4 – mannose - β-1,4 – glucose - α-1,3 – galactose glycosidic linkages.
- ☆ Starch molecule contains a mixture of two polysaccharides, amylase and amylopectin.
  - ❒ Amylase is unbranched polymer of glucose residues joined by α-1,4 linkage.
  - ❒ Amylopectin is the branched form in which most of the glucose residues are joined in α-1,4 linkages but additional α-1,6 bonds occur at every 25-30 residues creating the branch points.
- ☆ Glycogen molecule consists of glucose units which are linked in long chains by α-1,4 bonds. For every 10 units or so, the chain is branched by the formation of α-1,6 glycosidic bond.
- ☆ Chitin is a long chain polymer of a N- acetyl glucosamine, a derivative of glucose units linked by β-1,4 glycosidic linkage.
- ☆ Inulin is a heterogeneous collection of fructose polymers. It consists of chain-terminating glucosyl moieties and a repetitive fructosyl moiety, which are linked by β-(2,1) bonds.
- ☆ Pectins, also known as pectic polysaccharides, are rich in galacturonic acid. Homogalacturonans are linear chains of α-(1–4)-linked D-galacturonic acid

Conjugated Polysaccharides

Besides occurring in free state, the carbohydrates occur in nature in conjugation with other biomolecules like lipids and proteins to form glycolipids and glycoproteins. Mucopolysaccharides are glycoproteins characterized by the presence of amino sugars like glucosamine, galactosamine.

e.g. Hyaluronic acid and Heparin.

Functions of Carbohydrates

1. They are used as material for energy storage and production. *e.g.* Starch and glycogen, respectively in plants and animals, are stored carbohydrates from which glucose can be mobilized for energy production.
   - ☆ Glucose can supply energy both fueling ATP synthesis and in the form of reducing power as NADPH.
2. They exert a protein-saving action: If present in adequate amount in daily nourishment, the body does not utilize proteins for energy purpose, an

anti-economic and "polluting" fuel because it will need to eliminate nitrogen (ammonia) and sulfur present in some aminoacids.

3. Their presence is necessary for the normal lipid metabolism. More than 100 years ago Pasteur said: *"Fats burn in the fire of carbohydrates"*. This idea continues to receive confirmations from the recent scientific studies. Moreover, excess carbohydrates may be converted in fatty acids and triglycerides (processes that occur mostly in the liver).
4. Glucose is indispensable for the maintenance of the integrity of nervous tissue (some central nervous system areas are able to use only glucose for energy production) and red blood cells.
5. Two sugars, ribose and deoxyribose, are part of the bearing structure, respectively of the RNA and DNA and obviously find themselves in the nucleotide structure as well.
6. They take part in detoxifying processes. For example, at hepatic level glucuronic acid, synthesized from glucose, combines with endogenous substances, as hormones, bilirubin *etc.*, and exogenous substances, as chemical or bacterial toxins or drugs, making them atoxic, increasing their solubility and allowing their elimination.
7. They are also found linked to many proteins and lipids.
   - ✰ Within cells they act as signals that determine the metabolic fate or the intracellular localization of the molecules which are bound.
   - ✰ On the cellular surface their presence is necessary for identification processes between cells that are involved *e.g.* in the recognition between spermatozoon and oocyte during fertilization, in the return of lymphocytes in the lymph nodes of provenance or still in the leukocyte adhesion to the lips of the lesion of a blood vessel.
8. Two homopolysaccharides, cellulose (the most abundant polysaccharide in nature) and chitin (probably, next to cellulose, the second most abundant polysaccharide in nature), serve as structural elements, respectively, in plant cell walls and exoskeletons of nearly a million species of arthropods (*e.g.* insects, lobsters, and crabs).
9. Heteropolysaccharides provide extracellular support for organisms of all kingdoms: In bacteria, the rigid layer of the cell wall is composed in part of a heteropolysaccharide contained two alternating monosaccharide units while in animals the extracellular space is occupied by several types of heteropolysaccharides, which form a matrix with numerous functions, as hold individual cells together and provide protection, support, and shape to cells, tissues, and organs.

Define proteins? Explain in detail the different levels of protein structure?

Proteins are polymers of aminoacids linked together in straight chains to form polypeptides.

Proteins are traditionally looked upon as having four dfferent levels of structure. These levels are hierarchial, the protein being built up stage by stage with each level of structure.

Primary Structure

- ☆ The primary structure of the protein is formed by joining amino acids into a polypeptide.
- ☆ The amino acids are linked by peptide bonds that are formed by a condensation reaction between the carboxyl group of one amino acid and the amino group of a second amino acid.
- ☆ The direction of polypeptide can therefore be expressed as either N→C (left to right) or C→N (right to left).

$$ {}^{+}H_3N-\underset{R_1}{\overset{H}{C}}-COO^- + {}^{+}H_3N-\underset{R_2}{\overset{H}{C}}-COO^- \rightleftharpoons {}^{+}H_3N-\underset{R_1}{\overset{H}{C}}-\overset{O}{\overset{\|}{C}}-\underset{H}{N}-\underset{R_2}{\overset{H}{C}}-COO^- + H_2O $$

Peptide bond

Secondary Structure

- ☆ The secondary structure refers to the different conformations that can be taken up by the polypeptide.
- ☆ The two main types of secondary structure are the α- helix and β- pleated sheet.
- ☆ These are stabilized mainly by hydrogen bonds that form between different amino acids in the polypeptide.

α-*Helix*

- ☆ The α-helix is a common secondary structure encountered in proteins of the globular class.
- ☆ The formation of the α-helix is spontaneous and is stabilized by H-bonding between amide nitrogen and carbonyl carbons of peptide bonds spaced four residues apart.
- ☆ In this, the polypeptide backbone is tightly wound around an imaginary axis drawn longitudinally through the middle of the helix and R groups protrude outward from the helical back bone.
- ☆ Single turn of the helix has 5.4 Amino acids and it is called as the pitch of the helix.
- ☆ There are 3.6 amino acids residues per turn of the helix.
- ☆ Distance between the peptide bonds is 1.5 Å.
- ☆ The helix structure is maintained by the hydrogen bond formed between every first and the fourth amino acids of the helix

- The hydrogen bonds are intra molecular and are parallel to the central axis.
- There are three types of α- helices based on the direction and the nature *i.e.* left handed α- helix, right handed α- helix and triple helix

β-*pleated Sheet*

- Pauling and Corey also proposed a second ordered structure, the β- pleated sheet for polypeptide.
- This structure is a result of intermolecular hydrogen bonding between the polypeptide chains to form a sheet like arrangement.
- There are two ways in which proteins chains can form the pleated sheet structure.
- One is with the chains running in the same direction *i.e.* the -COOH or NH_2 ends of the polypeptide chains lying all at the top or all at the bottom of the sheet. This is called parallel pleated-sheet structure.
- In another type, known as antiparallel β- pleated sheet structure, the polypeptide chains alternate in such a way that the -COOH end of the one polypeptide is next to the -NH2 end of the other *i.e.* polypeptide chains run in opposite directions.

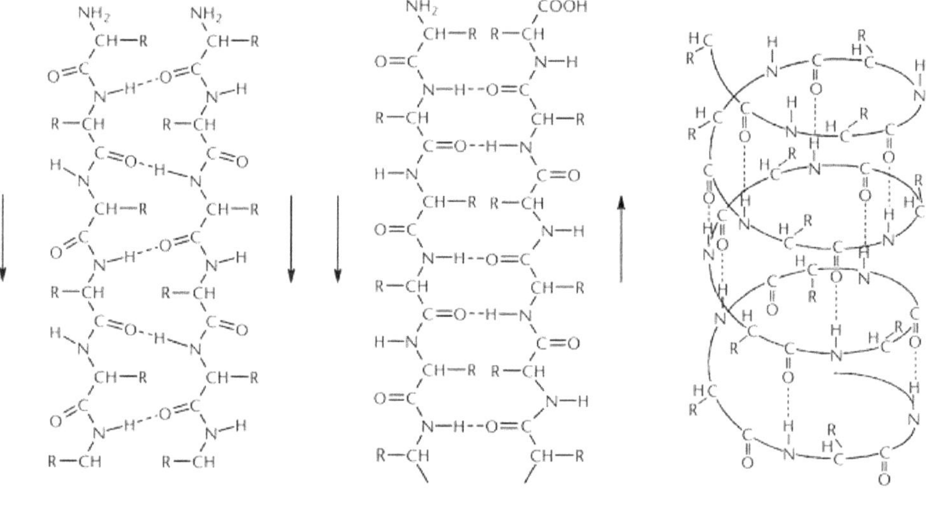

Parallel β pleated sheet

Antiparallel β pleated sheet

Right-handed α helix

Tertiary Structure

- The tertiary structure results from folding the secondary structural components of the polypeptide in to a three dimensional configuration.
- The tertiary structure is stabilized by various chemical forces *i.e.* hydrogen bonding between individual amino acids, electro static interaction between

the R groups of charged amino acids, and hydro phobic forces, which dictate that amino acids with non polar side-groups must be shielded from water by embedding within internal regions of the protein.

☆ There may also be covalent linkages called disulfide bridges between cysteine amino acid residues at various places in the polypeptide.

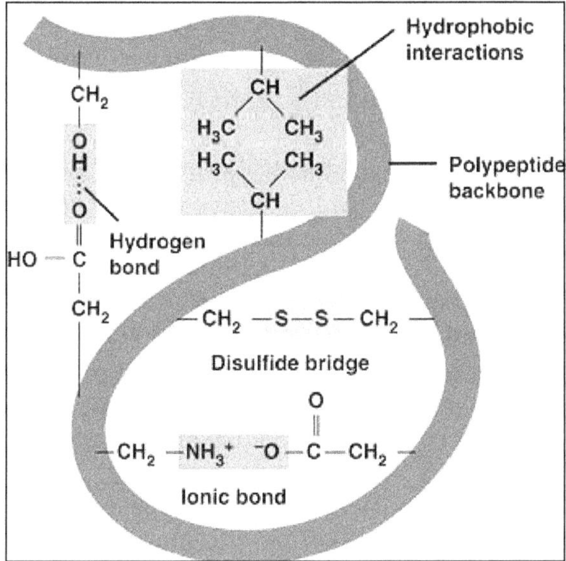

Quaternary Structure

☆ The quaternary structure involves the association of two or more polypeptides, each folded in to its tertiary structure, into a multi subunit protein.

☆ Proteins with multiple polypetide chains are oligomeric proteins. The structure formed by monomer-monomer interaction in an oligomeric protein is known as quaternary structure.

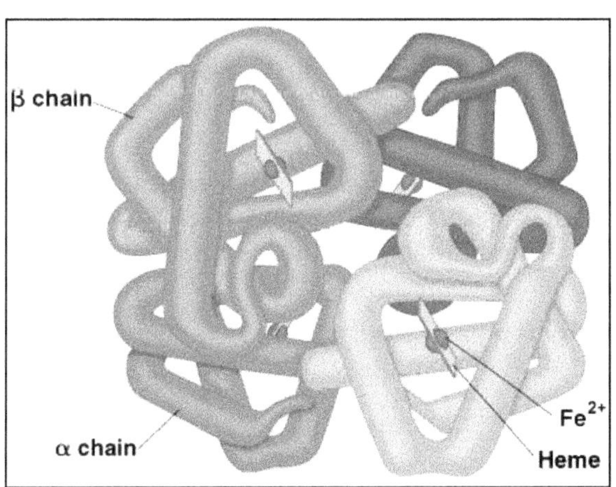

☆ *e.g.* Hemoglobin, the oxygen carrying protein of the blood, contains two α and two β subunits arranged with a quaternary structure in the form, $\alpha_2\beta_2$. Hemoglobin is, therefore, a heterooligomeric protein.

Explain in detail about the various types of proteins based on their functions?

1. **Contractile Proteins:** These proteins are responsible for the movement of muscles in the body. They are involved in the transport of nutrients in cells, the genetic make up, cell division, as well as muscular coordination.

   e.g. The proteins myosin and actin, together produce muscle contractions and relaxations.

2. **Defensive Proteins:** The antibodies produced by the body to fight diseases or prevent injury are called defensive proteins. Presence of an antigen or a foreign particle like bacteria, viruses, pollen or non-matching blood types, triggers the production of antibodies. It opposes the antigen and weakens it, so that it can be eradicated or destroyed by the white blood cells. Antibodies are also called immunoglobulins.

   e.g. Fibrinogen and thrombin are antibodies that facilitate blood clotting, and prevent the loss of blood following an injury. They also aid in the healing process, so that an individual recovers faster.

3. **Enzymatic Proteins:** Enzymes are the catalysts of biochemical reactions that occur in the body. They accelerate and alleviate these reactions, which otherwise may take years to complete. Thus, they increase the metabolic rate, and regulate various life processes like digestion, blood clotting, *etc.*

   e.g. The enzymes amylase and pepsin aid digestion by breaking down complex molecules like starch and proteins respectively, into simpler ones, so they can be absorbed by the small intestine.

4. **Hormonal Proteins:** Hormones are secretions that act as messengers to initiate or influence a function and coordinate certain metabolic processes in the body. These hormonal proteins help in regulating this action.

   e.g. In females, oxytocin is the hormone that stimulates contractions during childbirth. Insulin regulates glucose in the blood.

5. **Storage Proteins:** These proteins store amino acids and metal ions needed in the body. They also act as food reserves that provide energy as and when required by the body.

   e.g. The protein ferritin stores iron and controls the amount of iron present in the human body. Casein, found in milk, is another type of storage protein that provides certain amino acids, carbohydrates, calcium, and phosphorus.

6. **Structural Proteins:** These proteins help maintain structure and provide support to the human body. They give strength and protection to the human anatomy.

e.g. The protein collagen is the major component of tendons, cartilages, and bones. Hair and fingernails consist an insoluble protein called keratin.

7. **Transport Proteins:** These proteins help transport various molecules which include nutrients, gases, and all the essential chemicals that help maintain balance in the human body.

   e.g. Hemoglobin that carries oxygen to the lungs and various cells in the human body, and lipoproteins which help transport lipids or fats, are examples of transport proteins.

Other Functions of Proteins

- Proteins help regulate the fluid balance in the body, and control the movement of water and other fluids in the cells.
- They also release hydrogen ions to maintain the acid-base level in the body.
- Some proteins are receptor proteins, which act as binding sites for various enzymes, hormones, and nutrients. They regulate the flow of nutrients in the cells.

Explain in detail about structure, classification, properties and advantages of amino acids?

Amino Acid

A simple organic compound containing both a carboxyl (—COOH) and an amino (—NH_2) group.

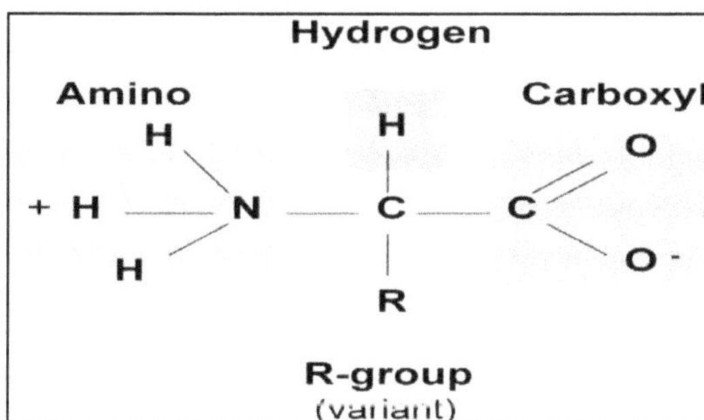

- Proteins (building block of life) represent the polymers of alpha-amino acids, which can exist in either dextro (D) or levo (L) form, also known as stereoisomers.
- Dextro and levo refer to the absolute confirmation of optically active components.
- Aside from glycine, all other amino acids can be described as mirror images not able to be superimposed.
- Major parts of them, which are found in nature, appear to be of the L-type. That's why eukaryotic proteins turn out to be always composed of levo

amino acids, despite the fact that dextro ones can be found in bacterial cell walls and a number of peptide antibiotics.

- Thus far, more than three hundred kinds of amino acids have been discovered in nature. However, only 20 of them are usually found as compounds of human peptides and proteins.
- Molecule of an amino acid contains carboxyl (COOH) and amino (NH_2) functional groups, and each molecule features a different side chain, also known as R group, which vary greatly in properties. Alanine is one of the standard amino acids.

Classifications

Experts classify amino acids based on lots of different features.

One of them is whether or not people can acquire them through the diet. According to this factor, scientists recognize 3 types: the nonessential, essential, and conditionally essential amino acids.

i. Those 8 called essential (or indispensable) can't be produced by the body and therefore should be supplied by food: Leucine, Isoleucine, Lysine, Threonine, Methionine, Phenylalanine, Valine, and Tryptophan.
ii. One more amino acid, Histidine, can be considered semi-essential, as the human body doesn't always need dietary sources of it. Meanwhile, conditionally essential amino acids aren't usually required in the human diet, but are able to become essential under some circumstances.
iii. Finally, nonessential ones are produced by the human body either out of the essential ones or from normal proteins breakdown. These include Asparagine, Alanine, Arginine, Aspartic acid, Cysteine, Glutamic acid, Glutamine, Proline, Glycine, Tyrosine, and Serine.

One more classification depends on the side chain structure, and experts recognize 5 types in this classification:

i. Containing sulfur (Cysteine and Methionine)
ii. Neutral (Asparagine, Serine, Threonine, and Glutamine)
iii. Acidic (Glutamic acid and Aspartic acid) and basic (Arginine and Lysine)
iv. Alphatic (these include Leucine, Isoleucine, Glycine, Valine, and Alanine)
v. Aromatic (these include Phenylalanine, Tryptophan, and Tyrosine)

Another classification based on structure of the side chain that divides the list of twenty into 4 groups, two of which are main groups and two are subgroups: non-polar, polar, acidic and polar, basic and polar.

i. Side chains having pure hydrocarbon alkyl or aromatic groups are considered non-polar, and their list includes Phenylalanine, Valine, Leucine, Alanine, Isoleucine, Proline, Methionine, and Tryptophan.

ii. If the side chain contains different polar groups like amides, acids, and alcohols, they are classified as polar. Their list includes Tyrosine, Glycine, Serine, Asparagine, Threonine, Glutamine, and Cysteine.

iii. Acidic-polar includes Aspartic Acid and Glutamic Acid *i.e.* if the side chain has a carboxylic acid (negatively charged R groups),

iv. And basic-polar includes Lysine, Arginine, and Histidine *i.e.* if the side chain contains an amino group (positively charged R groups).

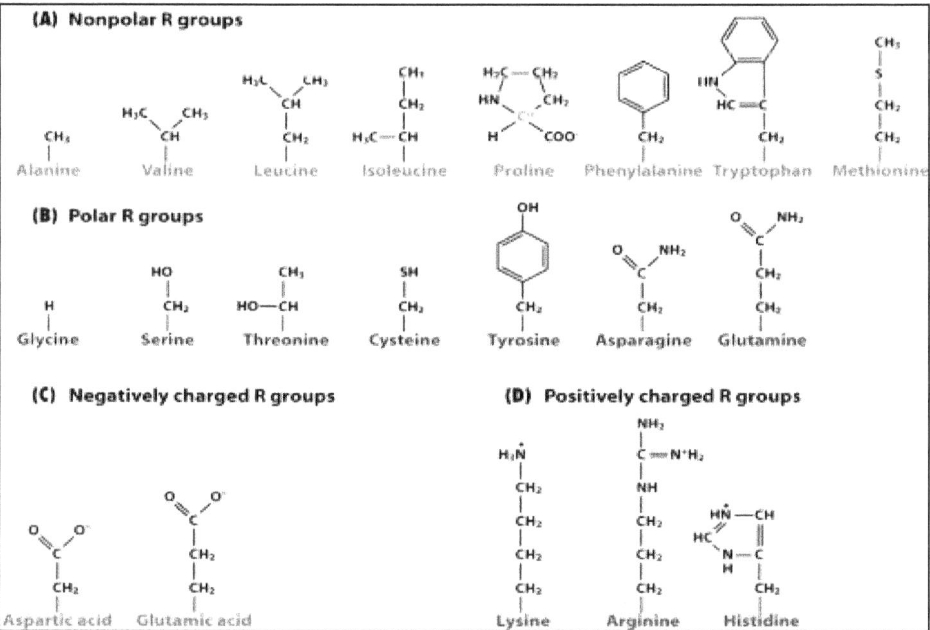

Advantages

- ☆ Aside from playing an important role in protein and enzyme synthesis, amino acids are considered very crucial for your good health, since they contribute considerably to the health of the human nervous system, hormone production, and muscular structure.
- ☆ In addition, they are needed for vital organs and cellular structure. If a person experiences low levels of the essential amino acids, this may cause hormonal imbalances, lack of concentration, irritability, and even depression.

Non-protein Functions

In humans, non-protein amino acids also have important roles as metabolic intermediates, such as in the biosynthesis of the neurotransmitter gamma-amino-butyric acid (GABA). Many amino acids are used to synthesize other molecules, for example:

- ☆ Tryptophan is a precursor of the neurotransmitter serotonin.
- ☆ Tyrosine (and its precursor phenylalanine) are precursors of the catecholamine neurotransmitters dopamine, epinephrine and norepinephrine.
- ☆ Glycine is a precursor of porphyrins such as heme.
- ☆ Arginine is a precursor of nitric oxide.
- ☆ Ornithine and S-adenosylmethionine are precursors of polyamines.
- ☆ Aspartate, glycine, and glutamine are precursors of nucleotides.
- ☆ Phenylalanine is a precursor of various phenylpropanoids, which are important in plant metabolism.
- ☆ Some non-standard amino acids are used as defenses against herbivores in plants. *e.g.* Canavanine is an analogue of arginine that is found in many legumes, and in particularly large amounts in *Canavalia gladiata* (sword bean). This amino acid protects the plants from predators such as insects and can cause illness in people if some types of legumes are eaten without processing.
- ☆ The non-protein amino acid Mimosine is found in other species of legume, in particular *Leucaena leucocephala*. This compound is an analogue of tyrosine and can poison animals that graze on these plants.

Enlist the metabolic precursor compounds for biosynthesis of amino acids? And outline the biosynthetic pathway of any two of the essential aminoacids?

Precursor	*Amino Acids*
α Ketoglutarate	Glutamine, Glutamic acid, Proline and Arginine
3 Phosphoglycerate	Serine, Glycine and Cysteine
Oxaloacetate	Aspartic acid, Asparagine, Metheonine, Threonine, Lysine and Isoleucine
Pyruvate	Alanine, Valine and Leucine
Phosphoenol pyruvate and Erythrose 4 phosphate	Tryptophan, Phenyl alanine and Tyrosine
Ribose 5 phosphate	Histidine

- ☆ Essential amino acids are Leucine, Isoleucine, Lysine, Threonine, Methionine, Phenylalanine, Valine, and Tryptophan.

Valine

Valine is produced by a four-enzyme pathway.

- ☆ It begins with the reaction of two pyruvate molecules catalyzed by Acetohydroxy acid synthase yielding α-acetolactate.
- ☆ Step two is the $NADPH^{+} + H^{+}$ - dependent reduction of α-acetolactate and migration of the methane groups to produce α, β-dihydroxyisovalerate.

This is catalyzed by Acetohydroxy isomeroreductase.

- The third reaction is the dehydration reaction of α, β-dihydroxyisovalerate catalyzed by Dihydroxy acid dehydrase resulting in α-ketoisovalerate.
- Finally, a transamination catalyzed either by an alanine-valine transaminase or a glutamate-valine transaminase results in valine.
- Valine performs feedback inhibition to inhibit the Acetohydroxy acid synthase used to combine the first two pyruvate molecules.

Leucine

The leucine synthesis pathway diverges from the valine pathway beginning with α-ketoisovalerate.

- α-Isopropylmalate synthase reacts with this substrate and Acetyl CoA to produce α-isopropylmalate.
- An isomerase then isomerizes α-isopropylmalate to β-isopropylmalate.
- The third step is the NAD^+-dependent oxidation of β-isopropylmalate via the action of a dehydrogenase to yield α-ketoisocaproate.
- Finally is the transamination via the action of a glutamate-leucine transaminase to result in leucine.
- Leucine, like valine, regulates the first step of its pathway by inhibiting the action of the α-Isopropylmalate synthase.
- Because leucine is synthesized by a diversion from the valine synthetic pathway, the feedback inhibition of valine on its pathway also can inhibit the synthesis of leucine.

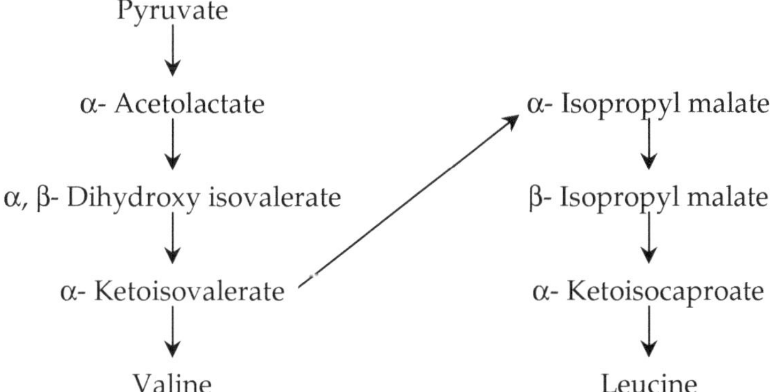

11

Biomolecules: Lipids and Nucleic Acids

Explain in detail about the classification, structure and functions of lipids?

Lipids are naturally occurring hydrophobic molecules. They are heterogenous group of compounds related to fatty acids. They include fats, oils, waxes, phospholipids, *etc.*

- ☆ Lipids are sparingly soluble in water and are soluble in organic solvents like chloroform, ether and benzene.

Structure of Lipids

Lipids has no single common structure. The most commonly occurring lipids are triglycerides and phospholipids.

- ☆ Triglycerides are fats and oils. Triglycerides have a glycerol backbone bonded to three fatty acids. If the three fatty acids are similar then the triglyceride is known as simple triglyceride. If the fatty acids are not similar then the fatty acids are known as mixed triglyceride.
- ☆ The second most common class of lipids is phospholipids. They are found in membranes of animal and plants. Phospholipids contains glycerol and fatty acids, they also contain phosphoric acids and a low-molecular weight alcohol. Common phospholipids are lecithins and cephalins.

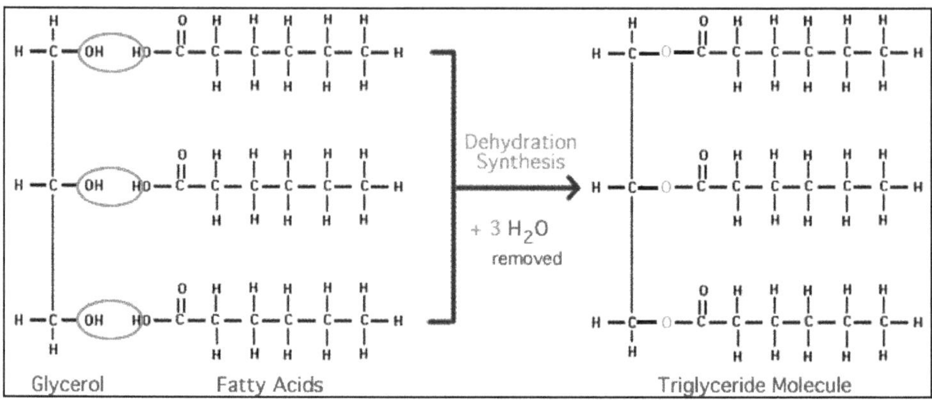

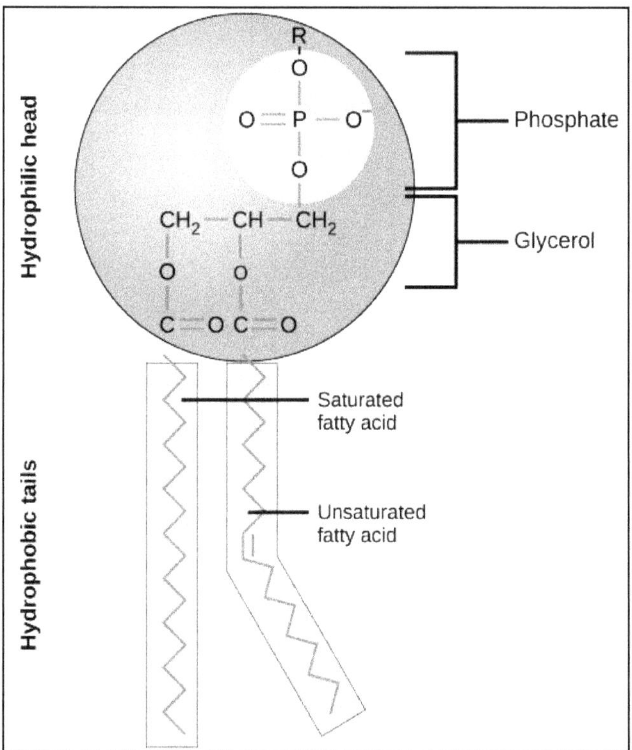

Classification of Lipids

1. *Fatty Acyls*

- Fatty acyls, a generic term are used for fatty acids and their derivatives.
- The fatty acid composed of one hydrocarbon chain which terminates with a carboxylic acid group.
- The carboxyl end is polar and soluble in water; hydrophilic.

- The long hydrocarbon chain is non-polar in nature and hydrophobic.
- The carbon chain in fatty acids can be saturated or unsaturated and can have some other functional groups.
- In case of unsaturated fatty acids, cis-trans isomers can be possible. Cis-fatty acids are found naturally while trans can be synthesized artificially by partial hydrogenation of fats and oils.

2. *Glycerolipids (triglycerides)*

- These are also called as neutral fats.
- They also known as triglycerides or triglycrols which composed of mono-, di- and tri-substituted glycerol.
- They formed by the esterification of glycerols with different fatty acids *i.e.* triglycerides have ester as functional group which comprised of one glycerol molecule (1,2,3 - trihydroxypropane) combined with three fatty acid molecules.
- During fat metabolism, glycerolipids releases glycerol and fatty acids from adipose tissue, thus they function as a food store comprise in the bulk as storage fat in animal tissues.
- Triglycerides molecules are considered for efficient energy storing as more energy could be stored in a pound of fat compared to a pound of carbohydrates like glycogen.
- Glycosylglycerols is a class of glycerolipids which composed of one or more sugar residues with glycerol through a glycosidic linkage and fatty acids.
- The plant membranes having Digalactosyldiacylglycerols and the mammalian sperm containing seminolipid are best example of glycosylglycerols.

3. *Glycerophospholipids*

- They also called as phospholipids, composed of fatty acids, glycerol with phosphate groups.
- Phospholidis are the main constituents of the lipid bilayer of cells and also are involved in metabolism.
- They also have one hydrophilic end with hydrophobic at another end and such type of molecules are known as amphipathic.
- In the aqueous solution of phospholipids; the hydrophobic end will tend to aggregate together away from water and the hydrophilic part that is polar phosphate group are oriented towards the molecules of water and tend to dissolve in it.

4. Sphingolipids

- ✩ These lipids are quite different from first two lipids and more complex compare to them.
- ✩ They composed of a sphingoid base backbone which is synthesized from serine; an amino acid and a fatty acyl CoA long-chain.
- ✩ This further converted into ceramides, glycosphingolipids, phosphosphingolipids and other compounds.
- ✩ Some common examples of sphingolipids are Ceramides, sphingomyelins (a phosphosphingolipids) are found in mammals.
- ✩ Whereas, ceramide phosphoethanolamines are found in insects and phytoceramide phosphoinositols, mannose-containing headgroups are found in fungi.
- ✩ Examples of glycosphingolipids are cerebrosides and gangliosides.

5. Sterol Lipids

- ✩ This class of lipids shows much similarity with triglycerides except the fact that they are hydrophobic in nature.
- ✩ These lipids are main components of membrane lipids along with sphingomyelins and the glycerophospholipids.
- ✩ The best common examples of sterol are cholesterol and its derivatives.
- ✩ Steroids are composed of fused four-ring core structure but show different biological roles such as hormones like estrogen, testosterone and androsterone.
- ✩ Some vitamins like vitamin-D also composed of one of the sterol; secosteroids.
- ✩ They are metabolic derivative of terpenes with tetracyclic skeleton of three fused six-member and one five member ring.
- ✩ Other examples of sterols are bile acids in mammals, phytosterols in plants like sitosterol, brassicasterol and stigmasterol.

6. Prenol Lipids

- ✩ These lipids are synthesized from the 5-carbon precursor dimethylallyl diphosphate and the isopentenyl diphosphate.
- ✩ The simple isoprenoids are classified according to the terpene units number.
- ✩ Carotenoids is a terpene act as antioxidants as well as precursors of vitamin A.
- ✩ However if there are more than 40 carbons present in lipids, they are called as polyterpenes.
- ✩ If an tail of isoprenoid gets attached to an quinonoid core of non-isoprenoid origin, they are exemplified by the quinones and hydro quinones.

7. Saccharolipids

- Saccharolipids describe compounds in which fatty acids are linked directly to a sugar backbone, forming structures that are compatible with membrane bilayers.
- In the saccharolipids, a monosaccharide substitutes for the glycerol backbone present in glycerolipids and glycerophospholipids.
- The most familiar saccharolipids are the acylated glucosamine precursors of the Lipid A component of thelipopolysaccharides in Gram-negative bacteria

8. Polyketides

- These lipids are synthesized by the polymerization method of acetyl subunit and propionyl subunit in the presence of enzymes which share mechanistic features with the fatty acid syntheses.
- Generally these molecules are cyclic in nature whose backbones are then further modified by hydroxylation, glycosylation, methylation, and oxidation.
- Polyketides are used as anti-microbial, anti-cancer and anti-parasitic agents.
- Some common examples of polyketides are tetracyclines, avermectins, erythromycins and anti-tumor epothilones.

Out of all lipid compounds; three types of lipids are more common.

- Triglycerides
- Phospholipids
- Steroids

Function of Lipids

- Lipids are storage compounds, triglycerides serve as reserve energy of the body.
- Lipids are important component of cell membranes structure in eukaryotic cells.
- Lipids regulate membrane permeability.
- They serve as source for fat soluble vitamins like A, D, E and K.
- They act as electrical insulators to the nerve fibres, where the myelin sheath contains lipids.
- Some lipids like prostaglandins and steroid hormones act as cellular metabolic regulators.
- Cholesterol is found in cell membranes, blood and bile of many organisms.
- As lipids are small molecules and are insoluble in water, they act as signaling molecules.

- ☆ Layers of fat in the sub cutaneous layer, provides insulation and protection from cold and Body temperature is maintained by brown fat.
- ☆ Polyunsaturated phospholipids are important constituents of phospholipids, they provides fluidity and flexibility to the cell membranes.
- ☆ Lipoproteins that are complexes of lipids and proteins occur in blood as plasma lipoprotein, they enable transport of lipids in aqueous environment, and their transport throughout the body.
- ☆ Cholesterol maintains fluidity of membranes by interacting with lipid complexes.
- ☆ Cholesterol is the precursor of bile acids, vitamin D and steroids.
- ☆ Essential fatty acids like linoleic and linolenic acids are precursors of many different types of ecosanoids including prostaglandins, thromboxanes. These play a important role

Answer the following questions.

1. **Define nucleic acids?**
2. **Explain in brief about the structure and types of DNA?**
3. **Explain in brief about the structure and types of RNA?**

Nucleic acids are biopolymers, or large biomolecules, essential for all known forms of life. Nucleic acids, which include DNA (deoxyribonucleic acid) and RNA (ribonucleicacid), are made from monomers known as nucleotides.

The salient features of double helix structure of DNA are:

1. The DNA molecule consists of two polynucleotide chains wound around each other in a right-handed double helix.
2. The two strands of a DNA molecule are oriented anti-parallel to each other *i.e.* the 5′ end of one strand is located with the 3′ end of the other strand at the same end of a DNA molecule.
3. Each polydeoxyribonucleotide strand is composed of many deoxyribonucleotides joined together by phosphodiester linkage between their sugar and phosphate residues.
4. The half steps of one strand extend to meet half steps of the other strand and the base pairs are called complementary base pairs. The adenine present in one stand of a DNA molecule is linked by two hydrogen bonds with the thymine located opposite to it in the second strand, and vice-versa.
5. Similarly, guanine located in one strand forms three hydrogen bonds with the cytosine present opposite to it in the second strand, and vice-versa *i.e.* complementary base pairing and these base pairs are called complementary base pairs or Watson and crick base pairs.

6. The pairing of one purine and one pyrimidine maintains the constant width of the DNA double helix. The bases are connected by hydrogen bonds.
7. Although the hydrogen bonds are weaker, the fact that so many of them occur along the length of DNA double helix provides a high degree of stability and rigidity to the molecule.
8. The diameter of this helix is 20°A, while its pitch (the length of helix required to complete one turn) is 34°A.
9. In each DNA strand, the bases occur at a regular interval of 3.4°A so that about 10 base pairs are present in one pitch of a DNA double helix.
10. The helix has two external grooves, a deep wide one, called major groove and a shallow narrow one, called minor groove. Both these groves are large enough to allow protein molecules to come in contact with the bases.

Main difference between A, B, C and Z form of DNA

Character	*A-DNA*	*B-DNA*	*C-DNA*	*Z-DNA*
Coiling	Right-handed	Right-handed	Right-handed	Left-handed
Pitch	28 Ao	34 Ao	31 Ao	60 Ao
Base pairs per turn	11	10	9.33	12
Diameter	26 Ao	20 Ao	19 Ao	18 Ao
Vertical rise per base pair	2.56 Ao	3.38 Ao	3.32 Ao	3.71 Ao
Sugar-phosphate back bone	Regular	Regular	Regular	Zig-zag
Major groove	Narrow and deep	Wide and deep	–	Flat
Minor groove	Wide and shallow	Narrow and deep	–	Narrow and deep

Structure of RNA

- ☆ RNA like DNA is a polynucleotide. RNA nucleotides have ribose sugar, which participate in the formation of sugar phosphate backbone of RNA.
- ☆ Thymine is absent and is replaced by Uracil.
- ☆ Usually RNA is a single stranded structure. Single stranded RNA is the genetic material in most plant viruses *e.g.* TMV.
- ☆ Double stranded RNA is also found to be the genetic material in some organisms. *e.g.* Plant wound and tumour viruses.
- ☆ RNA performs nongenetic function.

There are three main types or forms of RNA.

1. Messenger RNA (m-RNA)

- ☆ It constitutes about 5-10 per cent of the total cellular RNA.
- ☆ It is a single stranded base for base complimentary copy of one of the DNA strands of a gene.

☆ It provides the information for the amino acid sequence of the polypeptide specified by that gene.

☆ Generally a single prokaryotic mRNA molecule codes for more than one polypeptide; such a mRNA is known as polycistronic mRNA. All eukaryotic mRNAs are monocistronic *i.e.* coding for a single polypeptide specified by a single cistron.

2. *Ribosomal RNA (r – RNA)*

☆ Occur in association with proteins and is organized into special bodies of about 20°·A diameter called ribosomes.

☆ The size of ribosomes is expressed in terms of 'S' units, based on the rate of sedimentation in an ultracentrifuge.

☆ It constitutes about 80 per cent of the total cellular RNA.

☆ The function of rRNA is binding of mRNA and tRNA to ribosomes.

3. *Transfer RNA (t – RNA)*

☆ It is also known as soluble RNA(sRNA).

☆ It constitutes about 10-15 per cent of total RNA of the cell.

☆ It is a class of RNA which is of small size of 3S type and generally have 70 – 80 nucleotides. The longest t – RNA has 87 nucleotides.

☆ It's main function is to carry various types of amino acids and attach them to mRNA template for synthesis of protein.

☆ Each t – RNA species has a specific anticodon which base pairs with the appropriate m – RNA codon.

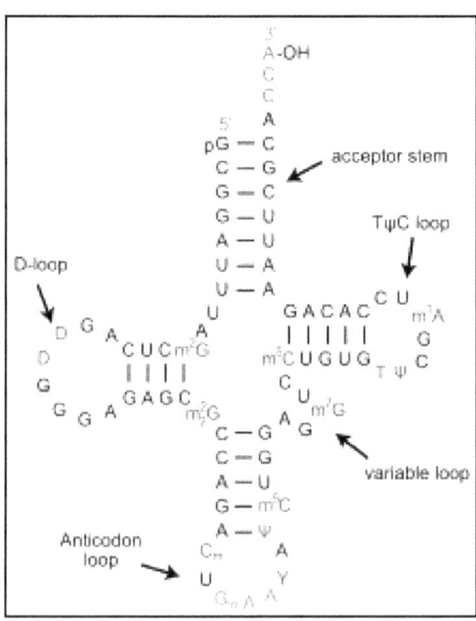

- ☆ The nucleotide sequence of the first tRNA (yeast alanyl tRNA) was determined by Robert Holley (1965), who proposed the clover leaf model of secondary structure of tRNA.
- ☆ Each tRNA is specific for each aminoacid.
- ☆ tRNA molecule contains the sequence of CCA at the 3'end, which is called amino acid attachment site. At the CCA end, it is joined to the single amino acid molecule for which that tRNA is specific. For example, a tRNA molecule specific for lysine cannot bind to the arginine.
- ☆ tRNA consists of three loops; (a) DHU-loop or D-loop (aminoacyl recognition region), (b) anticodon loop and (c) thymine loop (ribosome attachment region). Anticodon loop contains a short sequence of bases, which permits temporary complementary pairing with the codons of mRNA.

Write the difference between the following?

1. **Purines and Pyramidines**
2. **DNA and RNA**

Sl.No.	*Purines*	*Pyramidines*
1.	These are double ring (nine member) compounds.	These are single ring (six member) compounds.
2.	They are of two types, *viz.*, adenine and guanine.	They are of three types, *viz.*, cytosine, thymine and uracil.
3.	They occupy more space in DNA structure.	They occupy less space in DNA structure.
4.	Deoxyribose is linked at position 9 of purine.	Deoxyribose is linked at position 3 of pyrimidine.

Sl.No.	*Feature*	*DNA*	*RNA*
1.	Strands	Usually two, rarely one.	Usually one, rarely two.
2.	Sugar	Deoxyribose.	Ribose.
3.	Bases	Adenine, guanine, cytosine and thymine.	Adenine, guanine, cytosine and uracil
4.	Pairing	AT and GC	AU and GC
5.	Location	Mostly in chromosomes, some in mitochondria and chloroplasts.	In chromosomes and Ribosomes
6.	Replication	Self replicating	Formed from DNA. Self replication only in some viruses
7.	Size	Contains up to 4.3 million nucleotides	Contains up to 12,000 Nucleotides
8.	Function	Genetic role	Protein synthesis, genetic in some viruses

Sl.No.	*Feature*	*DNA*	*RNA*
9.	Origin	i. Replication of pre-existing DNA ii. In case of infection by RNA viruses, reverse transcription of genetic RNA	i. Genetic RNA either through transcription of DNA or ii. Through replication of RNA by RNA dependent RNA polymerase iii. Non-genetic RNA from transcription of DNA
10.	Native form	Double stranded DNA usually in B-form	Double stranded RNA usually in A-form
11.	Types	There are several forms of DNA like A, B, C and D.	Three types, *viz.*, mRNA, tRNA and rRNA

Answer the following questions.

1. **Define the replication of DNA?**
2. **Explain in brief about the semi conservative method of DNA replication?**
3. **How the replication of leading strand and lagging strand differ from each other?**
4. **Why the replication of lagging strand behind that of the leading strand?**
5. **Give the experimental proof of semi conservative method of DNA replication?**

The process by which a DNA molecule makes its identical copies is called DNA replication.

Modes of DNA Replication

There are three possible modes of DNA replication.

1. **Dispersive:** In dispersive mode of replication, the old DNA molecule would break into several pieces, each fragment would replicate and the old and new segments would recombine randomly to yield the progeny DNA molecule. Each progeny molecule would have both old and new segments along its length.
2. **Conservative:** According to conservative scheme, the two newly synthesized strands following the replication of a DNA molecule would associate to form one double helix, while the other two old strands would remain together as one double helix.
3. **Semi conservative:** In this model of DNA replication, each newly synthesized strand of DNA would remain associated with old strand against which it was synthesized. Thus, each progeny DNA molecule would consist of one old and one newly synthesized strand.

Method of Semi-conservative Replication

1. DNA replication was found to begin at various initiation points, called origin of replication, and proceed bi-directionally.

2. Two enzymes, DNA gyrase and DNA helicase induce unwinding of complementary strands of DNA.
3. Single-strand DNA binding (SSB) proteins bind to the single-stranded DNA, stabilizing it and preventing it from reannealing.
4. An enzyme, primase, initiates replication by synthesizing the primer.
5. DNA polymerases synthesize the complementary strand by progressively adding deoxyribonucleotides.
6. The DNA replication always proceeds in 5′ → 3′ direction.
7. During replication one strand of DNA can replicate continuously and the other strand discontinuously or in pieces. The continuously replicating strand is known as leading strand and the discontinuously replicating strand is known as lagging strand.
8. The replication of lagging strand generates small polynucleotide fragments called 'Okazaki fragments' which are later joined together by the enzyme, DNA ligase.

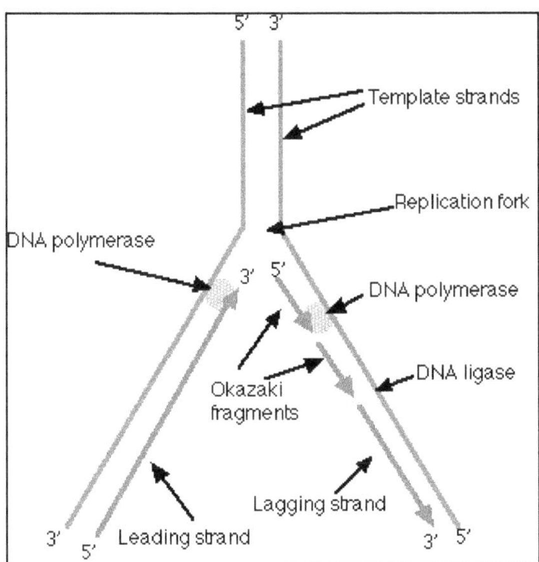

- ☆ In leading strand, DNA synthesized in the same direction as the growing replication fork on continuous basis while it is opposite to the direction of growing replication fork on discontinuous basis in lagging strand.
- ☆ And in leading strand, polymerase involved in DNA synthesis is DNA polymerase III in prokaryotes and polymerase in eukaryotes.
- ☆ But in lagging strand, enzyme primase reads the template DNA and initiates synthesis of a short complementary RNA primer. And DNA polymerase extends the primed segments, forming okazaki fragments. The DNA polymerase enzyme involved is DNA polymerase III in prokaryotes and polymerase in eukaryotes.

Sl.No.	Leading Strand	Lagging Strand
1.	It is replicated strand of DNA which grows continuously without any gap	It is replicated strand of DNA which is formed in short segments called okazaki fragments, its growth is discontinuous
2.	It does not require DNA ligase	DNA ligase is required for joining okazaki fragments
3.	The direction of growth of the leading strand is 5'-3'	The direction of growth of the lagging strand is 3'-5' through in each okazaki fragment it is 5'-3'
4.	Only a single RNA primer is required	Starting of each okazaki fragment requires a new RNA primer
5.	Formation of leading strand is quite rapid	Formation of lagging strand is slower
6.	Its template open in 3'-5' direction	Its template open in 5'-3' direction
7.	Formation of leading strand begins immediately at the beginning of replication	Formation of lagging strand begins a bit later than that of leading strand

- Later RNA primers are then removed by DNA polymerase I (with 5′-3′ exonuclease activity) in prokaryotes and polymerase in eukaryotes. And replaces the RNA nucleotides with DNA nucleotides and the fragments of DNA are joined together by DNA ligase.
- So the replication of lagging strand behind that of the leading strand.

Evidence for Semi Conservative Replication

The experiment conducted by Matthew Meselson and Franklin Stahl in 1958 on *E. coli* provided a conclusive proof that replication of DNA is by semi conservative model.

1. Meselson and Stahl labeled DNA of *E. coli* bacteria with heavy nitrogen *i.e.* ^{15}N by growing them on a medium containing ^{15}N for many generations to replace the normal nitrogen (^{14}N).
2. The density of normal and heavy nitrogen differs. The ^{14}N is lighter (1.710 g/cm) than ^{15}N (1.724 g/cm).
3. It is possible to detect such minute differences in density through density gradient centrifugation. Distinct bands are formed in the centrifuge tube for different density DNA.
4. DNA extracts of *E. coli* with ^{15}N gave a characteristic heavy band at one end of a tube that had been centrifuged at a high speed in an ultra-centrifuge.
5. These labeled cells were then grown on a normal unlabelled media containing ^{14}N for one generation.
6. DNA was again extracted and processed and it was found to consist of a hybrid DNA containing both ^{14}N and ^{15}N at the same time.
7. This indicated that the DNA had not replicated in two separate labeled and unlabelled forms.

8. The next generation of growth on unlabelled DNA was found to be in amounts equal to the partially labeled hybrid DNA.
9. Additional generation of growth on unlabelled media gave a relative increase in the amount of unlabelled DNA.
10. After two generations half the DNA was with intermediate density and half with light bands which further confirm semiconservative mode of DNA replication.
11. After third generation, ¾ DNA was found with ^{14}N and ¼ with hybrid nitrogen (^{14}N+^{15}N).
12. When the hybrid DNA was denatured by heating upto 100°C it was found to produce two separate single strands and in the ultracentrifuge density gradient, it was observed to form two separate bands; one band containing ^{15}N and the other ^{14}N. Thus it was concluded that DNA replication was by semi-conservative mode.

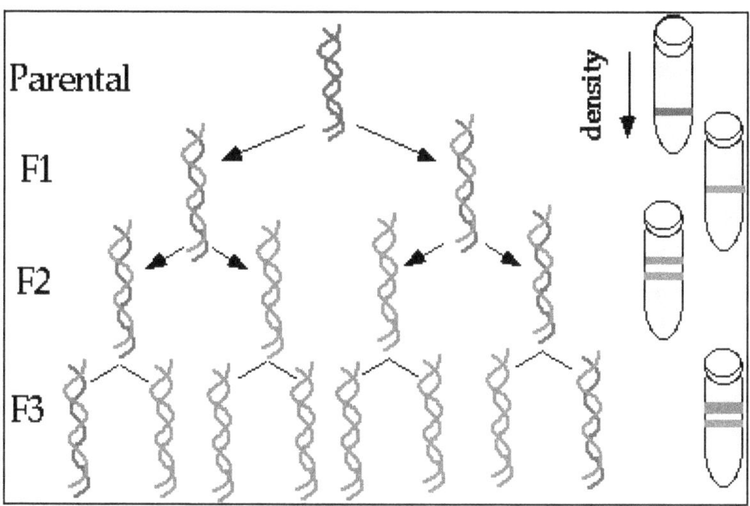

The major objection put forth for the semi-conservative replication was that the DNA molecule must unwind a number of times (1/10 of the total number of nucleotides) which cannot be accomplished without breaking within a short span of time say two minutes.

The presence of replication forks in *E. Coli* chromosomes first shown by J. Cairns in 1963 using the technique of autoradiography. He grew *E. Coli* cells in the presence of ^{3}H- thymidine (tritiated thymidine) since thymine is ordinarily found in DNA only. The *E. Coli* chromosome is a double stranded circular chromosome. It was shown that the two circular component strands separate during replication with each strand duplicating individually producing a è shaped structures during replication. This indicates that unwinding and replication proceed simultaneously.

12

Biomolecules – Enzymology

Define enzymes? And write a brief note on naming and classification of enzymes?

Enzymes are protein catalysts that increase the velocity of chemical reaction and are not consumed during the reaction they catalyze.

Naming and Classification

Except for some of the originally studied enzymes such as pepsin, rennin, and trypsin, most enzyme names end in "ase".

- ✰ The International Union of Biochemistry (I.U.B.) initiated standards of enzyme nomenclature which recommend that enzyme names indicate both the substrate acted upon and the type of reaction catalyzed.
- ✰ Under this system, the enzyme uricase is called urate: O_2 oxidoreductase, while the enzyme glutamic oxaloacetic transaminase (GOT) is called L-aspartate: 2-oxoglutarate aminotransferase.

Enzymes can be classified by the kind of chemical reactio catalyzed.

I. Addition or removal of water
   a) Hydrolases - these include esterases, carbohydrases, nucleases, deaminases, amidases, and proteases.
   b) Hydrases such as fumarase, enolase, aconitase and carbonic anhydrase.

II. Transfer of electrons
   a) Oxidases
   b) Dehydrogenases

III. Transfer of a radical

   a) Transglycosidases - of monosaccharides

   b) Transphosphorylases and phosphomutases - of a phosphate group

   c) Transaminases - of amino group

   d) Transmethylases - of a methyl group

   e) Transacetylases - of an acetyl group

IV. Splitting or forming a C-C bond

   a) Desmolases

V. Changing geometry or structure of a molecule

a) Isomerases

VI. Joining two molecules through hydrolysis of pyrophosphate bond in ATP or other tri-phosphate

   a) Ligases

Enzymes can accelerate reactions in several ways, all of which lower the activation energy.

1. By stabilizing the transition state: Creating an environment with a charge distribution complementary to that of the transition state to lower its energy.
2. By providing an alternative reaction pathway: Temporarily reacting with the substrate, forming a covalent intermediate to provide a lower energy transition state.
3. By destabilising the substrate ground state: Distorting bound substrate(s) into their transition state form to reduce the energy required to reach the transition state.
4. By orienting the substrates into a productive arrangement to reduce the reaction entropy change. The contribution of this mechanism to catalysis is relatively small.

Enzymes may use several of these mechanisms simultaneously. For example, proteases such as trypsin perform covalent catalysis using a catalytic triad, stabilize charge build-up on the transition states using an oxy anion hole, complete hydrolysis using an oriented water substrate.

Write the characteristics and chemical nature of enzymes? And also explain how do enzymes catalyze the chemical reactions within a cell?

Characteristics of Enzymes

1. Enzymes are biocatalysts. Hence they are not destroyed during an enzyme action. In other words, they are regenerated at the end of the reaction.
2. They can only speed up the reaction but can't initiate the reaction. They speed up the reaction by 10^7 to 10^{14} times.

3. These reactions occur even without enzymes but at a very slow pace.
4. Enzymes do not alter the equilibrium constant of a reaction but alters the rate at which equilibrium is reached.
5. A non enzymatic reaction may take several years to reach equilibrium, while an enzymatic reaction takes a fraction of a second.
6. Enzymes are very specific. Their specificity is with regard to substrate and the reaction they catalyse.

Chemical Nature of Enzyme

J.B Sumner explained that enzymes are protein in nature.

- Many enzymes require the presence of other compounds such as cofactors, coenzymes and metal ions for their catalytic activity.
- This entire active complex is referred to as the holoenzyme; *i.e.*, apoenzyme (protein portion) plus the cofactor (coenzyme, prosthetic group or metal ion or activator).

  Apoenzyme + Cofactor = Holoenzyme
- Apoenzyme is a protein substance which is thermo labile.

Write short notes on the following

1. **Coenzymes and Cofactors**
2. **Iso enzymes**
3. **Regulatory enzymes**
4. **Allosteric enzymes**

Coenzymes

The non protein component of the enzyme is called as coenzyme.

- They are loosely bound to the enzyme or non covalently bound.
- Whatever changes occur on the substrate the opposite changes occur on the co-enzyme.
- Hence co enzyme are also called as co- substrates.
- *e.g.* Glucose is converted to glucose 6 phosphate which is catalysed by hexokinase in presence of co enzyme ATP. Here glucose is gaining a phosphate molecule and ATP is loosing a phosphate and there by forming ADP.

Cofactors

A large number of enzymes require an additional non-protein component to carry out their catalytic functions. Generally these non-protein components are called as cofactors.

- The cofactors may be either one or more inorganic ions such as Fe^{2+}, Mg^{2+}, Mn^{2+} and Zn^{2+} or a complex organic molecules called coenzymes.
- Some enzymes require both coenzyme and one or more metal ions for their activity.
- Metals are required as cofactors in approximately two thirds of all enzymes.
- Metalloenzymes contain a definite quantity of functional metal ion that is retained throughout.
- Metal-activated enzymes bind metals less tightly but require added metals.
- The distinction between metalloenzymes and metal activated enzymes thus rests on the affinity of a particular enzyme for its metal ion.
- The iron-sulfur enzymes are unique class of metalloenzymes in which the active centre consists of one or more clusters of sulfur-bridged iron chelates.

Isoenzymes

Enzymes which exist in multiple forms within a single species of organism or even in a single cell are called isoenzymes or isozymes.

- Multiple forms of enzymes (isoenzymes) occur which have similar catalytic activities but different structures.
- Different isoenzymes are often organ-specific and their determination may improve the specificity of enzyme tests.
- The heterogeneity of some isoenzymes is due to different protein subunits which are coded for by separate genes.
- Since they are coded by genes, they differ in amino acid composition and thus in their isoelectric pH values.
- *e.g.* Lactate dehydrogenase is an example for the isoenzymes which occur as five different forms in the tissues of the human and other vertebrates.

Regulatory Enzymes

A regulatory enzyme is an enzyme in a biochemical pathway which, through its responses to the presence of certain other biomolecules, regulates the pathway activity.

- The enzymes which catalyse chemical reactions again and again are called regulatory enzymes.
- Regulatory enzymes exist at high concentrations (low V_{max}) so their activity can be increased or decreased with changes in substrate concentrations.
- Regulatory enzymes frequently catalyze thermodynamically irreversible reactions, that is, a large negative free energy change greatly favors formation of a given metabolic product rather than the reverse reaction.

- Regulatory enzymes require an extra activation process and need to pass through some modifications in their 3D in order to become functional, for instance, catalyzing enzymes (regulatory enzymes).
- The regulation of the activation of these catalyzing enzymes is needed in order to regulate the whole reaction speed, so that it is possible to obtain the amount of product required at anytime, that makes regulatory enzymes have a biological importance.
- Therefore, regulatory enzymes, by its controlled activation and are of two types: allosteric enzymes and covalently modulated enzymes.

Allosteric Enzymes

These are enzymes that change their conformational ensemble upon binding of an effector, which results in an apparent change in binding affinity at a different ligand binding site.

- This "action at a distance" through binding of one ligand affecting the binding of another at a distinctly different site, is the essence of the allosteric concept.
- Based on modulation, they can be classified in two different groups:
  - Homotropic allosteric enzymes: Substrate and effector play a part in the modulation of the enzyme, which affects the enzyme catalytic activity.
  - Heterotropic allosteric enzymes: only the effector performs the role of modulation.

Write short notes on the following?

1. **Enzyme active site**
2. **Enzyme specificity**
3. **Measures of enzyme activity**
4. **Multi enzyme complexes**

Active Sites

- Enzyme molecules contain a special pocket or cleft called the active site.
- The active site contains amino acid side chains that create a three-dimensional surface complementary to the substrate.
- The active site binds the substrate, forming an enzyme-substrate (ES) complex. ES is converted to enzyme-product (EP), which subsequently dissociates to enzyme and product.
- The substrate is bound in the active site by multiple weak forces like electrostatic interactions, hydrogen bonds, van der waal bonds, hydrophobic interactions and in some cases by reversible covalent bonds.

Enzyme Specificity

The specificity of an enzyme is determined by the functional groups of the substrate, the functional groups of the enzyme, and the physical proximity of these functional groups.

☆ Two theories have been proposed to explain the specificity of enzyme action.

a. **Lock and key theory:** The enzyme active site is complementary in conformation to the substrate, so that enzyme and substrate recognize one another.

b. **Induced-fit theory.** The enzyme changes shape on binding substrate, so that the conformation of substrate and enzyme is only complementary after binding.

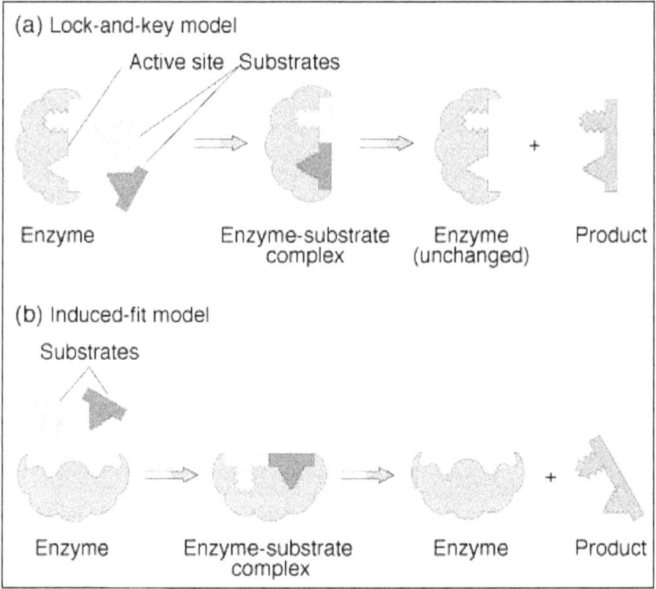

In general, there are four distinct types of specificity:

1. Absolute specificity: The enzyme will catalyze only one reaction by acting on a single type of substrate. *e.g.* Glucokinase which converts only glucose to glucose 6- phosphate.
2. Group specificity - The enzyme will act on a group of related molecules. *e.g.* Hexokinase which acts on all hexoses and converts to their hexose phosphates.
3. Linkage specificity - The enzyme will act on a particular type of chemical bond regardless of the rest of the molecular structure. *e.g.* Proteases acting on various proteins.

4. Stereochemical specificity - The enzyme will act on a particular steric or optical isomer. *e.g.* L-Amino acid oxidases acting only on L- Amino acids.

Measures of Enzyme Activity

Enzyme units: Amounts of enzymes can either be expressed as molar amounts, as with any other chemical, or measured in terms of activity, in enzyme units.

1. Turnover number, is the number of substrate molecules metabolized per enzyme molecule per unit time with units of min^{-1} or s^{-1}.
   - ☆ Enzyme activity is a measure of the quantity of active enzyme present and is thus dependent on conditions, which should be specified.
   - ☆ The SI unit is the katal, 1 katal = 1 mol s^{-1}, but this is an excessively large unit.
   - ☆ One katal of trypsin, for example is that amount of trypsin which breaks a mole of peptide bonds per second under specified conditions.
   - ☆ A more practical and commonly-used value is 1 enzyme unit (U) = 1 μmol min^{-1}. 1 U corresponds to 16.67 nanokatals.
2. Specific activity is usually expressed as μmol of substrate transformed to product per minute per milligram of enzyme under optimal conditions of measurement.
   - ☆ Specific activity gives a measurement of the purity of the enzyme, usually constant for a pure enzyme.
   - ☆ It is the amount of product formed by an enzyme in a given amount of time under given conditions per milligram of enzyme.
   - ☆ Specific activity is equal to the rate of reaction multiplied by the volume of reaction divided by the mass of enzyme.
   - ☆ The SI unit is katal kg^{-1}, but a more practical unit is μmol mg^{-1} min^{-1}

Multi Enzyme Complexes

- ☆ Multienzyme complexes are stable assemblies of more than one enzyme, generally involved in sequential catalytic transformations.
- ☆ A multienzyme complex is a protein possessing more than one catalytic domain contributed by distinct parts of a polypeptide chain ('domains'), or by distinct subunits, or both.
- ☆ *i.e.* a single enzyme is made up of one polypeptide chain, but when these polypeptide chain acts as a distinct part, and all act as a different enzyme having distinct catalytic domain, it is called multienzyme complex.
- ☆ These are distinct from a multienzyme polypeptide, in which multiple catalytic domains are found in a single polypeptide chain.
- ☆ Fatty-acyl-CoA Synthase is an example of six enzymes that assemble in a barrel shape to generate fatty acids.

What is meant by enzyme assay? Write its types and also explain the factors affecting enzyme assay?

Enzyme assays are laboratory methods for measuring enzymatic activity. They are vital for the study of enzyme kinetics and enzyme inhibition. *i.e.* the amount of enzyme protein can be determined (assayed) in terms of the catalytic effect it produces, that is the conversion of substrate to product.

Types of Enzyme Assay

Enzyme assays can be split into two groups according to their sampling method:

- Continuous assays, where the assay gives a continuous reading of activity, and
- Discontinuous assays, where samples are taken, the reaction stopped and then the concentration of substrates/products determined.

Continuous Assays

There are many different types of continuous assays.

1. *Spectrophotometric Method*

In spectrophotometric assays, an enzyme is most conveniently assayed by measuring the rate of appearance of product or the rate of disappearance of substrate.

- If the substrate absorbs the light at a specific wavelength, then changes in the concentration of these molecules can be measured by change in absorbance at this wavelength using a spectrophotometer.
- UV light is often used, since the common coenzymes NADH and NADPH absorb UV light in their reduced forms, but do not in their oxidized forms.
- An oxidoreductase using NADH as a substrate could therefore be assayed by following the decrease in UV absorbance at a wavelength of 340nm as it consumes the coenzyme.

2. *Fluorometric Method*

Fluorescence is when a molecule emits light of one wavelength after absorbing light of a different wavelength.

- Fluorometric assays use a difference in the fluorescence of substrate from product to measure the enzyme reaction.
- These assays are in general much more sensitive than spectrophotometric assays, but can suffer from interference caused by impurities and the instability of many fluorescent compounds when exposed to light.
- An example of these assays is again the use of the nucleotide coenzymes NADH and NADPH. Here, the reduced forms are fluorescent and the oxidised forms are non-fluorescent.
- Oxidation reactions can therefore be followed by a decrease in fluorescence and reduction reactions by an increase.

3. *Calorimetric Method*

Calorimetric is the measurement of the heat released or absorbed by chemical reactions.

- These assays are very general, since many reactions involve some change in heat and with use of a microcalorimeter, not much enzyme or substrate is required.
- These assays can be used to measure reactions that are impossible to assay in any other way.

4. *Chemiluminescent Method*

Chemiluminescent is the emission of light by a chemical reaction.

- Some enzyme reactions produce light and this can be measured to detect product formation.
- These types of assay can be extremely sensitive, since the light produced can be captured by photographic film over days or weeks, but can be hard to quantify, because not all the light released by a reaction will be detected.

Discontinuous Assays

Discontinuous assays are when samples are taken from an enzyme reaction at intervals and the amount of product production or substrate consumption is measured in these samples.

1. *Radiometric Method*

Radiometric assays measure the incorporation of radioactivity into substrates or its release from substrates.

- The radioactive isotopes most frequently used in these assays are ^{14}C, ^{32}P, ^{35}S and ^{125}I.
- Since radioactive isotopes can allow the specific labelling of a single atom of a substrate, these assays are both extremely sensitive and specific.
- They are frequently used in biochemistry and are often the only way of measuring a specific reaction in crude extracts.
- Radioactivity is usually measured in these procedures using a scintillation counter. This measures the ionizing radiation.

2. *Chromatographic Method*

Chromatographic assays measure product formation by separating the reaction mixture into its components by chromatography.

- This is usually done by high-performance liquid chromatography (HPLC), but can also use the simpler technique of thin layer chromatography.

- Although this approach can need a lot of material, its sensitivity can be increased by labelling the substrates/products with a radioactive or fluorescent tag.

Factors to Control in Assays

1. Salt Concentration

Most enzymes cannot tolerate extremely high salt concentrations.

- Typical enzymes are active in salt concentrations of 1-500 mM. As usual there are exceptions such as the halophilic (salt loving) algae and bacteria.

2. Effects of Temperature

All enzymes work within a range of temperature specific to the organism.

- Increases in temperature generally lead to increases in reaction rates.
- There is a limit to the increase because higher temperatures lead to a sharp decrease in reaction rates.
- This is due to the denaturating (alteration) of protein structure resulting from the breakdown of the weak ionic and hydrogen bonding that stabilize the three dimensional structure of the enzyme active site.
- The "optimum" temperature for human enzymes is usually between 35 and 40°C.
- The average temperature for humans is 37°C.
- Human enzymes start to denature quickly at temperatures above 40°C. Enzymes from thermophilic archaea found in the hot springs are stable up to 100°C.

3. Effects of pH

Most enzymes are sensitive to pH and have specific ranges of activity.

- All have an optimum pH. The pH can stop enzyme activity by denaturating (altering) the three dimensional shape of the enzyme by breaking ionic, and hydrogen bonds.
- Most enzymes function between a pH of 6 and 8; however pepsin in the stomach works best at a pH of 2 and trypsin at a pH of 8.

4. Substrate Saturation

Increasing the substrate concentration increases the rate of reaction (enzyme activity).

- However, enzyme saturation limits reaction rates. An enzyme is saturated when the active sites of all the molecules are occupied most of the time.
- At the saturation point, the reaction will not speed up, no matter how much additional substrate is added.

5. *Level of Crowding*

Large amounts of macromolecules in a solution will alter the rates and equilibrium constants of enzyme reactions, through an effect called macromolecular crowding.

Explain in detail about the regulation of enzyme activity?

Regulation of Enzyme Activity

Regulation of enzyme activity is important to coordinate the different metabolic processes.

It is also important for homeostasis *i.e.* to maintain the internal environment of the organism constant. Regulation of enzyme activity can be achieved by two general mechanisms:

1. **Control of enzyme quantity:** Enzyme quantity is affected by:
   i. Altering the rate of enzyme synthesis and degradation,
   ii. Induction
   iii. Repression
2. **Altering the catalytic efficiency of the enzyme by:** Catalytic efficiency of enzymes is affected by:
   i. Allosteric regulation
   ii. Feedback inhibition
   iii. Proenzyme (zymogen)
   iv. Covalent modification
   v. Protein – Protein interaction

1. *Control of Enzyme Quantity*

Control of the rates of enzyme synthesis and degradation: As enzymes are protein in nature, they are synthesized from amino acids under gene control and degraded again to amino acids after doing its work. Enzyme quantity depends on the rate of enzyme synthesis and the rate of its degradation.

- ☆ **Increased enzyme quantity** may be due to an increase in the rate of synthesis, a decrease in the rate of degradation or both.
- ☆ **Decreased enzyme quantity** may be due to a decrease in the rate of synthesis, an increase in the rate of degradation or both.
- ☆ For example, the quantity of liver arginase enzyme increases after protein rich meal due to an increase in the rate of its synthesis; also it increases in starved animals due to a decrease in the rate of its degradation.

Induction

- ☆ Induction means an increase in the rate of enzyme synthesis by substances called inducers.

☆ According to the response to inducers, enzymes are classified into:

a. **Constitutive enzymes,** the concentration of these enzymes does not depend on inducers.

b. **Inducible enzymes**, the concentration of these enzymes depends on the presence of inducers

c. For example, induction of lactase enzyme in bacteria grown on glucose media.

Repression

☆ Repression means a decrease in the rate of enzyme synthesis by substances called repressors.

☆ Repressors are low molecular weight substances that decrease the rate of enzyme synthesis at the level of gene expression.

☆ Repressors are usually end products of biosynthetic reaction, so repression is sometimes called feedback regulation.

☆ For example, dietary cholesterol decreases the rate of synthesis of HMG CoA reductase (β-hydroxy, β-methyl, glutaryl, CoA reductase), which is a key enzyme in cholesterol biosynthesis.

Derepression

Following removal of the repressor or its exhaustion, enzyme synthesis retains its normal rate.

Concentration of Substrates, Coenzymes and Metal Ion Activator

The susceptibility of the enzyme to degradation depends on its conformation. Presence of substrate, coenzyme or metal ion activator causes changes in the enzyme conformation decreasing its rate of degradation.

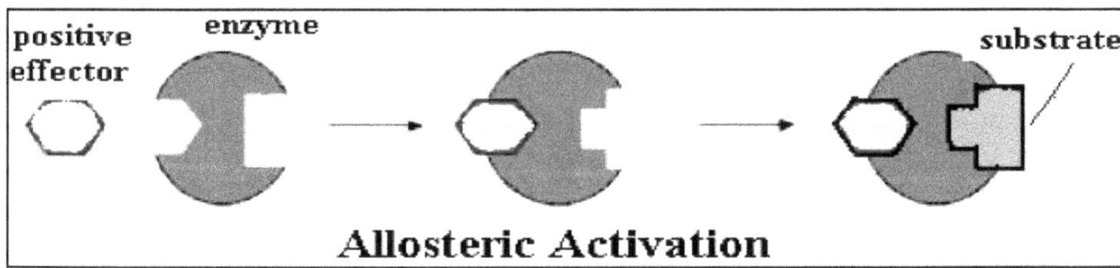

2. *Control of Catalytic Efficiency of Enzymes*

Allosteric Regulation

Allosteric enzyme is formed of more than one protein subunit. It has two sites; a catalytic site for substrate binding and another site (allosteric site), that is the regulatory site, to which an effector binds.

☆ Allosteric means another site, If binding of the effector to the enzyme increases it activity, it is called positive effector or allosteric activator *e.g.* ADP is allosteric activator for phosphofructokinase enzyme.

- ☆ If binding of the effector to the enzyme causes a decrease in its activity, it is called negative effector or allosteric inhibitor
- ☆ *e.g.* ATP and citrate are allosteric inhibitors for phosphofructokinase enzyme and Glucose-6-phosphate is allosteric inhibitor for hexokinase enzyme.

Mechanism of Allosteric Regulation

Binding of the allosteric effector to the regulatory site causes conformational changes in the catalytic site, which becomes more fit for substrate binding in positive effector (allosteric activator), and becomes unfit for substrate binding in negative effector (allosteric inhibitor).

Kinetics of Allosteric Enzymes

- ☆ One of the common characteristics of an allosteric enzyme is that it shows a sigmoid plot when velocity is plotted against substrate concentration
- ☆ Allosteric enzymes generally do not follow the Michaelis-Menten equation.
- ☆ The Lineweaver-Burk plot is concave upward.

Feedback Inhibition

In biosynthetic pathways, an end product may directly inhibit an enzyme early in the pathway.

- ☆ Such enzyme catalyzes the early functionally irreversible step specific to a particular biosynthetic pathway.
- ☆ Feedback inhibition may occur by simple feedback loop or by multiple feedback inhibition loops.

Proenzymes (Zymogens)

- ☆ Some enzymes are secreted in inactive forms called proenzymes or zymogens.
- ☆ Examples for zymogens include pepsinogen, trypsinogen, chymotrypsinogen, prothrombin and clotting factors.
- ☆ Zymogen is inactive because it contains an additional polypeptide chain that masks (blocks) the active site of the enzyme.
- ☆ Activation of zymogen occurs by removal of the polypeptide chain that masks the active site.

Biological Importance of Zymogens

- ☆ Some enzymes are secreted in zymogen form to protect the tissues of origin from auto digestion.
- ☆ Another biological importance of zymogens is to insure rapid mobilization of enzyme activity at the time of needs in response to physiological demands.

Protein-protein Interaction

- In enzymes that are formed from of many protein subunits, the enzyme may be present in an inactive form through interaction between its protein subunits.
- The whole enzyme, formed of regulatory and catalytic subunits, is inactive.
- Activation of the enzyme occurs by separation of the catalytic subunits from the regulatory subunits.
- Protein kinase A enzyme is an example for regulation of enzyme activity through protein interaction.
- This enzyme is formed of 4 subunits, 2 regulatory (2R) and 2 catalytic (2C) subunits. The whole enzyme (2R2C) is inactive. cAMP (cyclic adenosine monophosphate) activates the enzyme by binding to the 2 regulatory (2R) subunits releasing the 2 catalytic (2C) subunits and hence activating the enzyme.

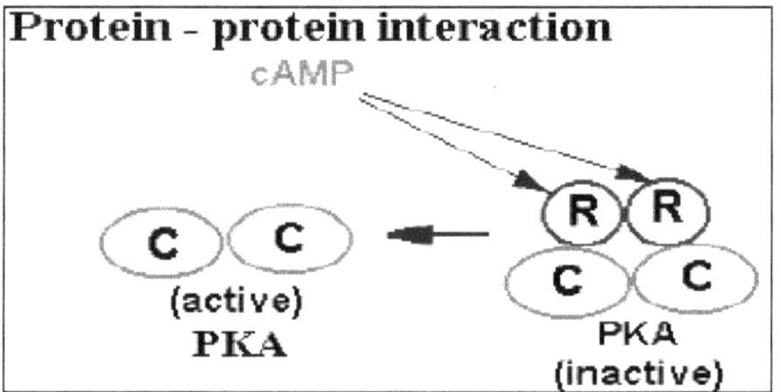

- Where PKA is protein kinase A enzyme, C is catalytic subunit and R is regulatory subunit.

Covalent Modification

It means modification of enzyme activity of many enzymes through formation of covalent bonds *e.g.*

a. Methylation (addition of methyl group).
b. Hydroxylation (addition of hydroxyl group).
c. Adenylation (addition of adenylic acid).
d. Phosphorylation (addition of phosphate group)
   - Phosphorylation is the most covalent modification used to regulate enzyme activity.
   - Phosphorylation of the enzyme occurs by addition of phosphate group to the enzyme at the hydroxyl group of serine, threonine or tyrosine. This occurs by protein kinase enzyme.
   - Dephosphorylation of the enzyme occurs by removal of phosphate group from the hydroxyl group of serine, threonine or tyrosine. This occurs by phosphatase enzyme.

- ☆ The phosphorylated form is the active form in some enzymes, while the dephosphorylated form is the active form in other enzymes.
- ☆ Examples of enzymes activated by phosphorylation. These are usually enzymes of degradative (breakdown) reactions *e.g.*
  - i. Glycogen phosphorylase that breaks down glycogen into glucose.
  - ii. Citrate lyase, which breaks down citrate.
  - iii. Lipase that hydrolyzes triglyceride into glycerol and 3 fatty acids.
- ☆ Examples of enzymes inactivated by phosphorylation. These are usually enzymes of biosynthetic reactions *e.g.*
  - i. Glycogen Synthetase, which catalyzes biosynthesis of glycogen.
  - ii. Acetyl CoA carboxylase, an enzyme in fatty acid biosynthesis.
  - iii. HMG CoA reductase, an enzyme in cholesterol biosynthesis.

What is meant by immobilized enzyme? Explain the different ways of immobilization of enzymes?

An immobilized enzyme is an enzyme that is attached to an inert, insoluble material such as calcium alginate (produced by reacting a mixture of sodium alginate solution and enzyme solution with calcium chloride).

- ☆ This can provide increased resistance to changes in conditions such as pH or temperature.
- ☆ It also allows enzymes to be held in place throughout the reaction, following which they are easily separated from the products and may be used again - a far more efficient process and so is widely used in industry for enzyme catalysed reactions.

Immobilization of an Enzyme

There are four different ways by which one can immobilize an enzyme.

1. *Affinity-tag Binding*

Enzymes may be immobilized to a surface, *e.g.* in a porous material, using non-covalent or covalent Protein tags.

- ☆ This technology has been established for protein purification purposes, and has recently been applied for biocatalysis applications by EziG™ with the His-tag.
- ☆ This technique is the only one which can be regarded as generally applicable, and can be performed without prior enzyme purification with a pure preparation as the result.
- ☆ Porous glass and derivatives thereof are used, where the porous surface can be adapted in terms of hydrophobicity to suit the enzyme in question.

2. *Adsorption on Glass, Alginate Beads or Matrix*

Enzyme is attached to the outside of an inert material.

- ☆ In general, this method is the slowest among those listed here.
- ☆ As adsorption is not a chemical reaction, the active site of the immobilized enzyme may be blocked by the matrix or bead, greatly reducing the activity of the enzyme.

3. *Entrapment*

The enzyme is trapped in insoluble beads or microspheres, such as calcium alginate beads. However, this insoluble substance hinders the arrival of the substrate, and the exit of products.

4. *Cross-linkage*

Enzyme molecules are covalently bonded to each other to create a matrix consisting of almost only enzyme.

- ☆ The reaction ensures that the binding site does not cover the enzyme's active site, the activity of the enzyme is only affected by immobility.
- ☆ However, the inflexibility of the covalent bonds precludes the self-healing properties exhibited by chemoadsorbed self-assembled monolayers.
- ☆ Use of a spacer molecule like polyethylene glycol helps reduce the steric hindrance by the substrate in this case.

Explain the different properties and applications of immobilized enzymes?

Properties of Immobilized Enzymes

- ☆ The stability of the immobilized enzymes is based on the temperature and time.
- ☆ The activity of the enzyme is retained throughout series of cycles.
- ☆ Due to immobilization, the properties of enzymes will be altered such as catalytic activity with respect to the support matrix.
- ☆ The change in the enzyme properties in the immobilized enzyme is due to the enzyme and the substrate reacts in the microenvironment which is different from the enzyme substrate reaction in the bulk solution environment.
- ☆ The change is also due to the change in the three dimensional conformation of the protein when linked with the support matrix.
- ☆ These conformational changes are to a lesser extent and these changes occur in the limited enzyme systems.
- ☆ Enzyme immobilization improves the operational stability and is also due to the increased enzyme loading which causes the controlled diffusion.
- ☆ The stabilization as a result of, number of bonds formed between the enzyme and the support matrix.
- ☆ When the immobilized enzyme acts on the macromolecular substrates, active site of the enzymes does not able to access with the substrates, hence the enzyme loses its activity.

Applications of Immobilized Enzymes

1. Biomedical Application

- ☆ Immobilized enzymes are used for diagnosis and treatment of diseases in the medical field.
- ☆ The inborn metabolic deficiency can be overcome by replacing the encapsulated enzymes (i.e, enzymes encapsulated by erythrocytes) instead of waste metabolites, the RBC acts as a carrier for the exogenous enzyme drugs and the enzymes are biocompatable in nature, hence there is no immune response.
- ☆ The enzyme encapsulation through the electroporation is a easiest way of immobilization in the biomedical field and it is a reversible process for which enzyme can be regenerated.
- ☆ The enzymes when combined with the biomaterials it provides biological and functional systems. The biomaterials are used in tissue engineering application for repair of the defect.
- ☆ The advantage of the enzyme immobilization in biomedical is that the free enzymes are consumed by the cells and not active for prolonged use, hence the immobilized enzymes remains stable, to stimulate the growth and to repair the defect.

2. Food Industry Application

- ☆ In food industry, the purified enzymes are used but during the purification the enzymes will denature. Hence the immobilization technique makes the enzymes stable.
- ☆ The immobilized enzymes are used for the production of syrups.
- ☆ Immobilized beta-galactosidase used for lactose hydrolysis in whey for the production of bakers yeast.

3. Biodiesel Production

Biodiesel is monoalkyl esters of long chain fatty acids.

- ☆ Biodiesel is produced through triglycerides (vegetable oil, animal fat) with esterification of alcohol (methanol, ethanol) in the presence of the catalyst.
- ☆ The production of catalyst is a drawback of high energy requirements, recovery of glycerol and side reaction which may affect the pollution.
- ☆ Lipase catalyses the reaction with less energy requirements and mild conditions required.
- ☆ But the production of lipase is of high cost, hence the immobilization of lipase which results in repeated use and stability.

4. *Wastewater Treatment*

The increasing consumption of fresh water and water bodies are mixed up with polluted industrial waste water and the waste water treatments are necessary at present.

- ✰ The sources of dye effluents are textile industry, paper industry, leather industry and the effluents are rich in dye colourants.
- ✰ These effluents are threat to the environment and even in low concentration it is carcinogenic.
- ✰ Nowadays enzymes are used to degrade the dye stuffs. The enzymes used in the wastewater treatments are preoxidases, laccase, azo reductases.
- ✰ These enzymes due to harsh conditions like extreme temperature, low or high pH and high ionic strength may lose its activity; to overcome this problem immobilized enzymes are used.
- ✰ The Horse radish Peroxidases are entrapped in calcium alginate beads, this method is still in lab scale research.

5. *Textile Industry*

The enzymes derived from microbial origin are of great interest in textile industry.

- ✰ The enzymes such as cellulase, amylase, liccase, pectinase, cutinase etc and these are used for various textile applications such as scouring,biopolishing, desizing, denim finishing, treating wools *etc.*
- ✰ The textile industries now turned to enzyme process instead of using harsh chemical which affects the pollution and cause damage to the fabrics.
- ✰ The processing of fabrics with enzymes requires high temperatures and increased pH, the free enzymes does not able to withstand the extreme conditions.
- ✰ Hence, enzyme immobilization for this process able to withstand at extreme and able to maintains its activity for more than 5-6 cycles.

6. *Detergent Industry*

- ✰ The detergent industry also employs enzymes for removal of stains. The enzymes used in detergent industry are
  - ❒ Protease which is used to remove the stains of blood, egg, grass and human sweat.
  - ❒ Amylase used to remove the starch based stains like potatoes, gravies, chocolate.
  - ❒ Lipase used to remove the stains of oil and fats and also used to remove the stains in cuffs and collars.
  - ❒ Cellulase is used for cotton based fabrics in order to improve softening, colour brightening and to remove soil stains.

- When compared to synthetic detergents the biobased detergents have good cleaning property because the enzymes based detergents can be used in low quantity when compared to the synthetic detergent, it has increased biodegradability, does not affect the environment works well in low temperature, and these are the advantages of enzymes in detergent industry.
- *e.g.* Protease directly cannot be used for wool and silk garments, protease loses its stability in the presence of surfactants and oxidizing agents, hence protease is immobilized by covalently linking with Eudragit S-100 using carbodiimide coupling. The immobilized protease treated with wool for 72hrs with 100U at 40°C the free enzyme was degraded the wool but the immobilized enzyme retained 76 per cent of the

What is meant by enzyme kinetics? And explain in detail about Michaelis – Menten Kinetics along with significance of K_M and V_{max} Values?

Enzyme kinetics is the study of the chemical reactions that are catalysed by enzymes. In enzyme kinetics, the reaction rate is measured and the effects of varying the conditions of the reaction are investigated.

According to Michaelis – Menten model, the enzyme catalysis reaction is

$$E+S \underset{k_2}{\overset{k_1}{\rightleftharpoons}} ES \xrightarrow{k_3} E+P$$

- It describes that, when enzyme (E), combines with its substrate (S), it leads to formation of enzyme-substrate complex (ES). The ES complex can dissociate again to form E+S, or can proceed chemically to form E and the product P.
- The rate constants k_1, k_2 and k_3 describes the rates associates with each step of the catalytic process.

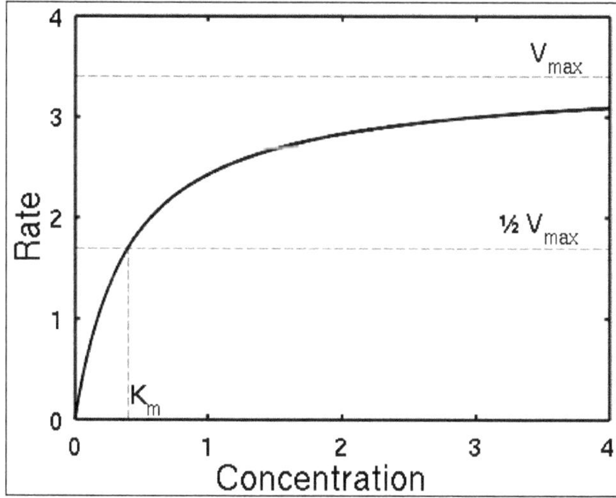

- At low substrate concentration [S], initial velocity (V_0) is directly proportional to substrate concentration [S], while at high substrate concentration [S], the velocity tends towards a maximum velocity (V_{max}). *i.e.* rate becomes independent of substrate concentration [S].
- The Michaelis – Menten equation is

$$V_0 = \frac{V_{max} \cdot [S]}{K_m + [S]}$$

  The equation describes a hyperbolic curve of V_0 against [S].
- Due to achievement of V_{max} at infinite substrate concentration, it is impossible to estimate V_{max} from a hyperbolic plot. But V_{max} and K_m can be determined experimentally by measuring V_0 at different substrate concentrations. It is called Lineweaver-Burk plot method. In this, plot of $1/V_0$ against $1/[S]$ is made.

  $1/V_0 = 1/V_{max} + K_m/Vmax. 1/[S]$

  It gives a straight line, with the intercept on the Y-axis equal to $1/V_{max}$, and the intercept on the X-axis equal to $-1/K_m$. The slope of the line is equal to K_m/V_{max}.

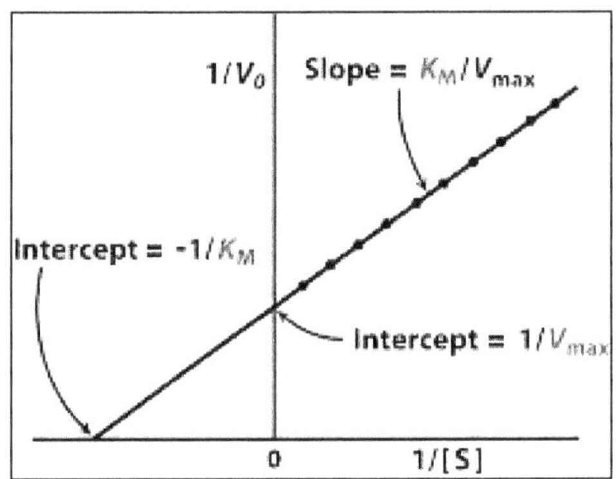

Significance of K_m and V_{max} Values

The Michaelis constant, K_m, has two meanings.

- First, K_m is the concentration of substrate at which half the active sites are filled. Thus, K_m provides a measure of the substrate concentration required for significant catalysis to occur. In fact, for many enzymes, experimental evidence suggests that K_m provides an approximation of substrate concentration in vivo. When the K_m is known, the fraction of sites filled, f_{ES}, at any substrate concentration can be calculated from

  $fES = V/V_{max} = [S]/[S] + K_m$
- Second, K_m is related to the rate constant of the individual steps in the catalytic scheme, *i.e.* K_m is equal to sum of the rates of breakdown of the enzyme-substrate complex over its rate of formation.

$K_m = (k_2 + k_3)/k_1$

And it also a measure of the affinity of an enzyme for its substrate. At high K_m, indicates weak substrate binding (k_2 predominant over k_1), a low km indicates strong substrate binding (k_1 predominant over k_2).

- ☆ K_m can be determined by fact that its value is equivalent to the substrate concentration at which the velocity equal to half of V_{max}.

The maximal rate, V_{max}, reveals the turnover number of an enzyme, which is the number of substrate molecules converted into product by an enzyme molecule in a unit time when the enzyme is fully saturated with substrate.

- ☆ It is equal to the kinetic constant k_3, which is also called $k_{cat.}$ The maximal rate, V_{max}, reveals the turnover number of an enzyme if the concentration of active sites $[E]_T$ is known, because

  $V_{max} = k_3[E]_T$

  and thus $k_3 = V_{max}/[E]_T$

Explain in brief about the following enzymes?

1. **RuBisCO**
2. **ATP synthase**
3. **PEP carboxylase**
4. **Nitrate reductase**
5. **Nitrogenase complex**
6. **Glutamine synthatase**

RuBisCO

Ribulose-1,5-bisphosphate carboxylase/oxygenase, is an enzyme involved in the first major step of carbon fixation, a process by which atmospheric carbon dioxide is converted by plants to energy-rich molecules such as glucose.

- ☆ The enzyme usually consists of two types of protein subunit, called the large chain and the small chain. The large-chain gene (rbcL) is part of the chloroplast DNA molecule in plants and small-chain genes in the nucleus of plant cells.
- ☆ The enzymatically active substrate (ribulose 1,5-bisphosphate) binding sites are located in the large chains that form dimers.
- ☆ A total of eight large-chains (4 dimers) and eight small chains assemble into a larger complex
- ☆ Magnesium ions (Mg^{2+}) are needed for enzymatic activity.
- ☆ During carbon fixation, the substrate molecules for RuBisCO are ribulose-1,5-bisphosphate and carbon dioxide. RuBisCO also catalyses a reaction between ribulose-1,5- bisphosphate and molecular oxygen (O_2) instead of carbon dioxide (CO_2).

- ☆ When carbon dioxide is the substrate, the product of the carboxylase reaction is two molecules of glycerate-3-phosphate.
- ☆ When molecular oxygen is the substrate, the products of the oxygenase reaction are phosphoglycolate and 3-phosphoglycerate. Phosphoglycolate is recycled through a sequence of reactions called photorespiration, which involves enzymes and cytochromes located in the mitochondria and peroxisomes.
- ☆ RuBisCO is usually only active during the day as ribulose 1,5-bisphosphate is not regenerated in the dark
- ☆ The rate of enzyme activity of RuBisCO is slow, being able to fix only 3-10 carbon dioxide molecules each second per molecule of enzyme.

ATP Synthase

ATP synthase is an important enzyme that provides energy for the cell to use through the synthesis of adenosine triphosphate (ATP).

- ☆ ATP synthase located within the thylakoid membrane and inner mitochondrial membrane.
- ☆ ATP synthase consists of 2 regions; the F_O portion is within the membrane. And the F_1 portion of the ATP synthase is above the membrane, inside the matrix of the mitochondria.
- ☆ ATP is the most commonly used "energy currency" of cells from most organisms. It is formed from adenosine diphosphate (ADP) and inorganic phosphate (P_i), and needs energy.
- ☆ The overall reaction sequence is: ADP + $P_i \rightarrow$ ATP, where ADP and P_i are joined together by ATP synthase.
- ☆ The mechanism of enzyme involves, a portion of the F_O (the ring of c-subunits) rotates as the protons pass through the membrane. The c-ring is tightly attached to the asymmetric central stalk (consisting primarily of the gamma subunit), which rotates within the alpha$_3$beta$_3$ of F_1 causing the 3 catalytic nucleotide binding sites to go through a series of conformational changes that leads to ATP synthesis.

PEP Carboxylase

Phosphoenolpyruvate carboxylase (PEP carboxylase, PEPCase, or PEPC) is an enzyme in the family of carboxy-lyases and catalyzes irreversible β-carboxylation of phospho*enol*pyruvate (PEP) in the presence of HCO_3^- and Mg^{2+} to yield oxaloacetate (OAA) and inorganic phosphate (Pi).

- ☆ This reaction is used for carbon fixation in CAM (crassulacean acid metabolism) and C4 organisms, as well as to regulate flux through the citric acid cycle.
- ☆ Phospho*enol*pyruvate carboxylase intrinsically located in the chloroplast and its expression was confined to mesophyll cells.

- PEPC is widely distributed in all photosynthetic organisms including vascular plants, algae, cyanobacteria, and photosynthetic bacteria, and also in nonphotosynthetic bacteria and protozoa.
- In vascular plants, the reaction catalyzed by PEPC is the primary fixation step of photosynthetic CO_2 assimilation in C_4 photosynthesis and crassulacean acid metabolism (CAM).
- In most nonphotosynthetic tissues and in the leaves of C_3 plants, the primary function of PEPC is anaplerotic, replenishing the tricarboxylic acid (TCA) cycle with intermediates that are withdrawn for a variety of biosynthetic pathways and nitrogen assimilation.
- PEPC also plays roles in the provision of malate in guard cells and legume root nodules.

Nitrate Reductase

Nitrate reductase (NR) is an iron-sulfur molybdenum flavoprotein which catalyses the reduction of nitrate to nitrite

- Nitrate reductase is a multi-domain enzyme comprising the prosthetic groups molybdopterin, Fe-heme, and FAD (flavin adenine dinucleotide) in a 1:1:1 stoichiometry that mediates an electron transfer from NAD(P) H to nitrate.
- The FAD and Mo-pterin domains function as the binding sites for NAD(P) H and NO_3^-, respectively, while the cytochrome b5-like heme domain facilitates the electron transfer from the FAD domain to the active-site Mo-pterin.
- NR serves plants, algae, and fungi as a central point for integration of metabolism by governing flux of reduced nitrogen by several regulatory mechanisms.

Nitrogenase Complex

Nitrogenases are a family of metalloenzymes that catalyse a key step in the global nitrogen cycle: the adenosine triphosphate (ATP)dependent reduction of dinitrogen (N_2) to ammonia (NH_3).

- It consists of two proteins: a reductase, which provides electrons with high reducing power, and nitrogenase, which uses these electrons to reduce N_2 to NH_3.
- The reductase consists of two identical subunits that contains one iron-sulfer cluster and two ATP binding sites.
- The nitrogenase consists of two α and two β subunits and contains an iron- molybdenum complex.
- The transfer of electrons from the reductase to the nitrogenase protein is coupled to the hydrolysis of ATP by the reductase.

- ☆ The reaction that this enzyme performs is:

  $N_2 + 8\,H^+ + 8\,e^- + 16\,ATP \longrightarrow 2\,NH_3 + H_2 + 16\,ADP + 16\,P_i$

- ☆ Nitrogenase has been found to catalyse the reduction of alternative substrates, such as H^+, N_3^-, CN^-, C_2H_2 and CO.

Glutamine Synthetase

Glutamine synthetase (GS) is an enzyme which catalyses the incorporation of ammonia in to glutamine, deriving energy from the hydrolysis of ATP.

Glutamate + ATP + NH_3 $\longrightarrow$ Glutamine + ADP + Phosphate

- ☆ This enzyme named as synthetase, rather than a synthase, because the reaction couples bond formation with the hydrolysis of ATP. In contrast, a synthase does not require ATP.

Terminology

Abiotic stress: The stress caused to plants due to herbicides, water deficiency, ozone, intense light *etc.*

Abscisic acid: A plant growth regulator involved in abscission, dormancy, stomatal opening/closure, and inhibition of seed germination. It also affects the regulation of somatic cell embryogenesis in some plant species.

Acetolactate synthase (ALS): An enzyme essential for amino acid production.

Acetyl co-enzyme A: acetyl CoA A compound formed in the mitochondria when an acetyl group (CH3CO-) – derived from breakdown of fats, proteins, or carbohydrates – combines with the thiol group (-SH) of co-enzyme A.

Activator: A substance or physical agent that stimulates transcription of a specific gene or operon or a compound that, by binding to an allosteric site on an enzyme, enables the active site of the enzyme to bind to the substrate. See gene expression.

Active site: The site on the enzyme at which the substrate binds.

Adaptor: A synthetic double-stranded oligonucleotide that has a blunt end, while the other end has a nucleotide extension that cans base pair with a cohesive end created by cleavage of a DNA molecule with a specific type II restriction endonuclease. The blunt end of the adaptor can be ligated to the ends of a target DNA molecule and the construct can be cloned into a vector by using the cohesive ends of the adaptor.

Adenine: A nitrogenous base, one member of the base pair A-T (adenine-thymine).

Adenosine diphosphate (ADP): High-energy phosphoric ester (nucleotide) of the nucleoside adenosine that functions as the principal energy-carrying compound in the living cell.

Adenosine triphosphate (ATP): High-energy phosphoric ester (nucleotide) of the nucleoside adenosine that functions as the principal energy-carrying compound in the living cell.

A-DNA: The dehydrated form of right-handed helical DNA obtained under nonphysiological conditions.

Aerobe: A microorganism dependent on oxygen for it's growth.

Affinity chromatography: A method for separating molecules by exploiting their ability to bind specifically to other molecules. There are several types of biological affinity chromatography. A biological molecule can be immobilized and a smaller molecule (ligand, q.v.,) to which it is to bind can be stuck to it, or the smaller ligand can be immobilized and the macromolecule stuck to it. A variant is to use an antibody as the immobilized molecule and use it to "capture" its antigen: this is often called immunoaffinity chromatography. A variation is pseudo-affinity chromatography, in which a compound which is like a biological ligand is immobilized on a solid material, and enzymes or other proteins are bound to it.

Agar: A polysaccharide solidifying agent used in nutrient media preparations and obtained from certain types of red algae (Rhodophyta). Both the type of agar and its concentration can affect the growth and appearance of cultured explants.

Agarose gel electrophoresis: A process in which a matrix composed of a highly purified form of agar is used to separate larger DNA and RNA molecules. See electrophoresis.

Agricultural Biotechnology: A range of tools, including traditional breeding techniques, that alter living organisms, or parts of organisms, to make or modify products; improve plants or animals; or develop microorganisms for specific agricultural uses. Modern biotechnology today includes the tools of genetic engineering.

Agrobacterium mediated transformation: Utilization of the tumor-inducing plasmid of bacterium species *Rhizobium radiobacter* to transfer a heritable piece of DNA to plant callus tissue.

***Agrobacterium rhizogenes*:** A species of Gram -negative, rod-shaped soil bacteria, often harboring large plasmids, called Ri plasmids; it can cause a tumorous growth known as hairy.

***Agrobacterium tumefaciens*:** A bacterium that causes crown gall disease in some plants; it infects a wound, and injects a short stretch of DNA into some of the cells around the wound; the DNA comes from a large plasmid the Ti (= tumor induction) plasmid.

Allele mining: An approach to access new and useful genetic variation in crop plant collections; it focuses on the detection of allelic variation in important genes and/or traits within a germplasm collection.

Allele: One of the different forms of a gene or DNA sequence that can exist at a single locus.

Allele-specific amplification (ASA): The use of polymerase chain reaction (PCR) at a sufficiently high stringency that only a primer with exactly the same sequence as the target DNA will be amplified.

Allele-specific oligonucleotide hybridization: The use of an oligonucleotide probe to determine which of the two alternative nucleotide sequences is contained in a DNA molecule.

Allelic exclusion: A phenomenon whereby only one functional allele of an antibody gene can be assembled in a given B lymphocyte. The "allele" on the other homologous chromosome in a diploid mammalian cell cannot undergo a functional re-arrangement, which would result in the production of two different antibodies by a single plasma cell.

Allosteric enzyme: An enzyme that has two structurally distinct forms, one of which is active and the other inactive. Active forms of allosteric enzymes tend to catalyse the initial step in a pathway leading to the synthesis of molecules. The end product of this synthesis can act as a feedback inhibitor, converting the enzyme to the inactive form, thus controlling the amount of product synthesized.

Allosteric regulation: A catalysis-regulating process in which the binding of a small effector molecule to one site on an enzyme affects the activity at another site.

Allosteric transition: A reversible interaction of a small molecule with a protein molecule, resulting in a change in the shape of the protein and a consequent alteration of the interaction of that protein with a third molecule.

Alternative splicing: Formation of diverse mRNAs through differential splicing of the same RNA precursor; it may result in proteins with different composition of amino acids, or it may involve just the length of 32 UTR; a reason for alternative/ differential splicing is base modification during RNA editing causing a change in slice sites.

Alu sequences: A family of 300-bp sequences occurring nearly a million times in the human genome.

Amber codon: Amber suppressor mutation that changes anticodon of amino acid-carrying tRNA.

Ameiosis: The failure of meiosis and its replacement by nuclear division without reduction of the chromosome number.

Amino acid: The building block of proteins. The messenger RNA tells the cell what amino acids are needed and what order they must be arranged in to build a particular protein. There are 20 different amino acids used in the human body.

Aminoacyl site (A-site): One of two sites on ribosomes to which the incoming aminoacyl tRNA binds.

Aminoacyl tRNA synthetase: Enzyme that attaches each amino acid to its specific tRNA molecule.

Aminoacyl-tRNA: A tRNA molecule covalently bound to an amino acid *via* an acyl bond between the carboxyl group of the amino acid and the 32-OH of the tRNA.

Amitosis: Cell division (cytokinesis), including nuclear division through constriction of the nucleus, without chromosome differentiation as in mitosis. The maintenance of genetic integrity and diploidy during amitosis is uncertain. This process occurs in the endosperm of flowering plants.

Amplification: An increase in the number of copies of a specific DNA fragment; can be *in vivo* or *in vitro*. See cloning, polymerase chain reaction.

Anaerobe: A micro-organism that can grow in the absence of O_2.

Amplified fragment length polymorphisms (AFLPs): Polymorphic DNA fragments are amplified through PCR procedure; their differences are used for genotype identification and linkage studies.

Anaerobic respiration: Respiration in which foodstuffs are partially oxidized, with the release of chemical energy, in a process not involving atmospheric oxygen, such as alcoholic fermentation, in which one of the end products is ethanol.

Annealing: The process of heating (de-naturing step) and slowly cooling (re-naturing step) double-stranded DNA to allow the formation of hybrid DNA or complementary strands of DNA or of DNA and RNA.

Anther culture: The aseptic culture of anthers for the production of haploid plants from microspores. See androgenesis; gynogenesis; parthenogenesis.

Anther: The terminal portion of a stamen of a flowering plant; the pollen sacs containing pollen are borne on the anther.

Antibiotic resistance: The ability of bacteria to tolerate an antibiotic and survive being exposed to it. Bacteria may develop this resistance naturally after being exposed to it over many years.

Antibiotic: A natural or synthetic chemical that is used to kill bacteria in order to treat diseases in humans and animals.

Antibody: A protein made by the immune system that is specific to an antigen. When an antibody detects this antigen in the body, it will start an immune response to rid the body of the antigen.

Anticoding strand: The strand of the DNA double helix that is actually transcribed. Also known as the antisense or template strand.

Anticodon: A triplet of nucleotides in a tRNA molecule that pairs with a complementary triplet of nucleotides, or codon, in an mRNA molecule during translation.

Antigen: A foreign substance that binds to an antibody and starts an immune response in the body.

Antimutator gene: Mutant genes that decrease the mutation rates.

Antiparallel orientation: The normal arrangement of the two strands of a DNA molecule, and of other nucleic-acid duplexes (DNARNA, RNA-RNA), in which the two strands are oriented in opposite directions so that the 5¢-phosphate end of one strand is aligned with the 3¢-hydroxyl end of the complementary strand.

Antisense DNA: The strand of chromosomal DNA that is transcribed or a DNA sequence that is complementary to all or part of an mRNA molecule.

Antisense gene: A gene construct placed in inverted orientation relative to a promoter.

Antisense RNA: An RNA sequence that is complementary to all or part of a functional mRNA molecule, to which it binds, blocking its translation.

Apoenzyme: Inactive enzyme that has to be associated with a specific organic molecule called a co-enzyme in order to function. The apoenzyme/co-enzyme complex is called a holoenzyme.

Apoptosis: The process of cell death, which occurs naturally as a part of normal development, maintenance and renewal of tissue in an organism. Apoptosis differs from necrosis, in which cell death is caused by a toxic substance.

ARS (autonomous replicating sequence) Any eukaryotic DNA sequence that initiates and supports chromosomal replication; they have been isolated in yeast cells.

Assay: A method for determining the presence or quantity of a component.

Association mapping: A population-based survey of molecular marker analysis in order to identify trait-marker relationships based on linkage disequilibrium; the association between a pair of linked markers is also called linkage disequilibrium (LD) or, less frequently, gametic disequilibrium.

Attenuation: A mechanism for controlling gene expression in prokaryotes that involves premature termination of transcription.

Autocatalysis: Catalysis in which one of the products of the reaction is a catalyst for the reaction. Usually the catalysis starts slowly and increases as the quantity of the catalyst increases, falling off as the product is used up.

Autoradiography: A technique that uses X-ray film to visualize radioactively labeled molecules or fragments of molecules; used in analyzing length and number of DNA fragments after they are separated by gel electrophoresis.

Autosomal dominant: Describes a type of inheritance where an individual with a mutation in only one copy of a gene will develop the associated trait or disorder.

Autosomal recessive: Describes a type of inheritance where an individual must inherit a mutation in both copies of a gene in order to develop the associated trait or disorder.

Autosome: A chromosome that is not a sex chromosome (X or Y); chromosomes 1 through 22.

Auxin: A group of plant growth regulators (natural or synthetic) which stimulate cell division, enlargement, apical dominance, root initiation, and flowering. One naturally produced auxin is indole-acetic acid (IAA).

***Bacillus thuringensis* (bt):** A naturally occurring soil bacterium that makes an endotoxin that is toxic to larvae of the European corn borer (Lepidoptera). The gene for this endotoxin has been incorporated into corn to produce a genetically modified corn plant that can defend itself against the European corn borer. The endotoxin is very specific in that it only affects the corn borer larvae. It is not toxic to people, domestic animals, fish or wildlife.

Back mutation: A reverse mutation in which a mutant gene reverts to the original standard.

Backcross-assisted selection (BCAS): A method that allows the selection of plants carrying a favorable recessive allele at each generation, limiting the need for a progeny test, which is common in traditional backcrossing; in cases where the traditional means of selection are limited by environmental conditions (*e.g.* the presence of an abiotic or a biotic stress such as drought) this selection strategy is superior to conventional ones; particularly in genetic transformation approaches, where the transgenes can be used as markers, BCAS may show a considerable advantage.

Backward selection: Selection of parent plants based on results from a progeny test.

Bacterial artificial chromosome (BAC): Pieces of plant DNA that have been cloned inside living bacteria; they can be used as probes to detect complementary DNA sequences within large pieces of DNA *via* hybridization techniques, or for marker-assisted selection by faster selection of segregant-bearing genes for a particular trait and to develop future crop varieties faster.

Bacteriophage: A virus that infects bacteria. Altered forms are used in DNA cloning work, where they are convenient vectors. The bacteriophages most used are derived from two "wild" phages, called M13 and lambda.

Base analogues: A purine or pyrimidine base that differs slightly in structure from a normal base, but that because of its similarity to that base may act as a mutagen when incorporated into DNA (*e.g.*uracil, 5-bromouracil, 5-fluorouracil, 5-methylcytosin, 5-bromocytosin, hypoxanthin).

Base pair (bp): The two strands that constitute DNA are held together by specific hydrogen bonding between purines and pyrimidines (A pairs with T; and G pairs with C). The size of a nucleic acid molecule is often described in terms of the number of base pairs (symbol: bp) or thousand base pairs (kilobase pairs; symbol: kb; a more convenient unit) it contains.

Base sequence analysis: A method, sometimes automated, for determining the base sequence.

Base substitution: Replacement of one base by another in a DNA molecule.

Batch culture: A suspension culture in which cells grow in a finite volume of liquid nutrient medium and follow a sigmoid pattern of growth.

Binary vector system: A two-plasmid system in *Agrobacterium tumefaciens* for transferring into plant cells a segment of T-DNA that carries cloned genes; one plasmid contains the virulence gene (responsible for transfer of the T-DNA), and another plasmid contains the T-DNA borders, the selectable marker, and the DNA to be transferred.

Bioassay: A method of determining the effect of a compound by quantifying its effect on living organisms or their component parts.

Biochemical oxygen demand (BOD): The oxygen required to meet the metabolic needs of aerobic organisms in water containing organic compounds.

Biodiversity: The variability among the living organism from all sources, soil, water, air, extreme habitat or associated with organisms.

Bioengineering: Engineering applied to biological and medical systems, such as biomechanics, biomaterials and biosensors. Bioengineering also includes biomedical engineering, as in the development of aids or replacements for defective or missing body organs.

Biofertilizer: Commercial preparation of microorganisms by using which the nitrogen nd phosphorus level and growth of plants increase.

Biofortification: The process of breeding food crops that are rich in bioavailable micronutrients; these crops fortify themselves that is, they load high levels of minerals and vitamins in their seeds and roots, which are then harvested and eaten by animals or humans; through biofortification, breeders can provide farmers with crop varieties that naturally reduce anemia, cognitive impairment, and other nutritionally related health problems, potentially for use in third-world countries.

Biofuel: A gaseous, liquid or solid fuel that contains energy derived from a biological source. For example, rapeseed oil or fish liver oil can be used in place of diesel fuel in modified engines.

Bioinformatics: It s an interdisciplinary field which addresses biological problems using computational techniques, and makes the rapid organization and analysis of biological data possible. The field may also be referred to as *computational biology*, and can be defined as, "conceptualizing biology in terms of molecules and then applying informatics techniques to.

Biolistics. The process of introducing DNA into plant cells by shooting DNA-coated pellets into the cell using a compressed air delivery system know as a gene gun.

Biological products/Biologicals/Biologics: Any virus, therapeutic serum, toxin, antitoxin, or analogous product used in the prevention, treatment or cure of diseases or injuries in humans.

Biomass: Any organic matter, particularly available on a renewable or recurring basis such as trees and plants (residues and fibers containing cellulose or lingo-cellulose), but also poultry litter and animal residues and waste, and industrial and municipal solid waste (for example, sawdust, wood chips, paper, grass and leaf compost).

Bioremediation: A process that uses living organisms to remove contaminants, pollutants or unwanted substances from soil or water.

Biosensor: A device that uses an immobilized agent (such as an enzyme, antibiotic, organelle or whole cell) to detect or measure a chemical compound. A reaction between the immobilized agent and the molecule being analysed is transduced into an electric signal.

Biotechnology: A general term used to describe the use of biological processes to make products, in contrast to purely chemical processes. Biotechnology has been in practice for centuries and includes such traditional applications as the use of yeast in making beer, as well as modern applications like recombinant DNA techniques to improve crops.

Biotic stress: The stress caused to plants by insects, pathogens (viruses, fungi, bacteria), wounds *etc.*

Blastocyst stage: Four to five days after the union of the sperm and the egg, before the embryo implants in the uterus.

Blue biotechnology: It is a term that has been used to describe the marine and aquatic applications of biotechnology, but its use is relatively rare.

Blunt end: The end of a DNA duplex molecule in which neither strand extends beyond the other. It is also called as flush end.

Blunt-end ligation: Ligation of DNA with blunt ends requires higher concentration of DNA ligase than sticky-end ligation; it is inhibited by ATP concentrations.

Bottleneck: A period when a population becomes reduced to only a few individuals; a form of genetic drift that occurs when a population is drastically reduced in size; some genes may be lost from the gene pool as a result of chance.

***Bt* crops:** Crops that are genetically engineered to carry a gene from the soil bacterium *Bacillus thuringiensis* (*Bt*). The bacterium produces proteins that are toxic to some pests but non-toxic to humans and other mammals. Crops containing the *Bt* gene are able to produce this toxin, thereby providing protection for the plant. *Bt* corn and *Bt* cotton are examples of commercially available *Bt* crops.

Bulked segregant analysis (BSA)**:** A rapid mapping strategy suitable for monogenic qualitative traits; when DNA of a certain number of plants is bulked into one pool, all alleles must be present; two bulked pools of segregants, differing for one trait, will differ only at the locus harboring that trait.

C value: The DNA quantity per genome (*i.e.*, per chromosome set); the content of diploids is referred to as the 2C; haploid cells contain the 1C amount of DNA.

CAAT box (also CAT box): A conserved sequence found within the promoter region of the protein-encoding genes of many eukaryotic organisms. It has the consensus sequence GGCCAATCT; it occurs around 75 bases prior to the transcription initiation site; and it is one of several sites for recognition and binding of regulatory proteins called transcription factors.

Callus culture: The *in vitro* culture of callus, often as first stage in the regeneration of whole plants in culture.

Callus induction: Undifferentiated plant tissue is produced at wound edges; callus can also be induced and grown *in vitro* by varying the ratio of hormones (*e.g.*auxin and cytokinin) in the growth medium.

Callus: Undifferentiated plant cells that are multiplied under tissue culture.

Candidate gene: A gene whose function suggests that it may be involved in the genetic variation observed for a particular trait (phenotype, disease, or condition); candidate genes can be divided into two categories: positional and functional.

Capsid: The protein coat of a virus. The capsid often determines the shape of the virus. See coat protein.

Carbohydrate: An organic compound based on the general formula Cx (H2O)y, performing many vital roles in living organisms. The simplest carbohydrates are the sugars (saccharides), including glucose and sucrose. Polysaccharides are carbohydrates of much greater molecular weight and complexity; examples are starch, which serves as energy store in plant seeds and tubers; cellulose and lignin that form the cell walls and woody tissue of plants of plants; glycogen, *etc.*

Carcinogen: Physical or chemical agent which induces cancer.

Carrier: (heterozygote) An individual who carries one copy of a recessive gene.

Cartagena Protocol on Biosafety: The first international treaty dealing with the movement of genetically modified organisms (GMOs) across country borders; the protocol was drawn up under the Convention of Biological Diversity and came into force in September 2003; more than 100 countries have ratified the agreement; although the biosafety protocol was pushed for by the South and drafted as a promise of legal protection against the introduction of GMOs, the weakness of its provisions means that the protocol and the national biosafety laws following its introduction are being steadily turned into tools to facilitate the introduction of GMOs.

Catabolite repression: Glucose-mediated reduction in the rates of transcription of operons that encode enzymes involved in catabolic pathways (such as the lac operon).

cDNA library: A collection of cDNA clones that were generated *in vitro* from the mRNA sequences isolated from an organism or a specific tissue or cell type or population of an organism.

cDNA: A duplex DNA where one strand is identical in sequence (except for T in place of U) and one is complementary to a particular RNA.

Cell culture: The culture of dispersed (or disaggregated) cells obtained from the original tissue, or from a cell line.

Cell cycle: The sequence of events that occurs between the formation of a cell and its division into daughter cells; it is conventionally divided into G_0, G_1, (G standing for gap), S (synthesis phase during which the DNA is replicated), G_2, and M (mitosis).

Cell division: Formation of two or more daughter cells from a single mother cell. The nucleus divides first, followed by the formation of a cell membrane between the daughter nuclei.

Cell fusion: Fusion of two previously separate cells, occurs naturally in fertilization; it can be induced artificially by the use of fusogens such as polyethylene glycol; fusion may be restricted to cytoplasm, nuclei may fuse as well; a cell formed by the fusion of dissimilar cells is referred to as a heterokaryon.

Cell: The basic structural and functional unit of a plant; it is a system surrounded by membranes and is compartmentalized into specific functional areas and/ or organelles with special tasks.

Centimorgan (cM): A unit of measure of recombination frequency. One cM is equal to 1 per cent chance that a marker at one genetic locus will be separated from a marker at a second locus due to crossing-over in a single generation.

Central dogma: Refers to F. CRICK's seminal concept that in nature genetic information generally flows from DNA to RNA to protein.

Centrifugation: Separating molecules by size or density using centrifugal forces generated by a spinning rotor. G-forces of several hundred thousand times gravity are generated in ultracentrifugation.

Chiasma: A cross-shaped structure forming the points of contact between nonsister chromatids of homologous chromosomes, first seen in the diplotene stage of meiotic prophase I.

Chimera: A tissue containing two or more genetically distinct cell types, or an individual composed of such tissue; chimeral plants may originate by grafting, spontaneous mutation, induced mutation, sorting-out from variegated seedlings, mixed callus cultures, or protoplast.

Chromatin remodeling: Refers to a reshaping (at molecular scale) of chromatin, that alters specific genes so that DNA subsequently gets expressed; it can be caused by short interfering RNA (siRNA) or certain transcription activators.

Chromatography: A technique used for separating and identifying the components from mixtures of molecules having similar chemical and physical properties; molecules are dissolved in an organic solvent miscible in water, and the solution is allowed to migrate through a stationary phase; since the molecules migrate at slightly different rates they are eventually separated.

Chromosome banding: The experimental production of differentially stained regions because of the distribution of different chromatin constituents along a chromosome.

Chromosome doubling: Induced or spontaneous doubling of chromosome sets leading to rediploidization or to polyploids.

Chromosome engineering: Manipulation of whole chromosome sets, individual chromosomes, or even chromosome segments, by different means, for scientific analysis or improvement of performance of crop plants.

Chromosome jumping: A technique that allows two segments of duplex DNA that are separated by thousands of base pairs (~200 kb) to be cloned together; after subcloning, each segment can be used as a probe to identify cloned DNA sequences that, at the chromosome level, are roughly 200 kb apart.

Chromosome map: A map showing the location of genes on a chromosome, deduced from genetic recombination and cytological experiments.

Chromosome painting: Fluorescent *in situ* hybridization of a specifically labeled DNA probe or probes that hybridizes to the entire chromosome of the probe's origin; using different fluorescent dyes and probes, a pattern of multicolored chromosomes or chromosome segments appear.

Chromosome reduplication: The synthesis of all compounds that result in an identical copy of the original chromosome.

Chromosome staining: The pretreatment and treatment of chromosomes with different dyes in order to make them more suitable for chromosome counting or specific analyses.

Chromosome substitution: The replacement of one or more chromosomes by others from another source by spontaneous events or a crossing scheme.

Chromosome walking: A technique that identifies overlapping cloned DNA fragments that form one continuous segment of a chromosome. These fragments can be generated either by random shearing or by partial digestion with a four-base-pair cutter such as Sau3A. A series of colony hybridizations is then carried out, starting with some cloned fragment which has already been identified and which is known to be in the region encompassed by the overlapping clones. This identified fragment is used as a probe to pick out clones containing adjacent sequences. These are then used as probes themselves to identify clones carrying sequences adjacent to them and so on. At each round of hybridization one "walks" further along the chromosome from the initial fragment.

Chromosome: The self-replicating genetic structure of cells, containing genes, which determines inheritance of traits. Chemically, each chromosome is composed of proteins and a long molecule of DNA.

Cis configuration: Two sites on the same molecule of DNA.

Cistron: A DNA sequence that codes for a specific polypeptide; a gene.

Cleaved amplified polymorphic sequences (CAPS): PCR-amplified DNA (STS, EST, or SCAR products) that is digested with restriction endonucleases to reveal polymorphisms in restriction sites.

Clone: A genetically identical copy of an organism or of a specific piece of DNA for use in research.

Cloning vector: DNA molecule originating from a virus, a plasmid, or the cell of a higher organism into which another DNA fragment of appropriate size can be integrated without loss of the vector's capacity for self-replication; vectors introduce foreign DNA into host cells, where it can be reproduced in large quantities. Examples are plasmids, cosmids, and yeast artificial chromosomes; vectors are often recombinant molecules containing DNA sequences from several sources.

Cloning: The process of asexually producing a group of cells (clones), all genetically identical, from a single ancestor. In recombinant DNA technology, the use of

DNA manipulation procedures to produce multiple copies of a single gene or segment of DNA is referred to as cloning DNA.

Closed continuous culture: A continuous culture in which inflow of fresh medium is balanced by outflow of corresponding volumes of spent medium. Cells are separated mechanically from outflowing medium and added back to the culture.

Coding sequence: The part of a gene that determines the sequence of amino acids of a protein, as opposed to noncoding sequences, such as promoter, operator, intron, or terminator regions.

Coding strand: The strand of duplex DNA that is transcribed into a complementary mRNA molecule.

Coding: The specification of a peptide sequence, by the code contained in DNA molecules.

Codon bias: Although several codons code for a single amino acid, a plant may have a preferred codon for each amino acid; this is called codon bias.

Codon: A triplet of nucleotides that represents an amino acid or a termination (STOP) signal.

Cohesive ends: Double-stranded DNA molecules with single-stranded ends which are complementary to each other, enabling the different molecules to join each other.

Colchicine: A poisonous alkaloid drug ($C_{22}H_{25}NO_6$) that is obtained from meadow saffron (*Colchicum autumnale*); it has a disruptive effect on microtubular activity; thus, it affects tissue metabolism generally, and mitosis and meiosis in particular; it is used for induction of polyploidy.

Colchiploidy: Polyploidy induced by application of colchicines.

Colony hybridization: A technique that uses a nucleic acid probe to identify a bacterial colony with a vector carrying a specific cloned gene or genes.

Comparative (gene) mapping: Localization (mapping) of a common set of DNA probes onto linkage maps of different species; the results of these comparisons indicate substantial conservation of blocks of genes and even large segments of chromosomes between species; this approach shows synteny of markers among related species or genera; such maps have been established for cereals (wheat, maize, oats, rye, rice, sorghum, millet) and Solanaceae, such as potato, tomato, and paprika.

Comparative genomic hybridization (CGH)**:** A molecular cytogenetic technique that allows detection of DNA sequence copy number changes throughout the genome in a single hybridization.

Comparative positional candidate gene: A gene that is likely to be located in the same region as a DNA marker that has been shown to be linked to a single-locus trait or to a quantitative trait locus (QTL), where the gene's likely location in the genome of the species in question is based on its known location in the map of another species (*i.e.* is based on the comparative map between the two species).

Complementary DNA (cDNA): DNA that is synthesised from a messenger RNA template; the single-stranded form is often used as a probe in physical mapping.

Complementary homopolymeric tailing: The process of adding complementary nucleotide extensions to different DNA molecules, for example, dG (deoxyguanosine) to the 32-hydroxyl ends of one DNA molecule and dC (deoxycytidine) to the 32-hydroxyl ends of another DNA molecule to facilitate, after mixing, the joining of the two DNA molecules by base pairing between the complementary extensions.

Complementary sequences: Nucleic acid base sequences that can form a double-stranded structure by matching base pairs; the complementary sequence to G-T-A-C is C-A-T-G.

Composite transposon:: A transposable element formed when two identical or nearly identical transposons insert on either side of a nontransposable segment of DNA, such as the bacterial transposon Tn5.

Conjugation: A process whereby organisms of identical species, but opposite mating types, pair and exchange genetic material (DNA); in molecular biology, natural process of DNA transfer between bacteria in which the DNA is never exposed.

Consanguineous matings: Matings between two individuals who share a common ancestor in the preceding two or three generations.

Conserved sequence: A base sequence in a DNA molecule (or an amino acid sequence in a protein) that has remained essentially unchanged throughout evolution.

Constitutive gene: A gene that is continually expressed in all cells of an organism.

Constitutive heterochromatin: The material basis of chromosomes or segments that exhibit heterochromatic properties under most conditions (*e.g.* centromeric or telomeric heterochromatin).

Constitutive mutation: Causes genes that usually are regulated to be expressed without regulation contain their own DNA which is separated from the cell's nuclear DNA; in transplastomic plants, the DNA in the chloroplasts has been genetically engineered; since only the nuclear DNA is inherited, the genetic modification in the chloroplasts will not be passed on to the next generation; therefore, transplastomic plants may be a solution for ensuring biocontainment.

Constitutive promoter: An unregulated promoter that allows for continual transcription of its associated gene.

Contig map: A map depicting the relative order of a linked library of small overlapping clones representing a complete chromosomal segment.

Contig: A set of overlapping clones that provide a physical map of a portion of a chromosome.

Contiguous map: The alignment of sequence data from large, adjacent regions of the genome to produce a continuous nucleotide sequence across a chromosomal region.

Continuous culture: An *in vitro* suspension culture continuously supplied with nutrients by the inflow of fresh medium; the culture volume is normally constant.

Corepressor: A metabolite that in conjugation with a repressor molecule binds to the operator gene present in an operon and prevents the synthesis of a repressible enzyme.

Cosmid: Artificially constructed cloning vector containing the cos gene of phage lambda. Cosmids can be packaged in lambda phage particles for infection into *E. coli*; this permits cloning of larger DNA fragments (up to 45 kb) than can be introduced into bacterial hosts in plasmid vectors.

Cosuppression: Silencing of a gene by addition of transgenic DNA copies or infection by a virus; this term, this can refer to silencing at the posttranscriptional (PTGS) or transcriptional.

CpG island: Repetitive CpG doublets creating a region of DNA greater than 200 bp in length with a G+C content of more than 0.5 and an observed/expected presence of CpG more than 0.6; usually associated with transcription-initiation regions of (housekeeping) genes transcribed at low rates that do not contain a TATA box; the CpG-rich stretch of 20–50 nucleotides occurs within the first 100–200 bases upstream of the start site region; a transacting transcription factor called SP1 recognizes the CpG islands.

Cross-protection: Plant protection conferred on a host by infection with one strain of, for example, a virus that prevents infection by a closely related strain.

Cry1A protein: Derived from the bacterium *Bacillus thuringiensis* that is toxic to some insects when ingested; this bacterium occurs widely in nature and has been used for decades as an insecticide although it constitutes less than 2 per cent of the overall insecticides used.

Cryopreservation: Storage and preservation at very low temperatures (-1960C).

Cryoprotectant: A chemical agent or a compound that can prevent damage to cells while they are frozen or defrosted.

Crystallography: The study of, for example, a protein structure by crystallizing the protein and examining the crystals using X-rays; the diffraction angles of the X-rays are used to compute the relative positions of components of the protein, and thus its structure.

Culture medium: The nutrients prepared in the form of a fluid (broth) or solid for the growth of cells/tissues in the laboratory.

Culture: A population of plant or animal cells/microorganisms that are grown under controlled conditions.

Cybrid: The hybrid formed from the fusion of a cytoplast and a whole cell; the cytoplast may transmit cytoplasmic components independently of the cell genome.

Cytogenetic map: A map showing the locations of genes on a chromosome (*i.e.* the visual appearance of a chromosome when stained and examined under a microscope).

Cytokines: Various chemicals produced in the body which mediate immunological responses.

Cytosine (C): A nitrogenous base, one member of the base pair G-C (guanine and cytosine).

Degeneracy: (of the genetic code) The specification of one amino acid by more than one codon. It arises from the inevitable redundancy resulting from 64 triplets in a triplet code (4 x 4 x 4 = 64) encoding only 20 amino acids.

Degenerate code: A genetic code in which some amino acids may be encoded by more than one codon each.

Deletion mutation: A mutation in which one or more bases are removed from the DNA sequence of a gene.

Denaturation: The separation of the two strands of a DNA double helix, or the severe disruption of the structure of any complex molecule without breaking the major bonds of its chains.

Density gradient centrifugation: The separation of macromolecules or subcellular particles by sedimentation through a gradient of increasing density under the influence of a centrifugal force; the density gradient may either be formed before the centrifugation run by mixing two solutions of different density (*e.g.* in sucrose density gradients), or it can be formed by the process of centrifugation itself (*e.g.* in $CsCl_2$ and Cs_2SO_4 density gradients).

Deoxyribonucleic acid (DNA): The molecule that carries the genetic information in most living organisms. It is a double-stranded helix held together by hydrogen bonds between pairs of nucleotides. The nucleotides in DNA (adenine, guanine, cytosine and thymine) are arranged in different combinations to represent each gene. The genes act like recipes in that they contain the information necessary for the cell to make the corresponding proteins.

derived from a common ancestor.

Differential centrifugation: A method of separating sub-cellular particles according to their sedimentation coefficients, which are roughly proportional to their size. Cell extracts are subjected to a succession of centrifuge runs at progressively faster rotation speeds. Large particles, such as nuclei or mitochondria, will be precipitated at relatively slow speeds; higher G forces will be required to sediment small particles, such as ribosomes.

Direct embryogenesis: Embryoids form directly in culture, without an intervening callus phase, on the surface of zygotic or somatic embryos or on explant tissues (leaf section, root tip, *etc.*).

Direct organogenesis: Formation of organs directly on the surface of cultured intact explants. The process does not involve callus formation.

Diversity arrays technology (DArT) marker: A DArT marker is a segment of genomic DNA, the presence of which is polymorphic in a defined genomic representation; DArT markers are diallelic and behave in a dominant (present versus absent) or codominant (two doses versus one dose versus absent) manner; to identify the polymorphic markers, a complexity reduction method is applied on the metagenome, a pool of genomes representing the germplasm of interest; the genomic representation obtained from this pool is then cloned and individual inserts are arrayed on a microarray resulting in a "discovery array"; labeled genomic representations prepared from the individual genomes included in the pool are hybridized to the discovery array.

DNA (deoxyribonucleic acid): The molecule that encodes genetic information. DNA is a double-stranded molecule held together by weak bonds between base pairs of nucleotides. the four nucleotides in DNA contain the bases stranded molecule held together by weak bonds between base pairs of nucleotides. The four nucleotides in DNA contain the bases: adenine (A), guanine (G), cytosine (C), and thymine (T). In nature, base pairs form only between A and T and between G and C; thus the base sequence of each single strand can be deduced from that of its partner.

DNA fingerprint(ing) (DFP**):** The unique pattern of DNA fragments identified by SOUTHERN hybridization (using a probe that binds to a polymorphic region of DNA) or by polymerase chain reaction (PCR) using primers flanking the polymorphic region.

DNA hybridization: Base pairing of DNA from two different sources; in biotechnology, a technique for selectively binding specific segments of single-stranded (ss) DNA or RNA by base pairing to complementary sequences of ssDNA molecules that are trapped on a nitrocellulose membrane.

DNA ligase (polynucleotide ligase)**:** An enzyme that creates a phosphodiester bond between the 52-PO_4 end of one polynucleotide and the 32-OH end of another, thereby producing a single, larger polynucleotide.

DNA marker: A DNA sequence that exists in two or more forms that can be used to genotype individual organisms.

DNA micro arrays: A powerful, versatile and economical molecular technique for screening of genetic aberrations; high-density gene sequences are printed onto glass slides; fluorephorelabeled genomic or complimentary DNA (cDNA) is hybridized to slides with fixed signature patterns and resolved using computer driven fluorescent images.

DNA polymerase: A group of enzymes mainly involved in copying a single-stranded DNA molecule to make its complementary strand.

DNA probe: A segment of DNA that is tagged with a label (*i.e.* isotope) so as to detect a complimentary base sequence in the DNA sample after a hybridization reaction.

DNA profiling: The term used to describe different methods for the analysis of DNA to establish the identity of an individual.

DNA repair: The biochemical processes that correct mutations occurring due to replication errors or as a consequence of mutagenic agents.

DNA replication: The use of existing DNA as a template for the synthesis of new DNA strands. In humans and other eukaryotes, replication occurs in the cell nucleus.

DNA sequencing Procedures for determining the nucleotide sequence of a DNA fragment. There are two common methods for doing this:.

DNA-amplified fingerprinting: A technology based on amplification of random genomic DNA sequences achieved by a single short (5–8 bases) oligonucleotide primer of arbitrary sequence; it produces a characteristic spectrum of short DNA pieces of varying complexity that are resolved on polyacrylamide gel (PAGE) following silver staining; it is used to detect genetic differences between genotypes as well as for detecting polymorphism even between organisms that are closely related, such as near isogenic lines.

DNA-DNA hybridization: The annealing of two complementary DNA strands to produce hybrid.

Dominant-negative mutation: A (heterozygous) dominant mutation on one allele blocking the activity of wild-type protein still encoded by the normal allele causing a loss-of-function phenotype; the phenotype is indistinguishable from that of homozygous dominant mutation.

Dosage compensation: A phenomenon whereby inactivation of all but one of the X chromosomes in female mammals results in males and females producing the same quantity of peptide from X-linked genes.

Dosage effect: The influence upon a phenotype of the number of times a genetic element is present.

Double helix: The shape that two linear strands of DNA assume when bonded together.

Drought hardening: Adapting plants to survive periods of time with little or no water by stepwise reducing water supply or germinating and/or growing under insufficient moisture conditions.

Drought resistance: the ability of a plant to withstand drought; this property can be very valuable in areas of uncertain rainfall, *e.g.* sorghum has greater drought resistance than maize, and is grown in many semi-arid areas for this reason.

EcoTILLING: A high throughput, low cost technique for rapid discovery of polymorphisms in natural populations of plants; it is similar to TILLING but it differs from TILLING in that natural polymorphisms are detected rather than polymorphisms induced through chemical mutagenesis; on the other hand, single nucleotide polymorphisms (SNPs), small insertions and deletions, and variations in microsatellite repeat number can be efficiently detected using the EcoTILLING technique.

Electrophoresis: A technique used to separate molecules such as DNA or proteins using an electric current. The mixture of molecules is added to one end of a gel-like medium. When a current is applied to it, the molecules will travel through the medium to the other end at different speeds depending on the charge and size of the molecule. Once the molecules are separated, the gel can be used in a blot (Southern, Northern and Western).

Embryo culture: A method of inducing the artificial growth of embryos by excising the young embryos under septic conditions and placing them on suitable nutrient media; the method is often applied when postgamous incompatibility exists (*e.g.* in wide crosses).

Embryo rescue: When cross-pollination is made or occurs between genetically widely different plants, the resulting embryo may be aborted because of parental mutual incompatibility; such embryos may be excised during an early stage and grown on a congenial medium such as nutrient.

Embryo: Defined in the *Assisted Human Reproduction Act* as a human organism during the first 56 days of its development following fertilization or creation, excluding any time during which its development has been suspended. It includes any cell derived from such an organism that is used for the purpose of creating a human being.

Endomitosis: A doubling of the chromosomes within a nucleus that does not divide, thus producing a polyploid; the doubling may be repeated a number of times, which leads to endopolyploidy.

Endonuclease: An enzyme that cleaves the phosphodiester bond within a nucleotide chain.

Enzyme: A protein that facilitates a biochemical reaction. Many essential reactions in the body require the help of enzymes and would not proceed on their own.

Enzyme-Linked Immuno Assays (EIA): Enzyme-Linked Immuno Assays (EIA) are use to measure the amount of a particular substance by virtue of its binding to a specific antibody. Examples of EIA include ELISA and Western blotting.

Enzyme-linked immunosorbent assay (ELISA): A technique using antibodies for detecting specific proteins. Used to test for the presence of a particular genetically engineered organism.

Epigenesis: The concept that an organism develops by the new appearance of structure and functions, as opposed to the hypothesis that an organism develops by the unfolding and growth.

Epigenetic regulation: Regulation of gene activity mediated by a reversible change in DNA modification or chromatin structure.

Epimutation: A heritable change in gene activity that is not due to a change in base sequence; generally applied to an abnormal change in gene activity (*e.g.* in a tumor or a cell culture.

Euchromatin: Genetic material that is stained less intensely by certain dyes during interphase, and that comprises many different kinds of genes.

Euchromatization: The induced or spontaneous change of heterochromatin into euchromatin.

Eugenic: Favorable to the genetic quality of a population, as opposed to dysgenic.

***Ex situ*:** Out of place or not in the original environment (*e.g.* seeds stored in a genebank).

Exogenous DNA: DNA originating outside an organism.

Exon: Any segment of an interrupted gene that is represented in the mature RNA product. The protein-coding sequences of a gene.

Exonuclease: An enzyme that cleaves nucleotides one at a time from an end of a polynucleotide chain.

Explant: The whole plants can be regenerated virtually from any plant referred to as explant.

Expressed sequence tag (EST): Short cDNA sequence. So-called because it represents part of the sequence (*i.e.*, the "tag" of a sequence) of an expressed gene.

Expression vector: A cloning vector that has been constructed in such a way that, after insertion of a DNA molecule, its coding sequence is properly transcribed and the RNA is translated.

F factor: A bacterial episome that confers the ability to function as a genetic donor in conjugation; the fertility factor in bacteria.

Facultative heterochromatin: Heterochromatin that is present in only one of a pair of homologues or not permanently present.

Feedback inhibition: The process by which the accumulated end product of a biochemical pathway stops synthesis of that product. A late metabolite of a synthetic pathway regulates synthesis at an earlier step of the pathway.

Fermentation: A process of growing microorganisms to produce various chemical or pharmaceutical compounds. Microbes are usually incubated under specific conditions in large tanks called fermenters. Fermentation is a specific type of bioprocessing.

Flow cytometry: Analysis of biological material by detection of the light-absorbing or fluorescing properties of cells or subcellular fractions (*i.e.*, chromosomes) passing in a narrow stream through a laser beam. An absorbance or fluorescence profile of the sample is produced. Automated sorting devices, used to fractionate samples, sort successive droplets of the analysed stream into different fractions depending on the fluorescence emitted by each droplet.

Fluorescence in situ hybridization (FISH): Hybridization of cloned DNA to intact chromosomes, where the cloned DNA has been labelled with a fluorescent dye. This is the major method of physical mapping of cloned DNA fragments on chromosomes.

Foot printing: A method used to determine the length of nucleotide chains that are close to a protein (which bind to DNA); for example, certain types of drugs act by binding tightly to certain DNA molecules in specific locations.

Foreign DNA: DNA that is not found in the normal genome concerned; usually it is directly or indirectly introduced into a recipient cell by several experimental means.

Forward mutation: A mutation that alters (usually inactivates) a wild-type allele of a gene.

Forward selection: Choosing good individuals out of a progeny test for possible use in seed orchards and/or subsequent generations of breeding.

Frame shift mutation: A mutation that is caused by a shift of the reading frame of the mRNA (usually by the insertion of a nucleotide) synthesized from the altered DNA template.

Functional genomics: The science of how the genes in organisms interact to express complex traits - that is, the field of research that aims to determine the function of newly discovered genes; attempts to convert the molecular information represented by DNA into an understanding of gene functions and effects; functional genomics also entails research on the protein function (proteomics) or, even more broadly, the whole metabolism (metabolics) of an organism.

Fusogen: An agent that induces fusion of protoplasts in somatic hybridization.

G banding: A special staining technique for chromosomes that results in a longitudinal differentiation by Giemsa stain, which is a complex of stains specific for the phosphate groups of DNA; the characteristic bands produced are called G bands; these bands are generally produced in AT-rich heterochromatic regions.

Gametic (phase) disequilibrium: In relation to any two loci, the occurrence of gametes with a frequency greater than or less than the product of the frequency of the two relevant alleles.

Gametic (phase) equilibrium: In relation to any two loci, the occurrence of gametes with a frequency equal to the product of the frequency of the two relevant alleles, *e.g.* loci *A* and *B* are in linkage equilibrium if the frequency of the gamete *AnBn* equals the product of the frequencies of alleles *An* and *Bn*.

Gametoclonal variation: Variation among regenerants obtained from pollen and/or anther culture.

Gel electrophoresis: An analytical method for separating molecules according to their size. Samples are put at one end of a slab of polymer gel; an electric field across the gel pulls the molecules through it; the smaller molecules pass more easily and so move towards the other end faster; the various sizes of molecules end up at different positions according their size.

Gene cloning: Insertion of a DNA fragment carrying a gene into a cloning vector; subsequent propagation of the recombinant DNA molecule in a host organism results in many identical copies of the gene (clones) in a form that is more easily accessible than the original chromosomal copy.

Gene discovery: A process whereby the individual base nucleotides in an organism's DNA are identified in order to learn more about the genome of the organism, and to learn about or identify specific areas of interest on chromosomes. This process allows for the identification of specific genes that code for a particular trait or traits.

Gene expression profiling: The analysis of gene expression level of many genes at the same time; technologies that can be used to obtain this information include Northern blot, DNA chips, high-density array spotted on glass or membranes, and quantitative techniques.

Gene expression: The phenotypic manifestation of a gene depending on the different levels of gene activation, or the process by which the information in a gene is used to produce a protein; in molecular genetics, the full use of the information in a gene *via* transcription and translation leading to production of a protein and hence the appearance of the phenotype determined by that gene; gene expression is assumed to be controlled at various points in the sequence leading to protein synthesis; this control is thought to be the major determinant of cellular differentiation.

Gene insertion: The addition of one or more copies of a normal gene into a defective chromosome.

Gene interaction: Modification of gene action by a nonallelic gene or genes, generally the interaction between products of nonallelic genes.

Gene library: In molecular genetics, a random collection of cloned DNA fragments in a number of vectors that ideally includes all genetic information of that species.

Gene map: A graphic presentation of the linear arrangement of a chromosome or segment; it shows the relative distance between loci gained in linkage experiments.

Gene mapping: Determining the relative physical locations of genes on a chromosome. Useful for plant and animal breeding.

Gene mutation: A heritable change of gene revealed by phenotypic modifications.

Gene silencing: The suppression of gene expression, for example, of the gene for polygalacturonase which causes fruit to ripen, or the gene for P34 protein in soybeans, *via* a variety of methods, for example, *via* RNA interference (RNAi), chemical genetics, effect of certain viruses, zinc finger proteins, or sense or antisense genes.

Gene splicing: The enzymatic attachment (joining) of one gene or part of a gene to another; also removal of introns and splicing of exons during mRNA synthesis.

Gene substitution: The replacement of one allele by another mutant allele in a population by natural or directed selection.

Gene tagging: The labeling of a gene by a marker gene or specific DNA sequence closely linked.

Gene therapy: Insertion of normal DNA directly into cells to correct a genetic defect.

Gene transfer: The physical transfer of a gene by crossing, chromosomal manipulation, and molecular means; in biotechnology, different methods are described, such as (1) microinjection, (2) insertion *via* microprojectiles (particle gun, particle bombardment) using silicon fibers as carriers of the DNA, (3) direct transfer, (4) electroporation, (5) liposome fusion, or (6) a vector-mediated transfer.

Gene translocation: The transfer or movement of a gene or gene fragment from one chromosomal location to another; often it alters or abolishes expression.

Gene: The fundamental physical and functional unit of heredity, which carries information from one generation to the next; a segment of DNA, composed of a transcribed region and a regulatory sequence that makes transcription possible.

Genetic (linkage) map: The linear arrangement of gene loci on a chromosome, deduced from genetic recombination experiments; a genetic map unit is defined as the distance between gene pairs for which one product of meiosis out of a hundred is recombinant (*i.e.* it equals a recombination frequency of 1 per cent).

Genetic code: The set of correspondences between base triplets in DNA and amino acids in protein; these base triplets carry the genetic information for protein synthesis.

Genetic engineering: Manipulation of an organism's genes by introducing, eliminating or rearranging specific genes using the methods of modern molecular biology, particularly those techniques referred to as recombinant DNA techniques.

Genetic library: A collection of clones representing the entire genome of an organism.

Genetic mapping: A research method that collects genetic information to determine the relative position of a gene or a phenotype in the genome.

Genetic maps: Maps giving relative distance and position of one gene with respect to the other, the distances are based on recombination values.

Genetic marker: A gene or DNA sequence at a known location on a chromosome that can be used to identify individuals or species. A genetic marker may be a short or long DNA sequence. Some common types of genetic markers are RFLPs (Restriction Fragment Length Polymorphism), SSLPs (Simple Sequence Length Polymorphism), SNPs (Single Nucleotide Polymorphism), and SSRs (Simple Sequence Repeat DNA marker). Genetic markers play a role in genetic engineering because they allow breeders to locate and utilize genes of interest for genetic gains.

Genetic modification: A general term which refers to any intentional change to the heritable traits of an organism. This includes both traditional breeding and recombinant DNA techniques.

G**enetic polymorphism:** Differences between DNA sequences.

Genetically modified organism (GMO): A plant or animal whose genetic material (DNA) has been altered through genetic engineering/biotechnology techniques (insertion/deletion of genes) to produce a genotype that possesses a modified trait that is not found in naturally occurring plants of that species. When genes are inserted, they usually come from a different species. The principle of producing a GMO is to add new genetic material into an organism's genome.

Genetics: The study of how traits are passed on in families and how genes are involved in health and disease.

Genome mapping: Determining a set of landmarks or genetic markers–genes or short DNA sequences–in the genome that will enable researchers to find new genes. A genome map is one-dimensional–a straight line with landmarks that stand for genes or DNA sequences. They guide a researcher toward a gene that is suspected to be involved in some process that is of interest.

Genome mutation: Spontaneous or induced changes in the number of complete chromosomes that result either in polyploids or aneuploids.

Genome projects: Research and technology development efforts aimed at mapping and sequencing some or all of the genome of human beings and other organisms.

Genome sequencing: Determining the order of every DNA base in the genome. More detailed than a genome map and usually done on short segments of the genome. Done to establish the distinctive order of genes of an organism.

Genome: The total genetic material of an organism, *i.e.* an organism's complete set of DNA sequences.

Genomic imprinting: The phenomenon whereby genes function differently depending on whether they are inherited from the maternal or paternal parent; this is thought to be caused by information superimposed on DNA sequences, which is different in male and female gametes; such information is transmitted, or inherited, in somatic cells but usually erased and reset in the germ line; it is due to methylation of one of the alleles depending on its origin.

Genomic *in situ* hybridization (GISH): An *in situ* hybridization technique that uses total genomic DNA of a given species as a probe and total genomic DNA of another species as a blocking DNA; it is based on fluorescence *in situ* hybridization; it is a useful method to detect interspecific or intergeneric genome differentiation, chromosome rearrangements (translocations), and substitutions or additions.

Genomic library: A collection of clones made from a set of randomly generated overlapping DNA fragments representing the entire genome of an organism.

Genomics: The study of the entire genome (chromosomes, genes and DNA) and how different genes interact with each other.

Genotype: The genetic make-up of an individual, usually referring to a particular pair of alleles for a gene that can be related to a particular phenotype of interest.

Genotyping: The process of identifying the genetic make-up of an organism, by using molecular markers, DNA sequencing, *etc.*

Germplasm: Germplasm refers to the sum total of all genes preset in a crop and its related species.

Golden rice: A biotechnology-derived rice (*Oryza sativa*) created in the 1990s by I. Potr ykus and P. Beyer, which contains large amounts of beta-carotene, *i.e.* precursor of vitamin A, in its seeds; the researchers utilized *Agrobacterium tumefaciens* bacteria to genetically engineer the rice plant, *i.e.* by inserting the following genes from daffodil and from the bacterium *Erwinia uredovora*: (a) phytoene synthase - from daffodil (narcissus), which converts geranylgeranyl-diphosphate into phytoene; (b) CRTL gene - from *Erwinia uredovora*, which codes for phytoene desaturase, which causes the rice plant to convert phytoene (a "light harvesting" carotenoid involved in photosynthesis) into lycopene (a carotenoid which is then utilized by the rice plant in the production of beta-carotene); and (c) lycopene beta-cyclase - from daffodil, which converts lycopene into beta-carotene.

Green biotechnology: It is biotechnology applied to agriculture processes. An example would be the selection and domestication of plants via micropropagation. Another example is the designing of Transgenic plants to grow under specific environments in the presence (or absence) of chemicals. Using geentic engineering, plants have been created that can express a pesiticide, thereby ending the need of external application of pesticides. eg Bt corn. Whether or not green biotechnology products such as this are ultimately more environmentally friendly is a topic of considerable debate.

Green fluorescent protein (GFP): Protein from the jellyfish, *Aequorea victoria* (Scyphozoa); the gene is used as a reporter gene; it fluoresces in UV light; several variants have been developed, which each exhibit characteristic spectra; significant advantages are that the protein.

Guanine (G): A nitrogenous base, one member of the base pair G-C (guanine and cytosine).

Guide RNA: An RNA molecule that contains sequences that function as a template during RNA editing.

GUS gene: a gene that codes for production of beta-glucuronidase (GUS protein) in *Escherichia.*

Haploid: A single set of chromosomes (half the full set of genetic material), present in the egg and sperm cells of animals and in the egg and pollen cells of plants.

Haploidization: The process whereby diploid somatic cells become haploid during a parasexual cycle or by experimental means, for example, in barley by crossing with wild (bulb) barley (*Hordeum bulbosum*), or in wheat by crossing with pearl millet (*Pennisetum glaucum*).

Heat shock protein: When certain plants are exposed to high temperature, heat shock proteins are synthesized; they provide thermal protection to subsequent heat stress.

Heredity: The transfer of genetic information from parents to children.

Heterochromatin: The darkly stained regions in chromosomes.

Heteroduplex analysis: A method of detecting gene mutation by mixing PCR-amplified mutant and wild-type DNA followed by denaturation and reannealing; the resultant products are resolved by gel electrophoresis, with single base substitutions detectable under optimal electrophoretic conditions and gel formulations; large base pair mismatches may also be analyzed by using electron microscopy to visualize heteroduplex regions.

Heteroduplex DNA: A double-stranded DNA molecule formed by the annealing of strands from two different sources, as opposed to homoduplex, which has homologous strands; as a result, there are regions noncomplementary and showing abnormalities in the form of extra loops.

Heterogeneous nuclear RNA (hnRNA): Large RNA molecules, which are unedited mRNA transcripts, or pre-mRNAs found in the nucleous of a eukaryotic cell.

Heterozygosity: The presence of different alleles at one or more loci on homologous chromosomes.

Heterozygote: An individual with two **different** alleles at a particular locus on a pair of chromosomes.

High-throughput DNA sequencing: The industrialized process of determining the exact order of the four letter code (bases, GAT or C) that composes all genetic material; because a typical higher organism contains > 20,000 genes, each with several thousand bases of DNA, data storage and analysis is one of the most important parts of this process.

Histone: One of a group of globular, simple proteins that have a high content of the amino acids arginine and lysine; it forms part of the chromosomal material of eukaryotic cells and appears to play an important role in gene regulation; highly conserved basic proteins that are involved in the packing of DNA;.

Homeotic mutation: A mutation that causes one body structure to be replaced by a different body structure during development.

Homoeologous: Partially homologous; chromosomes or genomes that are believed to have originated from ancestral homologous chromosomes.

Homologous chromosomes: Chromosomes that pair with each other at meiosis.

Homopolymer-tailing: Attachment of identical nucleotides to the 3' end of a DNA molecule that can be achieved with terminal deoxynucleotide transferase.

Homozygosity: The presence of identical alleles at one or more loci in homologous chromosomal segments.

Homozygote: An individual with two **identical** alleles at a particular locus on a pair of chromosomes.

***Hordeum bulbosum* procedure:** A method for producing zygotic haploids in barley by crossing *Hordeum bulbosum* with *Hordeum vulgare* genotypes; after formation of zygotes, the wild *H. bulbosum* chromosomes are subsequently eliminated during embryogenesis, which results in haploid *H. vulgare* plants.

Hormones: A chemical that is made by one type of cell in the body and acts on another. Hormones act as messengers to tell the target cell to stop or start certain cellular processes.

Housekeeping gene: Gene that is expressed in virtually all cells since it is fundamental to the any cell's functions.

Human Genome Project: An international research effort that aims to identify, map and sequence all human genes.

Hybrid DNA: DNA composed of sequences from two different organisms, also called as Recombinant DNA.

Hybridization: The creation of RNA-DNA hybrids by a heating process, so that the RNA becomes associated with the complementary DNA.

Hybridoma: A clone of hybrid cells produced by fusion of a myeloma cell with an antibody producing cell. Each hybridoma produces only one type of monoclonal antibody. ***In situ* hybridization:** The process of annealing a probe in order in order to screen a DNA library.

Idiogram: A diagrammatic representation of the karyotype of a plant.

Immobilized enzymes: An enzyme physically localized in a defined region enabling it to be reused in a continuous process.

Immuno-affinity chromatography: A purification technique in which an antibody is bound to a matrix and is subsequently used to bind and separate a protein from a complex mixture.

Map distance: The standard measure of distance between loci, expressed in centiMorgans (cM). Estimated from recombination fraction via a mapping function (q.v.). For small recombination fractions, map distance equals the percentage of recombination (recombination frequency) betwen two genes.

Immunoglobulins: The special group of proteins, commonly referred to as antibodies, produced by B-lymphocytes, and involved in humoral immunity.

Imprinting: Patterns of inheritance affected by whether the inheritance was from the mother or father.

***In silico*:** Modern term used to characterize biological experiments carried out entirely in a computer.

***In situ*:** in place; where naturally occurring.

***In vitro* culture:** The cell, organ, or tissue culture performed under artificial conditions in tubes, glasses, dishes, *etc.*

***In vitro* fertilization:** Pollination performed aseptically *in vitro* by direct application of the pollen to the ovule; it is used to overcome prezygotic incompatibility.

***In vitro* mutagenesis:** Methods for altering DNA outside the host cells; mutagenesis can be random or specific for the site and base change depending on the technique used.

***In vitro* selection:** Used to screen large numbers of plants or cells for a certain characteristic before growing them in the field or in glasshouses, *e.g.* salt tolerance.

In vitro - Literally means "in glass" refers to biological activities/reactions carried out in the test tube rather than the living cell or organism.

Independent assortment: The random distribution in the gametes of separate genes; if an individual has one pair of alleles *A* and *a*, and another pair *B* and *b* then it should produce equal numbers of four types of gametes: *AB*, *Ab*, *aB*, and *ab*; it is asserted in Mendel 's second law—the law of independent assortment.

Induced mutation: A change in a gene caused by a treatment.

Insulin: A hormone made by the pancreas that controls the level of sugar in the blood.

Interferons: A group of glycoproteins that resist viral infection and regulate immune responses.

Introgression library: Increasing the genetic diversity of elite breeding materials with exotic germplasm requires techniques that minimize negative side effects attributable to genetic interactions between recipient and donor; this seems achievable by an introgression library approach involving the systematic transfer of a limited and/or restricted number of short donor chromosome segments from an agricultural unadapted source (donor) into an elite line (recipient or recurrent parent); established introgression libraries represent a dynamic resource that can substantially foster breeding programs and provide an opportunity to proceed towards functional genomics.

Introgression: The incorporation of genes of one species into the gene pool of another.

Intron: A segment of DNA of unknown function within a gene; it may be transcribed in precursor RNA, but cannot be found in functional mRNA.

Karyotype: The entire chromosomal complement of an individual cell or individual, which may be observed during mitotic metaphase.

Karyotyping: The method of photographing the complete set of chromosomes for a particular cell type and organizing them into pairs based on size and shape.

Kinetochore: A dense, plaquelike area of the centromere region of a chromatid, to which the microtubules of the spindle attached during cell division.

Kilobase (kb): Unit of length for DNA fragments equal to 1000 nucleotides.

Labelling: Attaching radioactive or non radioactive molecules to specific substances in order to detect them.

***Lac* operon:** A cluster of structural genes specifying the enzymes acetylase, permease, and betagalactosidase.

Lagging strand: DNA strand growing in the 32 to 52 direction, synthesized discontinuously.

Leader sequence: The sequence at the 5' end of an mRNA that is not translated into protein.

Lethal doses: the concentration of a poison that kills a certain amount of cells, individuals, *etc.*; for example, LD50 = 50 per cent of the cells or individuals are killed.

Lethal gene: A gene whose expression results in the premature death of the organism carrying it; dominant alleles kill heterozygotes, whereas recessive alleles kill homozygotes only.

Lethal mutation: A gene mutation whose expression results in the premature death of the organism carrying it.

Library: A set of cloned fragments together representing the entire genome, created then placed into storage.

Ligase: DNA ligase; an enzyme that can rejoin a broken phosphodiester bond in a nucleic acid; requires a 5' phosphate and a 3' OH.

Linkage disequilibrium: The non-random association of alleles at two or more loci, *i.e.* they occur together more often than would be expected by chance.

Linkage drag: The inheritance of undesirable genes along with a beneficial gene due to their close linkage.

Linkage group: All genes in one chromosome form one linkage group.

Linkage map: A map of the relative positions of genetic loci on a chromosome, determined on the basis of how often the loci are inherited together. Distance is measured in centimorgans (cM).

Linkage, genetic linkage: When two chromosomal regions are located physically near each other such that there is a high likelihood they will be inherited together.

Lipids: Water-insoluble (fat) biomolecules that are highly soluble in organic solvents such as chloroform. Lipids serve as "fuel" molecules in organisms, highly concentrated energy stores, "signalling" molecules, and are basic components of cell membranes.

Locus: The position of a gene or a marker on a chromosome.

Lod score: The logarithm of the ratio of the odds that two loci are linked with a recombination fraction equal to or greater than 0 and less than 0.5 to the likelihood of independent assortment. Also called "Z.".

Lytic cycle: The replication cycle of bacteria that ultimately results in the lysis of host cells.

Macromutation: a mutation that results in a profound change in an organism, such as a change in a regulatory gene that controls the expression of many structural genes, as opposed to micromutation.

Map distance: The distance between any two markers on a genetic map, based on the percentage of crossing-over; the minimum distance between linked genes is 1 per cent and maximum 50 per cent.

Map-based cloning: The isolation of important genes by cloning the gene in question on the basis of molecular maps, where the biochemical function is unknown; it involves the identification of a mutant phenotype for the trait of interest (obtained by mutagenesis or from natural variation) and genetic fine mapping using many progeny plants; this map is then used for chromosome walking or landing, with the help of large-insert DNA libraries or physical maps to isolate the gene.

Mapping: The process of identifying the location of a gene or DNA segment along a chromosome. In *genetic* mapping, this is done by analyzing patterns of inheritance in segregating populations (measured in recombinational units, commonly centimorgans). In*physical* mapping, this describes the actual location of a sequence in a particular genomic region (measured in bp).

Marker gene: Genes that identify which plants have been successfully transformed.

Marker-assisted introgression: The use of DNA markers to increase the speed and efficiency of introgression of a new gene or genes into a population; the markers will be closely linked to the genes in question.

Marker-assisted selection (MAS): Use of genetic/DNA markers to guide in choosing specific plants or lines with targeted traits for new variety development.

Maternal effect: An effect attributable to some aspect of performance of the mother of the individual being evaluated.

Maternal inheritance: Phenotypic differences found between individuals of identical genotype due to an effect of maternal inheritance.

Megabase (Mb): Unit of length for DNA fragments equal to 1 million nucleotides and roughly equal to 1 cM.

Meiosis: A type of nuclear division that occurs at some stage in the life cycle of sexually reproducing organisms; by a specific mechanism, the number of chromosomes is halved to prevent doubling in each generation.

Melting: Denaturation of DNA.

Messenger DNA (mDNA): A single-stranded DNA that acts as a messenger of the protein biosynthesis.

Messenger RNA (mRNA): RNA that is complementary to the DNA of a gene and acts as a template to make the protein.

Metabolome: The quantitative complement of all the low molecular weight molecules present in cells in a particular physiological or developmental state.

Metabolomics: An extended discipline of biochemistry that involves the analysis (usually high throughput or broad scale) of small-molecule metabolites and polymers, such as starch; it involves descriptions of biological pathways and current metabolomic databases, such as the Kyoto Encyclopedia of Genes and Genomes (KEGG).

Microarray: A glass or plastic slide with many DNA spots attached to it, which allows researchers to study how many genes act and interact in different conditions.

Microinjection: The delivery of DNA or other compounds in to eukaryotic cells using fine microscopic needle.

Micropropagation: Miniaturized *in vitro* multiplication and/or regeneration of plant material under aseptic and controlled environmental conditions on specially prepared media that contain substances necessary for growth; used for three general types of tissue.

Microsatellite marker: Microsatellites or simple sequence repeats are a type of molecular marker; microsatellites consist of tandem repeats of 1–6 nucleotide motifs; the repeats usually are in units of ten or more, although repeats as small as six units have been found.

Microsatellite-primed PCR (MP-PCR): A technique resulting in RAPD-like patterns after agarose gel electrophoresis and ethidium bromide staining; the MP-PCR technique is more reproducible than RAPD analysis because of higher stringency.

Minichromosome: A very small chromosome, usually as a result of chromosome aberrations; engineered minichromosomes offer an opportunity to improve crop performance; unlike conventional gene transformation technologies, minichromosomes can be used simultaneously to transfer and to stably express (multiple) sets of genes; because they segregate independently of host chromosomes, they provide a platform for accelerating plant breeding; strategies for producing artificial chromosomes may consider engineering of endogenous chromosomes or *de novo* assembly from chromosomal constituents.

Minisatellite: Highly polymorphic DNA markers comprised of a variable number of tandem repeats that tend to cluster near the telomeric ends of chromosomes; the repeats often contain a repeat of 10 nucleotides; they are useful tools for genetic mapping.

Missense mutation: A single DNA base change which leads to a codon specifying a different amino acid.

Mitochondrial DNA (mtDNA): The genetic material found in the mitochondria, which is different from the cell's DNA in the nucleus. Mitochondria are passed on from one generation to the next in the cytoplasm of the egg, so they are inherited from the mother.

Mitosis: The process of cell division in most cells in the human body. Mitosis results in two daughter cells that are genetically identical to each other and to the original cell.

Mitotic recombination: The recombination of genetic material during mitosis and the process of asexual reproduction; the mechanism for the production of variation in heterokaryons.

Molecular biology: The study of the structure and function of proteins and nucleic acids in biological systems.

Molecular breeding: Plant breeding assisted by using DNA markers or protein markers.

Molecular farming: The application of biotechnology for the selected production of pharmaceutical compounds or other health or industrial compounds within a living organism (*e.g.* microbe or agricultural crop).

Molecular genetics: The study of genes, the base components of heredity. Genes are made up of DNA, whose bases are placed in a unique order that determines the hereditary traits in living organisms.

Molecular genetics: The study of the molecular structure and function of genes.

Molecular marker: A gene or DNA sequence that identifies a particular locus on the chromosome (whether the actual location is known or not) and whose inheritance can be followed.

Monoclonal antibodies: Antibodies derived from a single source (a group of cloned cells) and recognize only one kind of antigen. They are made in the laboratory from hybridoma cells, hybrids of antibody-producing cells and immortal cancer cells.

Morphogenesis: The growth and development of an undifferentiated structure to a differentiated structure or form.

Multiplexing: A sequencing approach that uses several pooled samples simultaneously, greatly increasing sequencing speed.

Mutagen: Any agent that is capable of increasing the mutation rate.

Mutagenisis: The formation or development of a mutation.

Mutant allele: An allele differing from the allele found in the standard, or wild type.

Mutant: A gene having undergone genetic change or mutation.

Mutation: Any heritable change in DNA sequence.

Nanotechnology: A precise molecule-by-molecule control of products and byproducts in the development of functional structures.

Natural selection: A complex process in which the total environment determines which members of a species survive to reproduce and so pass on their genes to the next generation.

Near-isogenic lines (NIL)**:** Not fully isogenic; for example, in maize, two distinct composites of F_3 lines from a single cross, one consisting of lines homozygous recessive and the other consisting of lines homozygous dominant for a certain gene (*i.e.* there is same genetic background), however, differing only in being homozygous dominant versus recessive for the genes; in wheat, near-isogenic lines have been produced for different *Rht* (reduced height) genes causing different straw length.

Nested association mapping (NAM): An integrated mapping strategy that allows genome-wide high-resolution mapping in a cost-effective way; it greatly facilitates complex trait dissection in many species; it is used as a large-scale maize mapping resource and as a genomics and bioinformatics tool; it requires large amounts of SNPs and/or individuals, and it is sensitive to genetic heterogeneity.

Nick translation: A procedure for labelling DNA. A DNA fragment is treated with DNase to produce single-stranded nicks. The nick is moved along the DNA molecule in the presence of labelled deoxyribonucleoside triphosphates by the concerted action of the 5¢-> 3¢ exonuclease and 5¢-> 3¢ polymerase activities of E. coli DNA polymerase I.

Nitrogen fixation: The process of conversion of atmospheric nitrogen to ammonia. Biological nitrogen fixation occurs in prokaryotes and is catalysed by the enzyme nitrogenase.

Northern blotting: The transfer of RNA from an electrophoresis gel to a membrane to perform Northern hybridization.

Nondisjunction: The failure of separation of paired chromosomes at metaphase, resulting in one daughter cell receiving both and the other daughter cell none of the chromosomes in question; it can occur both in meiosis and mitosis.

Nonhomologous chromosomes: The different chromosomes of a haploid chromosome set, which usually cannot pair with one another.

Nonsense codon: (also called STOP codon) Any one of three triplets (UAG, UAA, UGA) that cause termination of protein sysnthesis.

Non-sense mutation: A mutation that alters a gene so that a non-sense codon is inserted; such a codon is one for which no normal tRNA molecule exists, therefore it does not code for an amino acid; usually non-sense codons cause the termination of translation; several non-sense codons are recognized (*e.g.* amber, ochre, opal).

Nucleic acid: A large molecule composed of nucleotide subunits.

Nucleoid: A term used to represent the DNA containing region of a prokaryotic cell.

Nucleotide: A subunit of DNA or RNA consisting of a nitrogenous base (adenine, guanine, thymine, or cytosine in DNA; adenine, guanine, uracil, or cytosine in RNA), a phosphate molecule, and a sugar molecule (deoxyribose in DNA and ribose in RNA). Thousands of nucleotides are linked to form a DNA or RNA molecule.

Nucleus: The structure in eukaryotic cells (cells with a true nucleus) that contains the cellular DNA.

Null mutation: A mutation, which eliminates all enzymatic activity; usually deletion mutations.

Okazaki fragment: Since DNA can be replicated in only one direction, *i.e.*, nucleotides can be added only at the 3¢ end, only one of the two strands of a double helix can be replicated continuously. The other strand is replicated in small segments (Okazaki fragments) that are subsequently joined together by DNA ligase.

Oligonucleotide ligation assay (OLA): A PCR-based method for SNP typing; it is a ligase mediated gene detection system which uses exact 32 matching of a primer to one of the SNP allele; if this happens, the other labeled oligonucleotide that binds to the nucleotide immediately next to the SNP on the other side would be joined to the primer by ligase; the resulting sample can then be tested for the presence of the label.

Oligonucleotides: Small single-stranded segments of DNA typically 20-30 nucleotide bases in size which are synthesized *in vitro*.

Oncogene: A gene, one or more forms of which is associated with cancer. Many oncogenes are involved, directly or indirectly, in controlling the rate of cell growth.

Open reading frame (ORF): A sequence of nucleotides in a DNA molecule that has the potential to encode a peptide or protein: it starts with a start triplet (ATG), is followed by a string of triplets each of which encodes an amino acid, and ends with a stop triplet (TAA, TAG or TGA). This term is often used when, after the sequence of a DNA fragment has been determined, the function of the encoded protein is not known. The existence of open reading frames is usually inferred from the DNA (rather than the RNA) sequence.

Operator: The region of DNA that is upstream from a gene or genes and to which one or more regulatory proteins (repressor or activator) binds to control the expression of the gene(s).

Operon: A set of adjacent structural genes whose mRNA is synthesized in one piece, together with the adjacent regulatory genes that affect the transcription of the structural genes; it is under the control of an operator gene, lying at one end of it.

Organ culture: The growth in aseptic culture of plant organs, such as roots or shoots, beginning with organ primordia or segments and maintaining the characteristics of the organ.

Organogenesis: The process of morphogenesis that finally results in the formation of organs *e.g.* shoots, roots.

Overlapping DNA: A special type of gene organization; one DNA sequence may code for different proteins; it is performed by two open reading frames, which subsequently act.

Palindrome: Originally a word, sentence or verse that reads the same from right to left as it does from left to right. In biotechnology, the term is applied to a segment of DNA in which the base-pair sequence reads the same in both directions from a central point of symmetry.

Parallel mutation: A mutation that causes similar phenotypes but in different species or genotypes.

Paralogous genes: Homologous genes that have arisen through gene duplication and that have evolved in parallel with the same organism; as opposed to orthologous genes.

Paramutable allele: An unstable allele where the phenotypic consequences are enhanced by the wishbou allele.

Paramutagenic allele: An allele possessing the ability to cause paramutation.

Paramutation: A mutation in which one allele in the heterozygous condition permanently changes the partner allele.

Patent: A government issued document that provides the holder the exclusive rights to manufacture, use or sell an invention for a defined period usually 20 years.

Peptidyl-tRNA binding site (P-site): The site on a ribosome that hosts the tRNA to which an amino acid for the growing polypeptide chain is attached.

Phasmid: A cloning vector that has the capabilities to replicate as a plasmid or as a phage; the two modes of replication are usually functional in different bacterial species.

Phenotype: The visible appearance of an organism. The phenotype reflects the combined.

Physical map: A map showing physical locations on a DNA sequence, such as restriction sites and sequence-tagged sites. Also a diagram of a chromosome or a karyotype, showing the location of loci (genes and markers).

Phytohormone: A substance that stimulates growth or other processes in plants. They include auxins, abscissic acid, cytokinins, gibberellins and ethylene. Phytohormones are chemical messengers that may pass through cells, tissues and organs and stimulate biochemical, physiological and morphological responses.

Plant Molecular Farming (PMF): This technique involves using genetically modified plants to produce substances that the plants typically do not produce naturally, such as industrial compounds or therapeutics.

Plaque: In bacterial virus analysis, a clear area of a petri dish, devoid of bacterial cells, indicating the presence of viral particles.

Plasmid: Autonomously replicating, extra chromosomal circular DNA molecules, distinct from the normal bacterial genome and nonessential for cell survival under nonselective conditions. Some plasmids are capable of integrating into the host genome.

Plastid DNA: Organelle DNA that is present in a plastid.

Plastid inheritance: Non-Mendelian inheritance that is caused by hereditary factors present in the plastids plot with or without mass selection, generally followed by a single-plant selection; it is a procedure for inbreeding a segregating population until the desired level of homozygosity is achieved.

Point mutation: A change in a single base pair.

Poly(A) tail: A series of A- nucleotides attached to the 3′end of eukaryotic mRNA.

Polyacrylamide gel electrophoresis (PAGE): A method for separation of proteins and amino acids on the basis of their molecule size; the molecules move through the gel under the influence of an electric field.

Polyadenylation: Post-transcriptional addition of a polyadenylic acid tail to the 3¢ end of eukaryotic mRNAs. Also called poly-(A) tailing. The adenine-rich 3¢ terminal segments is called a poly (A) tail.

Polycistronic RNA: mRNA that codes for more than one polypeptide.

Polyclonal antibody: A serum sample that contains a mixture of immunoglobulin molecules secreted against a specific antigen, each recognizing a different epitope, some antibodies of which bind to different antigenic determinants of one antigen.

Polymerase chain reaction (PCR): A laboratory method used to make many copies of a DNA fragment in minutes using an enzyme called polymerase.

Polymerase, DNA or RNA: Enzymes that catalyse the synthesis of nucleic acids on preexisting nucleic acid templates, assembling RNA from ribonucleotides or DNA from deoxyribonucleotides.

POLYMORPHISM - The allelic variations in the genomes that results in different phenotypes.

Position effect: The situation in which a change in phenotype results from the change of the position of a gene or group of genes.

Positional candidate gene: A gene known to be located in the same region as a DNA marker that has been shown to be linked to a single-locus trait or to a quantitative trait locus and whose function suggests that it could be the source of genetic variation in the trait in question.

Positional cloning: The process where researchers obtain the clone of a gene without prior knowledge of its protein product or function; it uses large-scale physical and formal genetic linkage maps to find specific genes.

Post-transcriptional gene silencing (PTGS): Silencing of an endogenous gene caused by the introduction of a homologous dsRNA, transgene, or virus; in PTGS, the transcript of the silenced gene is synthesized but does not accumulate because it is rapidly degraded; this is a more general term than RNAi.

Pribnow box: Consensus sequence near the mRNA start-point of prokaryotic genes.

Primer: Short, pre-existing oligonucleotide or polynucleotide cahin to which new DNA can be added by DNA polymerase.

Primer-dimers: Occur when single stranded primer oligonucleotides bind to each other rather than the DNA template.

Primer walking: A method for sequencing long (>1 kb) cloned pieces of DNA. The initial sequencing reaction reveals the sequence of the first few hundred nucleotides of the cloned DNA. On the basis of these data, a primer containing about 20 nucleotides and complementary to a sequence near the end of sequenced DNA is synthesized, and is then used for sequencing the next few hundred nucleotides of the cloned DNA. This procedure is repeated until the complete nucleotide sequence of the cloned DNA is determined.

Prion: A protein particle found in brain cell membranes. Changes in its structure appear to be related to infectious diseases of the nervous system, such as Creutzfeld-Jakob disease in humans, bovine spongiform encephalopathy in cows, and sheep scrapie.

Probe: Single-stranded DNA or RNA molecules of specific base sequence, labeled either radioactively or immunologically, that are used to detect the complementary base sequence by hybridisation.

Processed pseudo-gene: A copy of a functional gene which has no promoter, no introns and which, consequently, is not itself transcribed. Pseudo-genes are thought to originate from the integration into the genome of cDNA copies synthesized from mRNA molecules by reverse transcriptase. Pseudo-genes therefore have a poly (dA) sequence at their 5¢ ends. Because they are not subject to any evolutionary pressure to maintain their coding potential, pseudo-genes accumulate mutations and often have stop codons in all three reading frames.

Prokaryote: An organism without a nucleus; eubacteria, archaebacteria, and blue-green algae.

Promoter: A region of DNA involved in binding of RNA polymerase to initiate transcription.

Proofreading: The enzymatic scanning of newly-synthesized DNA for structural defects, such as mis-matched base pairs. It is effected by DNA polymerase.

Protein engineering: Generating proteins with modified structures that confer properties such as higher catalytic specificity or thermal stability.

Protein targeting: The process of transport of proteins from one compartment to other with in a cell. Also called as protein sorting.

Protein: A large molecule composed of one or more chains of amino acids in a specific order; the order is determined by the base sequence of nucleotides in the gene coding for the protein. Proteins are required for the structure, function, and regulation of the bodys cells, tissues, and organs, and each protein has unique functions. Examples are hormones, enzymes, and antibodies.

Proteome: The complete protein complement of cells, tissues and organisms is referred to as its proteome.

Proteomics: Large scale characterization of the entire protein complement of cells, tissues, and organisms is called proteomics.

Protoclonal variation: Variability of somatic cells derived from protoplast culture.

Protoplast culture: The isolation and culture of plant protoplasts by mechanical means or by enzymatic digestion of plant tissue, organs, or cultures derived from these; protoplasts are utilized for selection or hybridization at the cellular level and for a variety of other purposes.

Protoplast fusion: A technique used in somatic hybridization experiments; it is used for overcoming crossing barriers; protoplasts are placed together and induced to fuse, applying fusogenic agents, such as polyethylene glycol or physical means; subsequent regeneration of the cell wall allows the propagation and regeneration of a somatic hybrid plant.

Protoplast: Cell without a cell wall; protoplasts are produced by enzymes, which digest the wall; they are used for production of hybrid cells by protoplast fusion or for injection of foreign DNA.

Pseudo-gene: A copy (duplicate) of a functional gene that, as a result of mutation following duplication, can no longer function.

pUC18 vector: A plasmid cloning vector; size: 2.7 kb; ~100 copies per chromosome.

Pulsed-field gel electrophoresis: An electrophoretic technique in which the gel is subjected to electrical fields alternating between different angles, allowing very large DNA fragments to move through the gel, and hence permitting efficient separation of mixtures of such large fragments.

Pyrimidine: A nitrogen-containing, double-ring, basic compound that occurs in nucleic acids. The pyrimidines in DNA are cytosine and thymine; in RNA, cytosine and uracil.

Quantitative trait loci (QTL): Regions of the chromosome associated with the inheritance of polygenic traits.

Random amplified polymorphic DNA (RAPD) technique: A comparative study (among individuals, populations, or species) of the DNA fragment length produced in controlled DNA synthesis reactions started with short sequences of DNA (primers); as a genetic mapping methodology, it utilizes as its basis the fact that specific DNA sequences (polymorphic DNA) are repeated (*i.e.*, appear in sequence) with a gene of interest; thus, the polymorphic DNA sequences are linked to that specific gene; their linked presence serves to facilitate genetic mapping within a genome.

Random mating: For a given population, where an individual of one sex has an equal probability.

RAPD - Randomly Amplified Polymorphic DNA - A PCR based method of DNA profiling. It basically involves the amplification of DNA sequences using random primers, and use of genetic fingerprints to identify individual organisms (mostly plants).

Reading frame: A sequence of sense codons such that each suceeding triplet generates the correct order of amino acids resulting in a polypeptide chain.

Recombinant DNA technology: Procedures used to join together DNA segments in a cell-free system (*e.g.* in a test tube outside living cells or organisms). Under appropriate conditions, a recombinant DNA molecule can be introduced into a cell and copy itself (replicate), either as an independent entity (autonomously) or as an integral part of a cellular chromosome.

Recombinant DNA: The DNA formed when DNA fragments from more than one organism are spliced together *in vitro*.

Recombinant inbred lines (RILs): A population of fully homozygous individuals that is obtained through the repeated selfing of an F_1 hybrid, and that comprises 50 per cent of each parental genome in different combinations; they derive from multiple inbred strains (either by selfing or by full-sib mating) and can serve as a powerful resource for the genetic dissection of complex traits; however, the use of such multiple-strain requires a detailed knowledge of the haplotype structure in such lines.

Red biotechnology: is applied to medical processes. Some examples are the designing of organisms to produce antibiotic, and the engineering of genetic cures through Genetic manipulation.

Redundant DNA: DNA that does not appear to be genetically active and hence is not translated or transcribed; it often consists of repeated sequences.

Redundant gene: A gene that is present in many functional copies, so that one copy can complement the loss of another copy.

Reduplication: Doubling of the genetic matter of a haploid chromosome set.

Regeneration: The embryonic development of shoots and roots from callus tissue, producing a plantlet.

Regulator gene: In the operon theory of gene regulation, a gene that is involved in switching on or off the transcription of structural genes; when transcribed, the regulator gene produces a repressor protein, which switches off an operator gene and hence the operon that this controls; the regulator gene is not part of the operon and may even be on a different chromosome.

Renaturation: The reassociation of denatured complementary single strands of a DNA double helix.

Restriction endonuclease: An enzyme that specifically cuts DNA molecule at specific nucleotide sequences.

Restriction enzyme cutting site: A specific nucleotide sequence of DNA at which a particular restriction enzyme cuts the DNA. Some sites occur frequently in DNA (*e.g.* every several hundred base pairs), others much less frequently (rare-cutter; *e.g.*, every 10,000 base pairs).

Restriction enzyme, endonuclease: A protein that recognises specific, short nucleotide sequences and cuts DNA at those sites. Bacteria contain over 400 such enzymes that recognise and cut over 100 different DNA sequences. See restriction enzyme cutting site.

Restriction enzyme: An enzyme used to cut DNA at specific sites. The resulting fragments can then be spliced together to form recombinant DNA, which can be separated out on a gel or inserted into a plasmid.

Restriction fragment length polymorphism (RFLP): Variation between individuals in DNA fragment sizes cut by specific restriction enzymes; polymorphic sequences that result in RFLPs are used as markers on both physical maps and genetic linkage maps. RFLPs are usually caused by mutation at a cutting site. See marker.

Restriction map: Representation of DNA with the position of restriction sites indicated.

Restriction site: A certain nucleotide sequence within the double-stranded DNA; it is recognized by a restriction endonuclease; the enzyme cuts the double strand within the recognition sequence; the restriction sites are usually composed of four to six base pairs and are bilaterally symmetric; both strands are cut either on exactly opposite positions (blunt ends) or alternated ones (sticky ends); the type of cutting depends on the enzyme used.

Retrotransposon: Retrotransposons are a ubiquitous and major component of plant genomes; those with long terminal DNA repeats (LTRs, Ty1-copia-like family) are widely distributed over the chromosomes of many plant species.

Reverse genetics: Using linkage analysis and polymorphic markers to isolate a disease gene in the absence of a known metabolic defect, then using the DNA sequence of the cloned gene to predict the amino acid sequence of its encoded protein; in general, a technology aiming at isolating mutants of a given sequence; it is also applied for identification of gene function.

Reverse mutation: The production, by further mutation, of a premutation gene from a mutant gene; it restores the ability of the gene to produce a functional protein; strictly, reversion is the correction of a mutation (*i.e.* it occurs at the same site).

Reverse transcriptase: An enzyme from retroviruses for the synthesis of a DNA complementary to a RNA molecule (*i.e.*, cDNA); it is used for (1) filling-in reactions, (2) for DNA sequencing, and (3) for cDNA synthesis.

Reverse transcription: The process of synthesis of DNA from RNA.

Ribonucleic acid (RNA): Like DNA, a type of nucleic acid. There are three major types: messenger RNA, transfer RNA, and ribosomal RNA. All are involved in the synthesis of proteins from the information contained in the DNA molecule. Synonyms: gene splicing,.

Ribosomal RNA (rRNA): A class of RNA found in the ribosomes of cells.

Ribosomes: Small cellular components composed of specialised ribosomal RNA and protein; site of protein synthesis.

RNA interference (RNAi): The use of double stranded RNA to interfere with gene expression; RNAi is usually mediated by approximately 21-nt small interfering RNAs.

RNA polymerase: An enzyme that transcribes an RNA molecule from the template strand of a DNA molecule; it adds to the 32 end of the growing RNA molecule one nucleotide at a time using ribonucleotide triphosphates (rNTPs) as substrates; RNA polymerase I is dedicated to the synthesis of only one type of RNA molecule (pre-rRNA); RNA polymerase II is required for general transcription reactions; RNA polymerase III produces small RNAs such as tRNAs and 5S rRNA.

RNA transcriptase: The enzyme responsible for transcribing the information encoded in DNA into RNA; it is also called transcriptase or RNA polymerase.

Robertsonian translocation: A chromosomal mutation due to centric fusion or centric fission (*i.e.*, a reciprocal translocation with breakpoints within the centromeric regions); in wheat, Robertsonian translocations arise from centric misdivision of univalents at anaphase/telophase I, followed by segregation of the derived telocentric chromosomes to the same nucleus, and fusion of the broken ends during the ensuing interkinesis.

Satellite chromosome: Each chromosome with a secondary constriction; this constriction divides a satellite part from the rest of the chromosome (arm).

Satellite DNA: A highly repetitive DNA, composed of repeated hepta- to deca-nucleotide sequences; DNA of different buoyant density; a minor DNA fraction that has sufficiently different base composition from the bulk of the DNA in order to separate distinctly during cesium chloride density gradient centrifugation; it can derive from nucleus, plastid, or mitochondrial.

Saturation mutagenesis: Induction and recovery of large numbers of mutations in one area of a genome, or in one biological function, in order to identify all the genes in that area, or affecting that function.

Secondary metabolite: A metabolite that is not required for the growth and maintenance of cellular functions.

Selectable marker: A gene, often encoding resistance to an antibiotic or an herbicide, introduced into a group of cells to allow identification of those cells that contain the gene of interest from the cells that do not. Selectable markers are used in genetic engineering to facilitate identification of cells that have incorporated another desirable trait that is not easy to identify in individual cells.

Selfish DNA: A segment of the genome with no apparent function other than to ensure its own replication.

Sense mutation: A mutation that changes a termination (stop) codon into one that codes for an.

Sense strand DNA: The DNA strand with the same sequence as mRNA.

Sequence characterized amplified region (SCAR): A molecular marker; the DNA amplified as a PCR derived marker (*e.g.* RAPD) is sequenced; the sequence is used to develop a new PCR marker comprising a single DNA species, unique to the target organism or genotype.

Sequence-tagged-microsatellites (STMs) marker: Primers constructed from the flanking regions of microsatellite DNA which can be used in PCR reactions to amplify the repeat region.

Sequencing: Determination of the order of nucleotides (base sequences) in a DNA or RNA molecule or the order of amino acids in a protein.

Sequence tagged site (STS): Short (200 to 500 base pairs) DNA sequence that has a single occurrence in the human genome and whose location and base sequence are known. Detectable by polymerase chain reaction, STSs are useful for localising and orienting the mapping and sequence data reported from many different laboratories and serve as landmarks on the developing physical map of the human genome. Expressed sequence tags (ESTs) are STSs derived from cDNAs.

Shine-Dalgarno sequence: A conserved sequence in prokaryotic mRNAs that is complementary to a sequence near the 5¢ terminus of the 16S ribosomal RNA and is involved in the initiation of translation.

Shotgun method: A way of preparing DNA for genetic studies; random-cut fragments of DNA are cloned into a vector; the fragments resemble a clone library; using this collection of clones, several molecular techniques are applicable.

Shuttle vector: A vector constructed in such a way that it can replicate in at least two different host species recombined into such a vector can be tested or manipulated in several cell types.

Signal transduction: The biochemical events that conduct the signal of a hormone or growth factor from the cell exterior, through the cell membrane, and into the cytoplasm. This involves a number of molecules, including receptors, ligands and messengers.

Silent mutation: Mutation in which the function of the protein product of the gene is unaltered.

Single cell protein (SCP): Cells or protein extracts of microorganisms produced in large quanities for use as human or animal protein supplement.

Single nucleotide polymorphism (SNP): Individual differences at a single nucleotide of DNA. This genotypic difference can cause a phenotypic difference in hair colour, height or response to a drug, depending on the gene.

Single sequence repeat (SSR) DNA marker technique: A genetic mapping technique that utilizes the fact that microsatellite sequences repeat (appear repeatedly in sequence within the DNA molecule) in a manner enabling them to be used as markers.

Single-strand conformational polymorphism (SSCP): This feature relies on secondary and tertiary structural differences between denatured and rapidly cooled amplified DNA fragments that differ slightly in their DNA sequences; different SSCP alleles are resolved on non-denaturing acrylamide gels, usually at low temperatures; the ability to resolve alleles depends on the conditions of electrophoresis and this requires DNA sequence data.

Site-directed mutagenesis: The technique used to produce a specified mutation at a predetermined position in a DNA molecule.

Site-specific mutagenesis: The use of recombinant DNA technology to create specific deletions, insertions, or substitutions *in vitro* in a particular gene; the technique allows the production of proteins having any desired amino acid at any position.

Site-specific recombination: A crossover event that requires homology of only a very short region and uses an enzyme specific for that recombination; recombination occurring between two specific sequences that need not be homologous, mediated by a specific recombination system.

Small interfering RNAs (siRNAs): Models of PTGS indicate that 21–23 nucleotide dsRNAs mediate PTGS; introduction of siRNAs can induce PTGS, *e.g.* in mammalian cells; siRNAs are apparently produced *in vivo* by cleavage of dsRNA introduced directly or *via* a transgene or virus; amplification by an RNA-dependent RNA polymerase (RdRP) may occur in some organisms; siRNAs are incorporated into the RNA-induced silencing complex (RISC), guiding the complex to the homologous endogenous mRNA where the complex cleaves the transcript.

Somaclonal variation: Variation among the tissues or plant derived from the *in vitro* somatic cell culture *i.e.*, callus or suspension cultures.

Somatic embryogenesis: Formation of embryos from asexual cells.

Somatic cells: All the cells of an organism except those of the germ line.

Somatic crossing-over: Crossing-over during mitosis of somatic cells such that parent cells heterozygous for a given allele, instead of giving rise to two identical heterozygous daughter cells, give rise to non-identical daughter cells, one of which is homozygous for one of these alleles, the other being homozygous for the other allele.

Somatic embryo: An organized embryonic structure morphologically similar to a zygotic embryo but initiated from somatic cells; somatic embryos develop into plantlets *in vitro* through developmental processes that are similar to those of zygotic embryos.

Somatic hybridization: The fusion of genetically different somatic cells by different means, usually to overcome natural crossing (incompatibility) barriers.

Somatic mutation: A mutation occurring in a somatic cell; if the mutated cell continues to divide, the individual will develop a patch of tissue with a genotype different from the cells of the rest of the body.

Southern blotting: Transfer by absorption of DNA fragments separated in electrophoretic gels to membrane filters for detection of specific base sequences by radiolabeled complementary probes. SPERM (abbreviation of Spermatozoon) The male reproductive cell carrying 23 chromosomes.

Spliceosomes: Organelles responsible for the removal of introns from mRNA by means of splicing.

Splicing: The removal of the intron (noncoding sequence) from an mRNA gene sequence and the fusion of the exons (coding sequence) during the mRNA processing.

Split genes: In eukaryotes, structural genes are typically divided up (split) by a number of non-coding regions called introns.

Spontaneous mutation: A naturally occurring mutation, as opposed to one artificially induced by chemicals or irradiation; usually such mutations are due to errors in the normal functioning of cellular enzymes.

Sub cloning: Transplantation of a piece of DNA from one vector to another.

Subculture: The aseptic transfer of a part of a stock culture to a fresh medium.

Suppression: Changes that eliminate the effects of a mutation without reversing the original change in DNA.

Suppressor mutation: A mutation that counteracts the effects of another mutation. A suppressor maps at a different site than the mutation it counteracts, either within the same gene or at a more distant locus. Different suppressors act in different ways.

Suspension cultures: Cells which do not attach to the surface of the culture vessel and grow in a suspended manner in the culture medium are called suspension cultures.

Tandem repeat: A chromosomal mutation in which two identical chromosome segments lie adjacent to each other, with the same gene order; the DNA, which codes for the rRNA, contains many tandem repeats.

Taq polymerase: A DNA-dependent RNA polymerase from phage T7, which recognizes a very specific promoter sequence; it is used in many expression vectors.

Targeting-induced local lesions in genomes (TILLING): A technique to screen a population that has either been deliberately mutagenized or possesses natural biodiversity (EcoTILLING) to identify those plants with a change in a specific gene of interest; the technique uses high-throughput methods useful to plant breeders, including a reverse molecular method that combines random chemical mutagenesis with PCR-based screening of gene regions of interest; it provides a range of allele types, including mis-sense and knockout mutations; by comparing the phenotypes of isogenic genotypes differing in single sequence motifs, TILLING provides direct proof of function of both induced and natural polymorphisms without the use of transgenic modifications.

TATA box: A canonic DNA sequence; part of a plant promoter; promotes the transcription of DNA.

T-DNA - The part of the Ti plasmid that is transferred to the plant DNA.

Telomerase: A reverse transcriptase (h*TERT*) containing an RNA molecule (h*TR*) that functions as the template for the tandem repeat at the telomere; it synthesizes the telomere to maintain its length after each cell division; it is active in young cells and gametes, inactive in differentiated somatic cells, and reactivated in malignant cells; telomerase can add one base at a time to the telomeric end of a chromosome; this maintenance work is required for cells to escape from replicative senescence.

Telomere: The ends of chromosomes. These specialized structures are involved in the replication and stability of linear DNA molecules. See DNA replication.

Temperate phage: A phage (virus) that invades but may not destroy (lyse) the host (bacterial cell). However, it may subsequently enter the lytic cycle.

Temperature-sensitive mutation: A class of conditional mutations; the mutant phenotype is observed in one temperature range and the wild-type phenotype is observed in another temperature range.

Template strand: The strand of the DNA double helix that is copied by base pair complementarity to make an RNA. The other, non-template strand of the DNA duplex has a sequence that is identical to the synthesized RNA (except in RNA, U replaces T).

Template: The DNA single strand, complementary to a nascent RNA or DNA strand, which serves to specify the nucleotide sequence of the nascent strand.

Terminalization: A progressive shift of chiasmata from their sites of origin to more terminal positions.

Termination codon: Stop codon.

Terminator genes: A set of transgenes designed to arrest the final stages of seed development in a crop plant; the resulting trait is the production of seeds that will not germinate; it is also called the technology protection system.

Terminator technology: Plants are genetically engineered so when the crops are harvested, all new seeds produced from these crops are sterile.

Thymine (T): A nitrogenous base, one member of the base pair A-T (adenine-thymine).

Ti plasmid: Established tumors in plants contain only a part of the total plasmid.

Tissue culture: A process where individual cells, or tissues of plants or animals are grown artificially.

Tissue engineering: The application of the principles of engineering to cell culture for the construction of functional anatomical units.

Ti plasmid: The large-sized tumour inducing plasmid found in *Agrobacterium tumefaciens*. It directs crown gall formation in certain plant species.

Topoisomerase: Enzyme that catalyzes the reduction of supercoiling of DNA; it is important during DNA replication to relieve the negative supercoiling that occurs when the two strands of the DNA double helix are separated.

Totipotency: The potential ability of a cell to express all its genetic information under appropriate conditions and to proceed through all the stages of development to produce a fully differentiated adult;.

Trans configuration: The configuration of two sites refers to their presence on two different molecules of DNA (chromosomes).

Transcription factors: Proteins that are directly involved in regulation of transcription initiation by binding to the control elements and allowing RNA polymerase to act; there are ubiquitous transcription factors as well as cell- and tissue-specific ones; several families have been identified including helix-loop-helix proteins, helix-turn-helix proteins, leucine zipper proteins and zinc finger proteins.

Transcription unit: The distance between sites of initiation and termination by RNA polymerase; may include more than one gene.

Transcription: A process in the cell where the DNA is used as a template to make the messenger RNA.

Transcriptome: DNA sequences of the expressed genome.

Transcriptomics: The study of global gene expression and changes caused by genetics, environmental effects, drugs or chemicals, or tissue location; it is the application of micro- or macroarrays and sequence-based methods to conduct expression profiling to determine gene expression at a global (genome wide) level.

Transfer RNA (tRNA): RNA molecules that bind to amino acids and carry them to the ribosomes where proteins are made.

Transformation efficiency: The number of cells that take up foreign DNA as a function of the amount of added DNA; expressed as transformants per microgram of added DNA.

Transformation A process by which the genetic material carried by an individual cell is altered by incorporation of exogenous DNA into its genome.

Transgene: A modified native gene or a gene from another species that is inserted into plants to bring about a desired change or introduce a trait not native to the natural genome of that species.

Transgenic organism: An organism resulting from the insertion of genetic material from another organism using recombinant DNA techniques.

Transgenics: The insertion or splicing of specific genetic sequences from one species into the functioning genome of an unrelated species to transfer desired properties for human purposes. This may be viewed as a more precise form of hybridization or plant/animal breeding, with the added consideration that genetic material from species significantly different from one another is involved (for example, the insertion of genetic material from an animal into a plant or vice versa). Another possibility is the transfer of genetically controlled properties between different animal species, such as the breeding of goats whose milk yields spider silk for possible development of new structural materials.

Transgressive segregation: The segregation of individuals in the F_2 or a later generation of a cross that shows a more extreme development of a character than either parent.

Translation The process in which the genetic code carried by mRNA directs the synthesis of proteins from amino acids.

Translocation: A change in the arrangement of genetic material, altering the location of a chromosome segment; the most common forms of translocation are reciprocal, involving the exchange of chromosome segments between two non-homologous chromosomes.

Transplastomic plants: Plants with genetic modifications in the DNA of their chloroplasts; plant cells contain chloroplasts, cell organelles responsible for photosynthesis; these chloroplasts.

Transposable element: A chromosomal locus that may be transposed from one spot to another within and among the chromosomes of the genome; it happens through breakage on either side of these loci and their subsequent insertion into a new position either on the same or a different chromosome.

Transposition: In molecular biology, the process of moving a transposon or other inserts from one position to another within a genome.

Transposon tagging: The insertion of a transposable element into or nearby a gene, thereby marking that gene with a known DNA sequence.

Transposon: Chromosomal loci capable of being transposed from one spot to another within and among the chromosomes of a complement.

Transverison: A mutation in which a purine is replaced with a pyrimidine or vice versa.

Ultracentrifugation: Centrifugation carried out at high rotor speeds (<100,000 rpm) and therefore under high centrifugal forces (<750,000 g).

Unique DNA sequence: A DNA sequence that is present only once per genome, with no repetitive nucleotide sequence.

Uracil: A nitrogenous base normally found in RNA but not DNA; uracil is capable of forming a base pair with adenine.

Variable number tandem repeat (VNTR): Genetic markers that consist of DNA segments that are duplicated end to end; the number of copies present at a locus can vary, giving rise to a large number of alleles; these genetic markers are commonly used for DNA fingerprinting.

Vector: A vehicle that carries foreign genes into an organism and inserts them into the organism's genome. Modified viruses are used as vectors for gene therapy.

Velocity density gradient centrifugation: A procedure used to separate macromolecules based on their rate of movement through a density gradient.

***V*max:** The maximal rate of an enzyme-catalysed reaction. *V*max is the product of *E*o (the total amount of enzyme) times the value of *K*eat (the catalytic rate constant).

Western blotting: A technique similar to Southern blotting but for the analysis of proteins instead of DNA; a technique in which proteins are separated by gel electrophoresis and transferred by capillary action to a nylon membrane or nitrocellulose sheet; a specific protein can be identified through hybridization to a labeled antibody.

White biotechnology, also known as industrial biotechnology, is biotechnology applied to industrial processes. An example is the designing of an organism to produce a useful chemical. Another example is the using of Enzymes as industrial Catalyst to either produce valuable chemicals or destroy hazardous/ polluting chemicals. White biotechnology tends to consume less in resources than traditional processes used to produce industrial goods.

Wobble hypothesis: An explanation of how one tRNA may recognize more than one codon. The first two bases of the mRNA codon and anticodon pair properly, but the third base in the anticodon has some wobble that permits it to pair with more than one base.

X-ray crystallography: A technique for deducing molecular structure by aiming a beam of X-rays at a crystal of the test compound and measuring the scatter of rays.

Yeast artificial chromosome (YAC): A vector used to clone DNA fragments (up to 400 kb); it is constructed from the telomeric, centromeric, and replication origin sequences needed for replication in yeast cells.

Z-DNA (zig-zag DNA): A form of DNA duplex in which the double helix is wound in a left-hand, instead of a right-hand, manner.

References

Ahlquist. P. 2002. RNA - dependent RNA polymerases, viruses, and RNA silencing. Science 296 (5571): 1270–1273.

Bahman Fazeli-nasab and Morteza Amozadeh. 2012. Molecular basis of plant harmones to stresses. *Technical journal of engineering and applied sciences,* 2(5): 128-130.

Bartlett J. M. S. and Stirling, D. 2003. A Short History of the Polymerase Chain Reaction. PCR Protocols. *Methods in Molecular Biology* 226 (2nd ed.). 3–6.

Beavis. W.D. 1998. QTL analyses: power, precision, and accuracy. In: Paterson AH (ed) *Molecular analysis of complex traits.* CRC Press, Boca Raton, 145–161.

Becker M. Wayne. 2009. The world of the cell. *Pearson Benjamin Cummings.* 480.

Bhojwani S. S and Razdan. M. K. 1996. Plant tissue culture: theory and practice (Revised ed.). Elsevier.

Bilgrami, K.S. and Pandey, A.K. 1992. Introduction to Biotechnology. *CBS Pub., New Dehli.*

Biofortification: Harnessing Agricultural Technology to Improve the Health of the Poor. 2002. *IFPRI and CIAT pamphlet.*

Biosafety in India. Rethinking GMO regulation. 2005. *Economical and political weekly,* 4284-4289.

Braman, Jeff, ed. 2002. In Vitro Mutagenesis Protocols. Methods in Molecular Biology 182 (2nd ed.). *Humana Press.*

Brown, Terry 2006. Gene cloning and DNA analysis: An introduction. *Cambridge, MA: Blackwell Pub.*

Butler, Jennifer, James T and Kadonaga. 2002. The RNA polymerase II core promoter: a key component in the regulation of gene expression. *Genes and Development.* 16 (20): 2583–2592.

Chahal G.S. and Gosal. S.S. 2002. Principles and Procedures of Plant Breeding – Biotechnological and Conventional Approaches. *Narosa Publishing House, New Delhi.*

Charlebois. R.L. (ed) 1999. Organization of the prokaryote genome. *ASM Press, Washington DC.*

Chawla. H.S. 2005. Introduction to Plant Biotechnology. *Oxford and IBH Publishing Co., New Delhi.*

Chie ishikawa, Hatanaka, Misoo. and Fukayama.2011. Functional incorporation of sorghum small subunit increases the catalytic turnover rate of Rubisco in transgenic rice. *Plant physiol.,* 156: 1603-1611.

Collard B.C. and Mackill. D.J. 2007. Marker-assisted selection: an approach for precision plant breeding in the twenty-first century. *Philos Trans R Soc Lond B Biol Sci.*

Cooper. G.M. 2000. "Chapter 14: The Eukaryotic Cell Cycle". *The cell: a molecular approach (2nd ed.). Washington, D.C: ASM Press.*

David L. Nelson and Michael M. Cox. 2010. Principles of Biochemistry, *Freeman publications* Delhi.

Devlin R.M. and Witham. F.H. 1986. Plant Physiology (1st ed.) *Delhi: CBS publishers and distributors.*

Donald Voet and Judith G. Voet. 2003. Biochemistry, *Wiley publishers*

Ehud Gazit. 2007. Plenty of room for biology at the bottom: An introduction to bionanotechnology. *Imperial College Press,*

Enelope Nestel, Howarth E. Bouis, J. V. Meenakshi and Wolfgang Pfeiffer. 2006. 'Biofortification of Staple Food Crops'. *The Journal of Nutrition,* 136 (4): 1066.

Farooq M., Wahid A., Kobayashi N, Fujita D and Basra S.M.A. 2009. Plant drought stress: effects, mechanisms and management. *Agron. Sustain. Dev.* 29: 185–212.

Fisher R.P. and Morgan. D.O. 1994. A novel cyclin associates with MO15/CDK7 to form the CDK-activating kinase. *Cell, 78*: 713–724.

Flint-Garcia S., Thornsberry J. M and Buckler ESIV. 2003. Structure of linkage disequilibrium in plants. *Annu Rev Plant Biol,* 54: 357–374.

Freeze. H. H. 1993. Preparation and analysis of glycoconjugates. Current Protocols in Molecular Biology. Chapter 17: 1771–1778.

Gage. D. J. 2004. Infection and invasion of roots by symbiotic, nitrogen-fixing rhizobia during nodulation of temperate legumes. *Microbiol Mol Biol Rev.* 68: 280-300.

Galas D. and Schmitz. A. 1978. DNAse footprinting: a simple method for the detection of protein-DNA binding specificity. *Nucleic Acids Research* 5 (9): 3157–70.

Gamborg and Phillips. 2005. Plant cell, Tissue and Organ culture, *Narosa publications.*

Gardner, Snustard and Simmonds. 1991. Principles of Genetics, *Wiley publishers.*

Glick B. R. and Pasternak J. J. 2005. Molecular Biotechnology Principles and Applications of Recombinant DNA (3rd ed.). *ASM Press.*

Goodman. M.M. 2004. Plant Breeding Requirements for Applied Molecular Biology. *Crop Sci.* 44: 6.

Griffiths J.F, Gelbart W.M, Lewontin R.C, Wessler, S.R, Suzuki D.T and Miller J.H. 2005. Introduction to Genetic Analysis. *W.H. Freeman and Co.* 34–40, 473–476, 626–629.

Gupta. P.K. 1994. Elements of Biotechnology. *Rastogi and Co., Educational Publishers, Meerut.*

Hablieb C. M. and Luden. P. W. 2000. Regulation of biological nitrogen fixation. *J. Nutr.* 130: 1081-1084.

Hamer B.D. and Hooper. N.M. 2007. Biochemistry, *Viva books Pvt. Ltd.*

Hand Book of Plant Tissue Culture, *ICAR publications.*

Heffner, E.L. Sorrells M.E. and Jannink. J.L. 2009. Genomic selection for crop improvement. *Crop Sci* 49: 1-12.

Hirayama T. and Shinozaki. K.2010. Research on plant abiotic stress responses in the post-genome era: past, present and future. *Plant Journal*, 61(6): 1041–1052.

Hoffman, B. M. Lukoyanov D, Dean D. R. and Seefeldt. L. C. 2013. "Nitrogenase: A Draft Mechanism". *Acc. Chem. Res.* 46: 587.

Humphreys M. and John; Chapple, Clint. 2000. Molecular 'pharming' with plant P450s. *Trends Plant Sci.* 5 (7): 271–272.

Indra K. Vasil and Trevor A. Thorpe. 2005. Plant cell and Tissue culture, *Springer publications.*

Ines Ben Rejeb, Victoria Pastor and Brigitte Mauch-Mani. 2014. Plant Responses to Simultaneous Biotic and Abiotic Stress: Molecular Mechanisms. *Plants,* 3: 458-475.

Jaspers P. and Kangasjärvi. J. 2010. Reactive oxygen species in abiotic stress signaling, *Physiologia Plantarum*, 138(4): 405–413.

Jennifer L. Reed, Ryan S. Senger, Maciek R. Antoniewicz, and Jamey D. Young. 2010. Computational approaches in metabolic engineering. *Journal of Biomedicine and Biotechnology,* 1-7.

Jha, T.B. and Ghosh, B. 2005. Plant Tissue Culture. *University Press, Hyderabad.*

Karas M. and Bahr. U. 1990. Laser Desorption Ionization Mass Spectrometry of Large Biomolecules. *Trends in Analytical Chemistry* 9 (10): 321–325.

Karp, Gerald 2009. Cell and Molecular Biology: Concepts and Experiments. John Wiley & Sons.

Karp.2010. Cell Biology, *Wiley publishers.*

Kevin J. Scanlon. 2004. Anti-Genes: siRNA, Ribozymes and Antisense. *Current Pharmaceutical Biotechnology,* 5(5): 1-6.

Khosro Mohammadi and Yousef Sohrabi. 2012. Bacterial biofertilizers for sustainable crop production. A review. *ARPN journal of Agricultural and Biological sciences,* 7(5): 307-316.

Kidwell M.G. and Lisch. D.R. 2000. Transposable elements and host genome evolution. *Trends in ecology and evolution* 15 (3): 95–99.

Kumar. H. D. 2004. Genomics and Cloning, *Oxford and IBH publishers.*

Kurien and Scofield. 2006. Western Blotting. *Methods* 38 (4): 283–293.

Lamour, Kurt and Tierney, Melinda. An Introduction to Reverse Genetic Tools for Investigating Gene Function. *APSnet. The American Phytopathological Society.*

Lee R Lynd, Mark S Laser, David Bransby, Bruce E Dale, Brian Davison, Richard Hamilton, Michael Himmel, Martin Keller, James D McMillan, John Sheehan and Charles E Wyman. 2008. How biotech can transform biofuels. *Nature biotechnology,* 26: 169-172.

Lewin. B. 2007. Genes IX, *Jones and Bartlett publishers. Burlington.*

Lindsey, Keith; Casson, Stuart; Chilley, Paul. 2002. Peptides: new signalling molecules in plants. *Trends in Plant Science* 7 (2): 78–83.

Liu, J. and Kipreos, E.T. 2000. Evolution of cyclin-dependent kinases (CDKs) and CDK-activating kinases (CAKs): differential conservation of CAKs in yeast and metazoa. *Mol. Biol. Evol. 17:* 1061–1074.

Luttman, Bratke and Kupper. 2006. Immunology. *Academic Press.*

Madhava Rao, K.V, Raghavendra A.S. and Janardhan Reddy. K. 2006. Physiology and Molecular Biology of Stress Tolerance in Plants. *Springer publications.*

Matthew A. Jenks and Paul M. Hasegawa. 2005. Plant abiotic stress. *Blackwell publishing.*

Matthews, C. E, Van Holde K. E. and Ahern. K. G. 1999. Biochemistry. 3rd edition. *Benjamin Cummings.*

Mattick, J. S. 2001. Non-coding RNAs: the architects of eukaryotic complexity. *EMBO Reports 2 (11): 986–991.*

Meagher, R. B. 2000. Phytoremediation of toxic elemental and organic pollutants. *Current Opinion in Plant Biology* 3(2): 153–162.

Meng-Huei Lin, Chia-Wei Lin, Jo-Chu Chen, Yu-Chung Lin, Shu-Yun Cheng, Tzung-Hua Liu, Fuh-Jyh Jan, Shu-Tu Wu, Fu-Sheng Thseng and Hsin-Mei Ku. 2007. Tagging Rice Drought related QTL with SSR DNA Markers, *Crop, Environment and Bioinformatics,* 4: 65-76.

Meyer C.A, Jacobs H.W, Datar S.A., Du W., Edgar B.A., and Lehner C.F. 2000. *Drosophila* Cdk4 is required for normal growth and is dispensable for cell cycle progression. *EMBO J. 19*: 4533–4542.

Miyao. M. 2003. Molecular evolution and genetic engineering of C_4 photosynthetic enzymes. *J. Exp. Bot,* 54: 179–89.

Mocellin, S. and Provenzano. M. 2004. RNA interference: learning gene knock-down from cell physiology. *Journal of translational medicine,* 2 (1): 39.

Mohan P. Arora. 2005. Molecular genetics, *Himalaya publishing house.*

Nathans D. and Smith. H.O. 1975. Restriction endonucleases in the analysis and restructuring of DNA molecules. *Annu. Rev. Biochem.* 44: 273–93.

NBTC Homepage, *Nanobiotechnology Center.*

Nguyen Thi Lang and Buu Chi Buu. Quantitative trait loci influencing drought tolerance in rice. *Omnorice,* 17: 22-28.

Osborne, Daphné J; Mc Manus and Michael T. 2005. Hormones, signals and target cells in plant development. *Cambridge University Press,* 158.

Papworth C. J, Bauer C. Braman J and Wright D. A. 1996. Site-directed mutagenesis in one day with >80% efficiency. *Strategies* 9 (3): 3–4.

Peoples M. B. and Herridge. D. F. 1990. Nitrogen fixation by legumes in tropical and sub-tropical agriculture. *Adv. Agron.* 44: 155-224.

Pettersson E. J, Lundeberg and Ahmadian. A. 2009. Generations of sequencing technologies. *Genomics* 93 (2): 105–11.

Pevsner, Jonathan. 2009. Bioinformatics and functional genomics (2nd ed.). *Wiley-Blackwell.*

Phundan singh. 2005. *Genetics,* Kalyani publishers

Pollack J.R., Perou C.M, Alizadeh A. A, Eisen M.B, Pergamenschikov A, Williams C.F, Jeffrey S.S, Botstein D. and Brown P.O. 1999. Genome-wide analysis of DNA copy-number changes using cDNA microarrays. *Nat Genet* 23 (1): 41–46.

Postgate, J. 1998. Nitrogen Fixation, 3rd Edition. *Cambridge University Press*

Primrose S.B. and Twyman. R.M. 2003. Principles of Genome Analysis and Genomics (3rd ed.). *Malden, MA: Blackwell Publishing.*

Rastogi. S.C. 2005. Cell Biology. *New Age publishers.*

Razdan. M. K. 2002. Introduction to Plant Tissue Culture. *Oxford and IBH Publishing Co., New Delhi.*

Redberry, Grace. 2006. Gene silencing: new research. New York: Nova Science Publishers.

Ribaut, J.M and Hoisington. D. A. 1998. Marker assisted selection: new tools and strategies. *Trends Plant Sci.* 3: 236–239.

Richard G.F, Kerrest A and Dujon. B. 2008. Comparative Genomics and Molecular Dynamics of DNA Repeats in Eukaryotes. *Microbiology and Molecular Biology Reviews* 72 (4): 686–727.

Ridley. M. 2006. Genome. *New York, NY: Harper Perennial.*

Ritika bhattacharjee and Utpal dey. 2013. Biofertilizer, a way towards organic agriculture: A Review. *African journal of microbial research,* 8(24): 2332-2342.

Roeder. R. 1996. The role of general initiation factors in transcription by RNA polymerase II. *Trends in Biochemical Sciences* 21 (9): 327–335.

Rosa M. Pérez-Clemente, Vicente Vives, Sara I. Zandalinas, María F. López-Climent, Valeria Muñoz, and Aurelio Gómez-Cadenas. 2013. Biotechnological Approaches to Study Plant Responses to Stress. *BioMed Research International,* 1-10.

Roy, Scott William. 2003. Recent evidence for the exon theory of genes. *Genetica* 118 (2–3): 251–66.

Russell, W. David and Sambrook, Joseph. 2001. Molecular cloning: a laboratory manual. *Cold Spring Harbor, N.Y: Cold Spring Harbor Laboratory.*

Salt D.E, Smith R.D. and Raskin. I. 1998. Phytoremediation. *Annu Rev Plant Physiol Plant Mol Biol.* 49: 643-668.

Sanger F.and Coulson. A. R. 1975. A rapid method for determining sequences in DNA by primed synthesis with DNA polymerase. *J. Mol. Biol.* 94 (3): 441–8.

Sherr, C.J. and Roberts. J.M. 1999. CDK inhibitors: positive and negative regulators of G1-phase progression. *Genes Dev. 13*: 1501–1512.

Shortle D, Dimaio D. and Nathans D. 1981. Directed Mutagenesis. *Annual Review of Genetics* 15: 265–294.

Singh. B. D. 2004. Fundamentals of Genetics (ed), *Kalyani publishers*

Singh. B.D. 2006. Plant Biotechnology (ed). *Kalyani Publishers, Ludhiana.*

Snustard and Simmonds. 2013 Principles of Genetics (6th ed.). *Wiley publishers.*

Southern, Edwin Mellor. 1975. Detection of specific sequences among DNA fragments separated by gel electrophoresis. *Journal of Molecular Biology.* 98 (3): 503–517.

Srivastava. L. M. 2002. Plant growth and development: hormones and environment. *Academic Press,* 140.

Stein. L. 2001. Genome annotation: from sequence to biology. *Nature Reviews Genetics* 2 (7): 493–503.

Stein. R. A 2008. Next-Generation Sequencing Update. *Genetic Engineering and Biotechnology News* 28 (15).

Stevaux O.and Dyson. N.J. 2002. A revised picture of the E2F transcriptional network and RB function. *Curr. Opin. Cell Biol. 14*: 684–691.

Stojanovic, edited by Nikola. 2007. Computational genomics: current methods. *Wymondham: Horizon Bioscience.*

Swanson and Webster.1989. The cell, *Princes of Hall of India Pvt Ltd.*

Taiz and Zeiger. 2010. Plant Physiology, *Panima Publishing Corporation.*

Tamarind. 2002. Principles of Genetics. *Tata McGraw hill publishers.*

Taylor, S. John, Raes and Jeroen. 2004. Duplication and divergence: the evolution of new genes and old ideas. *Annual review of genetics* 38: 615–43.

Towbin, Staehelin T. and Gordon. J. 1979. Electrophoretic transfer of proteins from polyacrylamide gels to nitrocellulose sheets: procedure and some applications. *PNAS* 76 (9): 4350–4.

Wahab. S. 1995. India's biotechnology policies and biosafety regulations. *African crop science journal,* 3(3): 319-326.

Wang Q. L. and Li. Z.H. 2007. The functions of microRNAs in plants. Front. Biosci. 12: 3975–82.

Warren Lau A, Michael. Fischbach, Anne Osbourn and Elizabeth S. Sattely. 2014. Key applications in plant metabolic engineering. *PLOS Biology,* 12(6): e1001879.

Watson J.D. and Crick. F.H. 1953. The structure of DNA. *Cold Spring Harb. Symp. Quant. Biol.* 18: 123–31.

Weier, Thomas Elliot, Rost, L. Thomas, T. Weier and Elliot. 1979. Botany: a brief introduction to plant biology. *New York: Wiley.* 155–170.

Wessler. S.R. 2006. Eukaryotic Transposable Elements and Genome Evolution Special Feature: Transposable elements and the evolution of eukaryotic genomes. *Proceedings of the National Academy of Sciences* 103 (47): 17600–17601.

Work, T.S, Hinton R, Work E, Dobrota M and Chard. T. 1980. Laboratory Techniques in Biochemistry and Molecular Biology.

Wrrcombe J.R, Parr L.B and Atlin. G.N 2002. Breeding rainfed rice for drought prone areas: Integrating conventional and participatory plant breeding in south and southeast asia. *Proceedings of a DFID plant sciences research programme/ IRRI conference.*

Xin-Guang Zhu, Stephen, Long. and Donald, R. Ort. 2010, Improving photosynthetic efficiency for greater yield. *Ann. Rev. Plant Biol.,*61: 235-261.

Yang Y.T, Bennet G. N, San K.Y, 1998. Genetic and Metabolic Engineering, *Electronic Journal of Biotechnology.*

Yu, J., Holland, J.B., McMullen, M.D., Buckler, E.S. 2008. Genetic design and statistical power of nested association mapping in maize. *Genetics* 178 (1): 539–551.

Zenobi, R and Knochenmuss, R. 1998. Ion formation in MALDI mass spectrometry. *Mass Spectrometry Reviews* 17(5): 337–366.

Zhu M, Zhao S. 2007. Candidate Gene Identification Approach: Progress and Challenges. *Int J Biol Sci.* 3(7): 420-427.

www.ingramcontent.com/pod-product-compliance
Ingram Content Group UK Ltd.
Pitfield, Milton Keynes, MK11 3LW, UK
UKHW021530300726
14060UKWH00011B/189